# THE IMPACT OF CLIMATE CHANGE MITIGATION ON INDIGENOUS AND FOREST COMMUNITIES

The international legal framework for valuing the carbon stored in forests, known as Reducing Emissions from Deforestation and Forest Degradation (REDD+), will have a major impact on indigenous peoples and forest communities. The REDD+ regime contains many assumptions about the identity, tenure and rights of indigenous and local communities who inhabit, use or claim rights to forested lands. The authors bring together expert analysis of public international law, climate change treaties, property law, human rights and indigenous customary land tenure to provide a systemic account of the laws governing forest carbon sequestration and their interaction. Their work covers recent developments in climate change law, including the Agreement from the Conference of the Parties in Paris that came into force in 2016. *The Impact of Climate Change Mitigation on Indigenous and Forest Communities* is a rich and much-needed new contribution to contemporary understanding of this topic.

MAUREEN F TEHAN is Principal Fellow at Melbourne Law School. Her scholarship has centred on indigenous land rights, property and land and resource management.

LEE C GODDEN is Professor of Melbourne Law School and Director of the Centre for Resources, Energy and Environmental Law. Her research focuses on the intersection of indigenous peoples' rights to land and waters, and environmental law. She recently served with the Australian Law Reform Commission in an inquiry into the Native Title Act.

MARGARET A YOUNG is Associate Professor of Melbourne Law School and was Director of Studies at the Hague Academy of International Law in 2016. Her research interests are public international law, international trade law, climate change law and the law of the sea.

KIRSTY A GOVER is Associate Professor of Melbourne Law School. She writes on the law, policy and political theory of indigenous rights in settler states and in international law.

# THE IMPACT OF CLIMATE CHANGE MITIGATION ON INDIGENOUS AND FOREST COMMUNITIES

## International, National and Local Law Perspectives on REDD+

MAUREEN F TEHAN
*University of Melbourne*

LEE C GODDEN
*University of Melbourne*

MARGARET A YOUNG
*University of Melbourne*

KIRSTY A GOVER
*University of Melbourne*

CAMBRIDGE
UNIVERSITY PRESS

**CAMBRIDGE**
**UNIVERSITY PRESS**

University Printing House, Cambridge CB2 8BS, United Kingdom

One Liberty Plaza, 20th Floor, New York, NY 10006, USA

477 Williamstown Road, Port Melbourne, VIC 3207, Australia

314–321, 3rd Floor, Plot 3, Splendor Forum, Jasola District Centre, New Delhi – 110025, India

79 Anson Road, #06–04/06, Singapore 079906

Cambridge University Press is part of the University of Cambridge.

It furthers the University's mission by disseminating knowledge in the pursuit of education, learning, and research at the highest international levels of excellence.

www.cambridge.org
Information on this title: www.cambridge.org/9781107074262

DOI: 10.1017/9781139696609



First published 2017

Printed in the United Kingdom by Clays, St Ives plc

*A catalogue record for this publication is available from the British Library.*

*Library of Congress Cataloging-in-Publication Data*
Names: Tehan, Maureen, author.
Title: The impact of climate change mitigation on indigenous and forest communities : international, national and local law perspectives on REDD+ / Maureen Frances Tehan, Lee Carol Godden, Margaret Anne Young, Kirsty Ann Gover.
Description: Cambridge, United Kingdom ; New York, NY : Cambridge University Press, 2017. | Includes bibliographical references.
Identifiers: LCCN 2017023167 | ISBN 9781107074262 (hardback)
Subjects: LCSH: Reducing Emissions from Deforestation and Forest Degradation (Program) | Forestry law and legislation–Developing countries. | Greenhouse gas mitigation–Law and legislation–Developing countries. | Indigenous peoples–Legal status, laws, etc.–Developing countries. | Forest degradation–Control. | Greenhouse gas mitigation. | Indigenous peoples–Land tenure.
Classification: LCC K3884 .T44 2017 | DDC 305.8009172/4–dc23 LC record available at https://lccn.loc.gov/2017023167

ISBN 978-1-107-07426-2 Hardback

CONTENTS

# FIGURES

# ACKNOWLEDGEMENTS

This book is part of a broader Australian Research Council–funded project, 'Climate Change Law and Mitigation: Forest Carbon Sequestration and Indigenous and Local Communities' (DP110100259). The project involved a study of the global climate change mitigation mechanism of REDD+ ('reducing emissions from deforestation and forest degradation, and the role of conservation, sustainable management of forests and enhancement of forest carbon stocks in developing countries'), established by the United Nations Framework Convention on Climate Change (UNFCCC), the Paris Agreement, and related guidance and decisions. In the book, we have sought to analyse the intersections amongst REDD+, the cultural, economic, land and resource rights of indigenous and local (forest dwelling) communities and associated international law regimes.

Throughout the life of the project the chief investigators have been engaged in an ongoing scholarly exchange, each bringing their specific knowledge and research expertise to lively debates about the fundamental issues addressed in the project. The book is a result of this enterprise and it has extended this scholarly engagement and exchange in its intellectual framing and in the detail of its completion. The methodology used to write the book is a direct result of the general and specific critical engagement of each author in the research and writing of their co-authors. In the book we have each written in our selected area of expertise and should be considered the sole author of the chapters as described here, while recognising the inputs from the other authors and the research team across the project as a whole. In addition, each of the authors has benefitted from the critical commentary and editing of the other authors. Thus, the book is the result of the collective research, intellectual engagement and integration of the authors' specific expertise in relation to the operation of REDD+ and its implications for indigenous peoples and local communities. In the context of this methodology, Chapters 2, 3 and 10 are the work of Margaret Young; Chapters 4 and 5 are the work of Kirsty Gover; Chapters 6, 7 and 8 are the work of Lee Godden; and Chapter 9 is the work of Maureen Tehan. The introduction and conclusion reflect collective views and are jointly authored.

We gratefully acknowledge the Australian Research Council's support in allowing us to research and examine the many dimensions of climate change law as well as the specifics of REDD+ at international, national and local levels. We thank the Centre for Resources, Energy and Environmental Law and the Melbourne Law School for ongoing support of our project and the book. The Melbourne Law School Academic Research Service has been integral to the successful completion of the book, particularly manager Robin Gardner and researcher Cate Read.

We have benefitted greatly from the expertise and support of a range of people. We are extremely grateful to Professor Fadzilah Majid Cooke, who provided invaluable assistance with both fieldwork in Malaysia and in developing links with government and other agencies in Sabah. She gave us the benefit of her extensive knowledge of land tenure and human rights issues in forests, land claims and resource management in Malaysia. She provided guidance for the evolution of the Malaysia chapter, including comments on earlier drafts. Sam Johnson provided a wealth of knowledge and insight into REDD+, indigenous peoples and savanna-burning projects both internationally and locally, and Tim Clairs and Julian Fox were generous in sharing their expertise. We are immensely grateful to staff members at the United Nations Food and Agriculture Organization (FAO), United Nations Development Programme (UNDP), United Nations Environment Programme (UNEP), World Bank, government agencies, nongovernmental organizations (NGOs) and our academic colleagues across the globe whose input, through workshops, conferences and generally, is an integral part of this book.

The publication has benefitted from the invaluable work of a number of project managers and researchers over the life of the project. In particular, we owe enormous gratitude to our long-serving research fellow, Stephanie Niall, whose planning, management, enthusiasm and intellectual curiosity contributed both to the project and to the book. The project's first research fellow, Jessica Rae, contributed significantly to early discussions about the shape and content of the book and the early research. Carly Godden has contributed to the project as a project manager and researcher for this book. We thank her for her commitment to the project, her intellectual contribution to it, and her flexibility and responsiveness to our many requests for research assistance. Special thanks are due to Lauren Pavli for her hard work on our behalf in the last stages of the book's production. Other researchers and research assistants who have assisted us at various stages include: Tim Baxter, Dr Kathleen Birrell, Lisa Caripis, Dr Angus Hervey, Rachel MacLeod, Dr Elizabeth McPherson, Cobi Smith and Sean Mulcahy. Faye Yik-Wei provided invaluable transcription of interviews. We thank all of them for their particular contributions.

We also thank the editors at Cambridge University Press, particularly Elizabeth Spicer, for their patience and assistance with the publication of the book, and acknowledge the generosity of the reviewers of our publication proposal, whose insightful and helpful comments assisted greatly in refining the book's key themes. Judith Belzer produced the wonderful painting that is on the cover of the book; we are very grateful and thank her for allowing us to use her work. We note that some parts of Gover's and Young's contributions build on their respective chapters to the collection edited by Christina Voigt, *Research Handbook on REDD+ and International Law* (2016), and are reproduced here in adapted and extended form.

Thanks are due, as always, to our family and friends for their moral support and good humour. Finally, the authors share mutual thanks for the intellectual challenges and collegiality enjoyed throughout, in an ambitious project that brought together diverse perspectives to enrich the understanding of REDD+ and its ramifications for indigenous communities.

# ABBREVIATIONS

| | |
|---|---|
| CANZUS | Canada, Australia, New Zealand and the United States |
| CBD | Convention on Biological Diversity |
| CDM | Clean Development Mechanism |
| CEDAW | Convention on the Elimination of All Forms of Discrimination Against Women |
| CERD | Convention on the Elimination of All Forms of Racial Discrimination |
| CFM | community forest management |
| CIFOR | Center for International Forestry Research |
| CIFOR-GCS | CIFOR's Global Comparative Study on REDD+ |
| CITES | Convention on the International Trade in Endangered Species of Wild Flora and Fauna |
| COP | Conference of the Parties |
| FAO | Food and Agriculture Organization |
| FCPF | Forest Carbon Partnership Facility |
| FPIC | free, prior and informed consent |
| FSC | Forest Stewardship Council |
| GHG | greenhouse gas |
| HRC | Human Rights Council |
| ILO | International Labour Organization |
| MRV | measurement, reporting and verification |
| NCMC | National Carbon Monitoring Centre |
| NGO | nongovernmental organization |
| PDD | project design document |
| PES | payment for environmental (or ecosystem) services |
| PFM | participatory forest management |
| REDD+ | [policy approaches and positive incentives for] reducing emissions from deforestation and forest degradation, and the role of conservation, sustainable management of forests and enhancement of forest carbon stocks |
| R-PP | Readiness Preparation Proposal |
| SBSTA | Subsidiary Body for Scientific and Technological Advice |

| | |
|---|---|
| SEPC | Social and Environmental Principles and Criteria |
| SIS | Safeguards Information System |
| SFM | sustainable forest management |
| UNDP | United Nations Development Programme |
| UNEP | United Nations Environment Programme |
| UNFCCC | United Nations Framework Convention on Climate Change |
| UNDG | United Nations Development Group |
| UNDRIP | United Nations Declaration on the Rights of Indigenous Peoples |
| UN-REDD Programme | United Nations Collaborative Programme on Reducing Emissions from Deforestation and Forest Degradation in Developing Countries |

# CASES

*Adong bin Kuwau v Kerajaan Negeri Johor* [1997] 1 MLJ 418
*Akiba v Commonwealth* (2013) 250 CLR 209
*Akiba v Queensland [No 2]* (2010) 204 FCR 1
*Attorney-General v Brown* (1847) 1 Legge 312
*Commonwealth v Tasmania* (1983) 158 CLR 1
*Cooper v Stuart* (1889) 14 App Cas 286
*Delgamuukw v British Columbia* [1997] 3 SCR 1010
*Gaya Nomgui & Others v The Administration of the Territory of Papua New Guinea* [1974] PNGLR 349
*Gerhardy v Brown* (1985) 159 CLR 70
*Grand River Enterprises Six Nations Ltd v United States of America (Award)* (Arbitral Tribunal constituted under Chapter 11 of the North American Free Trade Agreement, 12 January 2011), www.state.gov/documents/organization/156820.pdf
*Guerin v The Queen* [1984] 2 SCR 335
*Harris v Great Barrier Reef Marine Park Authority* (2000) 98 FCR 60
*Ifira Trustees Ltd v Family Kalsakau* [2006] VUCA 23 (6 October 2006)
*Kerajaan Negeri Selangor v Sagong Tasi* [2005] 4 CLJ 169
*Koperasi Kijang Mas v Kerajaan Negeri Perak* [1991] CLJ 486
*Kuna Indigenous People of Madungandi and Embera Indigenous People of Bayano and Their Members v Panama*, IACHR Report No 58/09, Petition 12.354, Admissibility (21 April 2009)
*Kuna Indigenous People of Madungandi and Embera Indigenous People of Bayano and Their Members v Panama*, IACHR Report No 125/12, Merits (13 November 2012)
*Legality of the Use by a State of Nuclear Weapons in Armed Conflict* [1966] ICJ Reports 66
*Mabo v Queensland* (1988) 166 CLR 186
*Mabo v Queensland (No 2)* (1992) 175 CLR 1
*Manitoba Metis Federation v Canada (Attorney General)* [2013] 1 SCR 623
*Maya Indigenous Communities of the Toledo District v Belize*, IACHR Report No 40/04, OEA/Ser.L/V/II.122 Doc 5 Rev 1 at 727 (12 October 2004)
*Mayagna (Sumo) Awas Tingni Community v Nicaragua*, IACHR Series C No 79 (31 August 2001)

*McGlade v Native Title Registrar* [2017] FCAFC 10
*Members of the Yorta Yorta Aboriginal Community v Victoria* (2002) 214 CLR 422
*Milirrpum v Nabalco Pty Ltd* [1972–1973] ALR 65
*Moiwana Communities v Suriname*, IACHR Series C No 124 (15 June 2005)
*New Zealand Maori Council v Attorney-General* [1987] 1 NZLR 641 (HC & CA)
*Nor Anak Nyawai v Borneo Pulp Plantations* [2001] 2 CLJ 769
*Pulp Mills on the River Uruguay (Argentina v Uruguay)* (2010) 2010 ICJ Reports 14
*Reparations for Injuries Suffered in the Service of the United Nations* [1949] ICJ Reports 174
*Roqara v Takau* [2005] VUCA 5 (3 May 2005)
*Saramaka People v Suriname*, IACHR Series C No 172 (28 November 2007)
*Sawhoyamaxa Indigenous Community v Paraguay*, IACHR Series C No 146 (29 March 2006)
*Solomon v Turquoise* [2008] VUSC 64 (8 August 2008)
*Superintendent of Land & Surveys, Bintulu v Nor Anak Nyawai* [2006] 1 MLR 256
*Te Runanga o Wharekauri Rekohu Inc v Attorney-General* [1993] 2 NZLR 301
*Trail Smelter Arbitration (United States v Canada)* (1941) 3 UN Rep Int'l Arb Awards 1905
*Tsilhqot'in Nation v British Columbia*, 2014 SCC 44
*United States Diplomatic and Consular Staff in Tehran (United States of America v Iran)* [1980] ICJ Rep 3
*Valele Family v Toura* [2002] VUCA 3 (26 April 2002)
*Western Australia v BP (deceased)* (2014) 223 FCR 488
*Western Australia v Ward* (2002) 213 CLR 1
*Whaling in the Antarctic (Australia v Japan)* ICJ GL No 148 (31 March 2014)
*Willis on behalf of the Pilki People v State of Western Australia (No 1)* [2014] FCA 714 (4 July 2014)
*Willis on behalf of the Pilki People v State of Western Australia (No 2)* [2014] FCA 1293 (2 December 2014)
*Yakye Axa v Paraguay*, IACHR Series C No 125 (17 June 2005)

# CONVENTIONS, DECLARATIONS AND DECISIONS OF CONFERENCES OF THE PARTIES

*Articles of Agreement of the International Monetary Fund*, adopted 22 July 1944, 2 UNTS 39 (entered into force 27 December 1945)

*Convention (No 169) Concerning Indigenous and Tribal Peoples in Independent Countries*, opened for signature 27 June 1989, 1650 UNTS 383 (entered into force 5 September 1991)

*Convention on Biological Diversity*, opened for signature 5 June 1992, 1760 UNTS 79 (entered into force 29 December 1993)

*Convention on the Elimination of All Forms of Discrimination against Women*, opened for signature 18 December 1979, 1249 UNTS 13 (entered into force 3 September 1981)

*Convention on the Rights of the Child*, opened for signature 20 November 1989, 1577 UNTS 3 (entered into force 2 September 1990)

*International Convention on the Elimination of All Forms of Racial Discrimination*, opened for signature 7 March 1966, 660 UNTS 195 (entered into force 4 January 1969)

*International Covenant on Civil and Political Rights*, opened for signature 19 December 1966, 999 UNTS 171 (entered into force 23 March 1976)

*International Covenant on Economic, Social and Cultural Rights*, opened for signature 19 December 1966, 993 UNTS 3 (entered into force 3 January 1976)

International Labour Organization, *Convention Concerning Indigenous and Tribal Peoples in Independent Countries*, opened for signature 27 June 1989, 1650 UNTS 383 (entered into force 5 September 1991)

*International Tropical Timber Agreement*, opened for signature 18 November 1983, 1393 UNTS 119 (provisionally entered into force 1 April 1985)

*International Tropical Timber Agreement*, opened for signature 1 April 1994, 33 ILM 1014 (entered into force 1 January 1997)

*Nagoya Protocol on Access to Genetic Resources and the Fair and Equitable Sharing of Benefits Arising from Their Utilization to the Convention on Biological Diversity*, opened for signature 2 February 2011, UN Doc UNEP/CBD/COP/DEC/X/1, annex I (entered into force 12 October 2014)

*United Nations Convention to Combat Desertification in Those Countries Experiencing Serious Drought and/or Desertification, Particularly in Africa*, opened for signature 14 October 1994, 1954 UNTS 3 (entered into force 26 December 1996)

*United Nations Declaration on the Rights of Indigenous Peoples*, GA Res 61/295, UN GAOR, 61st sess, 107th plen mtg, Agenda item 68, Supp No 49, UN Doc A/RES/61/295 (2 October 2007, adopted 13 September 2007)

*United Nations Framework Convention on Climate Change*, opened for signature 4 June 1992, 1771 UNTS 107 (entered into force 21 March 1994)

*Vienna Convention on the Law of Treaties*, opened for signature 23 May 1969, 1155 UNTS 331 (entered into force 27 January 1980)

*2200 (XXI). International Covenant on Economic, Social and Cultural Rights, International Covenant on Civil and Political Rights and Optional Protocol to the International Covenant on Civil and Political Rights*, UN GAOR, 21st sess, 1496th plen mtg, UN Doc A/RES/21/2200 (16 December 1966)

*63/278. International Mother Earth Day*, GA Res 63/278, UN GAOR, 63rd sess, 80th plen mtg, Agenda Item 49(d), UN Doc A/RES/63/278 (1 May 2009)

Ad Hoc Working Group on Long-Term Cooperative Action under the Convention, UNFCCC, *Compilation of Information on Nationally Appropriate Mitigation Actions to Be Implemented by Parties Not Included in Annex I to the Convention*, UN Doc FCCC/AWGLCA/2011/INF.1 (18 March 2011)

Commission on Human Rights, Draft Declaration on the Rights of Indigenous Peoples, UN Doc E/CN.4/Sub.2/1994/2/Add.1 (20 April 1994)

Committee on the Elimination of Racial Discrimination (CERD), *General Recommendation VIII Concerning the Interpretation and Application of Article 1, Paragraphs 1 and 4 of the Convention*, 38th sess, 884th mtg (21 August 1990) contained in *Report of the Committee on the Elimination of Racial Discrimination*, UN GAOR, 45th sess, Supp No 18 (A/45/18) (30 January 1991)

*Compilation of General Comments and General Recommendations Adopted by Human Rights Treaty Bodies*, UN Doc HRI/GEN/1/Rev.9 (vol. II) (27 May 2008)

Conference of the Parties, Convention on Biological Diversity, *XI/19. Biodiversity and Climate Change Related Issues: Advice on the Application of Relevant Safeguards for Biodiversity with Regard to Policy Approaches and Positive Incentives on Issues Relating to Reducing Emissions from Deforestation and Forest Degradation in Developing Countries; and the Role of Conservation, Sustainable Management of Forests and Enhancement of Forest Carbon Stocks in Developing Countries*, UN Doc UNEP/CBD/COP/DEC/XI/19 (5 December 2012)

Conference of the Parties, Convention on Biological Diversity, *X/33. Biodiversity and Climate Change*, UN Doc UNEP/CBC/COP/DEC/X/33 (29 October 2010)

Conference of the Parties, Convention on International Trade in Endangered Species of Wild Fauna and Flora (CITES), *Cooperation and Synergy with the Convention on Biological Diversity*, Res Conf 10.4 (Rev. CoP14) (June 1997)

Conference of the Parties, Convention on International Trade in Endangered Species of Wild Fauna and Flora (CITES), 'Cooperation with Other Organizations' (Seventeenth Meeting of the Conference of the Parties, Johannesburg, South Africa, 24 September–5 October 2016)

Conference of the Parties Serving as the Meeting of the Parties to the Kyoto Protocol, United Nations Framework Convention on Climate Change, *Report of the Conference of the Parties Serving as the Meeting of the Parties to the Kyoto Protocol on Its Eighth Session, Held in Doha from 26 November to 8 December 2012 – Addendum – Part Two: Action Taken by the Conference of the Parties Serving as the Meeting of the Parties to the Kyoto Protocol at Its Eighth Session*, UN Doc FCCC/KP/CMP/2012/13/Add.1 (28 February 2013) Decision 1/CMP.8 ('Amendment to the Kyoto Protocol Pursuant to Its Article 3, Paragraph 9 (the Doha Amendment)') (not yet in force)

Conference of the Parties, United Nations Framework Convention on Climate Change, *Adoption of the Paris Agreement – Proposal by the President – Draft Decision -/CP.21*, UN Doc FCCC/CP/2015/L.9 (12 December 2015)

Conference of the Parties, United Nations Framework Convention on Climate Change, *Aggregate Effect of the Intended Nationally Determined Contributions: An Update – Synthesis Report by the Secretariat*, UN Doc FCCC/CP/2016/2 (2 May 2016)

Conference of the Parties, United Nations Framework Convention on Climate Change, *Organizational Matters: Adoption of the Rules of Procedure – Note by the Secretariat*, UN Doc FCCC/CP/1996/2 (22 May 1996) ('Draft Rules of Procedure of the Conference of the Parties and Its Subsidiary Bodies')

Conference of the Parties, United Nations Framework Convention on Climate Change, *Reducing Emissions from Deforestation in Developing Countries: Approaches to Stimulate Action – Submission by the Governments of Papua New Guinea and Costa Rica*, UN Doc FCCC/CP/2005/Misc.1 (11 November 2005) (Submission of Papua New Guinea and Costa Rica)

Conference of the Parties, United Nations Framework Convention on Climate Change, *Report of the Conference of the Parties on Its Thirteenth Session, Held in Bali from 3 to 15 December 2007 – Addendum – Part Two: Action Taken by the Conference of the Parties at Its Thirteenth Session*, UN Doc FCCC/CP/2007/6/Add.1 (14 March 2008) Decision 1/CP.13 ('Bali Action Plan')

Conference of the Parties, United Nations Framework Convention on Climate Change, *Report of the Conference of the Parties on Its Fifteenth Session, Held in Copenhagen from 7 to 19 December 2009 – Addendum – Part Two: Action Taken by the Conference of the Parties at Its Fifteenth Session*, UN Doc FCCC/CP/2009/11/Add.1 (30 March 2010) Decision 2/CP.15 ('Copenhagen Accord')

Conference of the Parties, United Nations Framework Convention on Climate Change, *Report of the Conference of the Parties on Its Fifteenth Session, Held in Copenhagen from 7 to 19 December 2009 – Addendum – Part Two: Action Taken by the Conference of the Parties at Its Fifteenth Session*, UN Doc FCCC/CP/2009/11/Add.1 (30 March 2010) Decision 4/CP.15 ('Methodological Guidance for Activities Relating to Reducing Emissions from Deforestation and Forest Degradation and the Role of Conservation, Sustainable Management of Forests and Enhancement of Forest Carbon Stocks in Developing Countries')

Conference of the Parties, United Nations Framework Convention on Climate Change, *Report of the Conference of the Parties on Its Sixteenth Session, Held in Cancun from 29 November to 10 December 2010 – Addendum – Part Two: Action Taken by the Conference of the Parties at Its Sixteenth Session*, UN Doc FCCC/CP/2010/7/Add.1 (15 March 2011) Decision 1/CP.16 ('The Cancun Agreements: Outcome of the Work of the Ad Hoc Working Group on Long-Term Cooperative Action under the Convention')

Conference of the Parties, United Nations Framework Convention on Climate Change, *Report of the Conference of the Parties on Its Seventeenth Session, Held in Durban from 28 November to 11 December 2011 – Addendum – Part Two: Action Taken by the Conference of the Parties at Its Seventeenth Session*, UN Doc FCCC/CP/2011/9/Add.1 (15 March 2012) Decision 3/CP.17 ('Launching the Green Climate Fund'), Annex ('Governing Instrument for the Green Climate Fund')

Conference of the Parties, United Nations Framework Convention on Climate Change, *Report of the Conference of the Parties on Its Seventeenth Session, Held in Durban from 28 November to 11 December 2011 – Addendum – Part Two: Action Taken by the Conference of the Parties at Its Seventeenth Session*, UN Doc FCCC/CP/2011/9/Add.2 (15 March 2012) Decision 12/CP.17 ('Guidance on Systems for Providing Information on How Safeguards Are Addressed and Respected and Modalities Relating to Forest Reference Emission Levels and Forest Reference Levels as Referred to in Decision 1/CP.16')

Conference of the Parties, United Nations Framework Convention on Climate Change, *Report of the Conference of the Parties on Its Nineteenth Session, Held in Warsaw from 11 to 23 November 2013 – Addendum – Part Two: Action Taken by the Conference of the Parties at Its Nineteenth Session*, UN Doc FCCC/CP/2013/10/Add.1 (31 January 2014) Decisions 9–15/CP.19 ('Warsaw Framework for REDD-Plus')

Conference of the Parties, United Nations Framework Convention on Climate Change, *Report of the Conference of the Parties on Its Nineteenth Session, Held in Warsaw from 11 to 23 November 2013 – Addendum – Part Two: Action Taken by the Conference of the Parties at Its Nineteenth Session*, UN Doc FCCC/CP/2013/10/Add.1 (31 January 2014) Decision 9/CP.19 ('Work Programme on Results-Based Finance to Progress the Full Implementation of the Activities Referred to in Decision 1/CP.16, paragraph 70')

Conference of the Parties, United Nations Framework Convention on Climate Change, *Report of the Conference of the Parties on Its Nineteenth Session, Held in Warsaw from 11 to 23 November 2013 – Addendum – Part Two: Action Taken by the Conference of the Parties at Its Nineteenth Session*, UN Doc FCCC/CP/2013/10/Add.1 (31 January 2014) Decision 10/CP.19 ('Coordination of Support for the Implementation of Activities in Relation to Mitigation Actions in the Forest Sector by Developing Countries, including Institutional Arrangements')

Conference of the Parties, United Nations Framework Convention on Climate Change, *Report of the Conference of the Parties on Its Nineteenth Session, Held in Warsaw from 11 to 23 November 2013 – Addendum – Part Two: Action Taken by the*

*Conference of the Parties at Its Nineteenth Session*, UN Doc FCCC/CP/2013/10/Add.1 (31 January 2014) Decision 12/CP.19 ('The Timing and the Frequency of Presentations of the Summary of Information on How All the Safeguards Referred to in Decision 1/CP.16, Appendix I, Are Being Addressed and Respected')

Conference of the Parties, United Nations Framework Convention on Climate Change, *Report of the Conference of the Parties on Its Nineteenth Session, Held in Warsaw from 11 to 23 November 2013 – Addendum – Part Two: Action Taken by the Conference of the Parties at Its Nineteenth Session*, UN Doc FCCC/CP/2013/10/Add.1 (31 January 2014) Decision 13/CP.19 ('Guidelines and Procedures for the Technical Assessment of Submissions from Parties on Proposed Forest Reference Emission Levels and/or Forest Reference Levels')

Conference of the Parties, United Nations Framework Convention on Climate Change, *Report of the Conference of the Parties on Its Nineteenth Session, Held in Warsaw from 11 to 23 November 2013 – Addendum – Part Two: Action Taken by the Conference of the Parties at Its Nineteenth Session*, UN Doc FCCC/CP/2013/10/Add.1 (31 January 2014) Decision 15/CP.19 ('Addressing the Drivers of Deforestation and Forest Degradation')

Conference of the Parties, United Nations Framework Convention on Climate Change, *Report of the Conference of the Parties on Its Twenty-First Session, Held in Paris from 30 November to 13 December 2015 – Addendum – Part Two: Action Taken by the Conference of the Parties at Its Twenty-First Session*, UN Doc FCCC/CP/2015/10/Add.1 (29 January 2016) Decision 1/CP.21, annex ('Paris Agreement')

Conference of the Parties, United Nations Framework Convention on Climate Change, *Report of the Conference of the Parties on Its Twenty-First Session, Held in Paris from 30 November to 13 December 2015 – Addendum – Part Two: Action Taken by the Conference of the Parties at Its Twenty-First Session*, UN Doc FCCC/CP/2015/10/Add.3 (29 January 2016) Decision 18/CP.21 ('Methodological Issues Related to Non-Carbon Benefits Resulting from the Implementation of the Activities Referred to in Decision 1/CP.16, paragraph 70')

Inter-American Commission on Human Rights (IACHR), *American Declaration on the Rights of Indigenous Peoples*, AG/RES 2888 (XLVI–O/16), 46th reg sess, 3rd plen sess (adopted 15 June 2016)

International Law Commission, *Fragmentation of International Law: Difficulties Arising from the Diversification and Expansion of International Law*, UN GAOR, 58th sess, UN Doc A/CN.4/L.682 (13 April 2006), UN Doc A/CN.4/L.682/Add.1 (2 May 2006), and UN Doc A/CN.4/L.682/Corr.1 (11 August 2006) ('ILC Fragmentation Analytical Study')

International Law Commission, *Responsibility of International Organizations*, UN Doc A/CN.4/L.778 (30 May 2011)

International Law Commission, *Report of the International Law Commission on the Work of Its Sixty-Sixth Session*, UN GAOR, 69th sess, Supp No 10, UN Doc A/69/10 (5 May–6 June and 7 July–8 August 2014)

*Kyoto Protocol to the United Nations Framework Convention on Climate Change*, opened for signature 16 March 1998, 2303 UNTS 148 (entered into force 16 February 2005)

*Non-Legally Binding Instrument on All Types of Forests*, GA Res 62/98, UN GAOR, 62nd sess, 74th plen mtg, UN Doc A/RES/62/98 (31 January 2008)

*Permanent Sovereignty over Natural Resources*, GA Res 1803 (XVII), UN GAOR, 17th sess, 1194th plen mtg, UN Doc A/RES/1803 (XVII) (14 December 1962)

*Permanent Sovereignty over Natural Resources*, GA Res 3171 (XXVIII), UN GAOR, 28th sess, 2203rd plen mtg, UN Doc A/RES/3171 (XXVIII) (17 December 1973)

*Programme of Action for the Second International Decade of the World's Indigenous People*, UN GAOR, 60th sess, Agenda Item 68, UN Doc A/RES/60/142 (7 February 2006)

*Report of the Committee on the Elimination of Racial Discrimination*, UN GAOR, 52nd sess, Supp No 18, UN Doc A/52/18 (26 September 1997) annex V ('*General Recommendation on the Rights of Indigenous Peoples*') ('General Recommendation 23')

*Report on the Fourth Session of the Intergovernmental Forum on Forests*, UN ESCOR, 46th mtg, UN Doc 2000/35 (18 October 2000)

*Report of the United Nations Conference on Environment and Development*, UN Doc A/CONF.151/26 (vol. I) (12 August 1992) annex I ('Rio Declaration on Environment and Development')

*Report of the United Nations Conference on Environment and Development*, UN Doc A/CONF.151/26 (vol. III) (14 August 1992) annex III ('Non-Legally Binding Authoritative Statement of Principles for a Global Consensus on the Management, Conservation and Sustainable Development of All Types of Forests')

*Report of the United Nations Conference on Environment and Development*, UN Doc A/CONF.151/26 (vol. IV) (28 September 1992)

*Transforming Our World: The 2030 Agenda for Sustainable Development*, GA Res 70/1, UN GAOR, 70th sess, UN Doc A/RES/70/1 (21 October 2015)

# LEGISLATION

Aboriginal Land Rights (Northern Territory) Act 1976 (Cth)
Aboriginal Peoples Act 1954 (Revised 1974) (Malaysia)
Aboriginal Land Rights Act 1983 (NSW)
Aboriginal Land Act 1991 (Qld)
Anangu Pitjantjatjara Yankunytjatjara Land Rights Act 1981 (SA)
Canada Act 1982 (UK) c 11, Sch B (Constitution Act 1982) (Canada)
Canada Act 1982 (UK) c 11, Sch B Pt I (Canadian Charter of Rights and Freedoms) (Canada)
Carbon Credits (Carbon Farming Initiative) Act 2011 (Cth)
Carbon Credits (Carbon Farming Initiative – Emissions Abatement through Savanna Fire Management) Methodology Determination 2015 (Cth)
Carbon Farming Initiative Amendment Act 2014 (Cth)
Carbon Rights Act 2003 (WA)
Clean Energy Act 2011 (Cth)
Clean Energy Legislation (Carbon Tax Repeal) Act 2014 (Cth)
Climate Change Act 2010 (Vic)
Climate Change Act 2015 (Vic)
Climate Change Authority Act 2011 (Cth)
Climate Change (Management) Act 2015 (PNG)
Commonwealth of Australia Constitution Act 1900 (Imp) 63 & 64 Vict, c 12
Conservation Forests and Lands Act 1987 (Vic)
Constitution of the Independent State of Papua New Guinea 1975
Constitution of the Republic of Vanuatu 1980
Conveyancing Act 1919 (NSW)
Corporations (Aboriginal and Torres Strait Islander) Act 2006 (Cth)
Custom Land Management Act 2013 (Vanuatu)
Environmental Management and Conservation Act 2002 (Cap 283) (Vanuatu)
Explanatory Memorandum, Carbon Farming Initiative Amendment Bill 2014 (Cth)
Federal Constitution (Malaysia)
Forest Property Act 2000 (SA)
Forestry Act 1959 (Qld)
Forestry Act 1991 (PNG)

Forestry Act (Cap 276) 2003 (Vanuatu)
Forestry Rights Registration Act 1990 (Tas)
Forestry Rights Registration and Timber Harvest Guarantee Act (Cap 265) 2000 (Vanuatu)
Human Rights Act 1993 (NZ)
Land Act 1994 (Qld)
Land Act 1996 (PNG)
Land Administration Act 1997 (WA)
Land Code 1958 (Sarawak) (Malaysia)
Land Disputes Settlement Act 1975 (PNG)
Land Fund and Indigenous Land Corporation (ATSIC Amendment) Act 1995 (Cth)
Land Groups Incorporation Act 1974 (PNG)
Land Groups Incorporation (Amendment) Act 2009 (PNG)
Land Leases Act 1983 (Cap 163) (Vanuatu)
Land Leases Act 1984 (Vanuatu)
Land Ordinance 1930 (Revised 1996) (Sabah) (Malaysia)
Land Reform Act 1980 (Vanuatu)
Land Reform Regulation 1980 (Vanuatu)
Land Registration (Amendment) Act 2009 (PNG)
Land Registration (Customary Land) (Amendment) Act 2007 (PNG)
Land Title Act 1994 (Qld)
Maralinga-Tjarutja Land Rights Act 1984 (SA)
National Land Code 1965 (Malaysia)
Native Court Ordinance 1993 (Malaysia)
Native Title Act 1993 (Cth)
Native Title (Prescribed Bodies Corporate) Regulations 1999 (Cth)
New Zealand Bill of Rights Act 1990 (NZ)
Pitjantjatjara Land Rights Act 1981 (renamed Anangu Pitjantjatjara Yankunytjatjara Land Rights Act 1981) (SA)
Racial Discrimination Act 1975 (Cth)
Replacement Explanatory Memorandum, Carbon Credits (Carbon Farming Initiative) Bill 2011 (Cth)
Sabah Forest Enactment 1968 (Malaysia)
Sarawak Forest Ordinance 1954 (Malaysia)
Torres Strait Islander Land Act 1991 (Qld)
Traditional Owner Settlement Act 2010 (Vic)
Transfer of Land Act 1893 (WA)
Underlying Law Act 2000 (PNG)

# 1

# Introduction

The reduction of greenhouse gas emissions is an internationally agreed objective that has become increasingly attentive to the volume of carbon stored in forests.[1] One of the most ambitious and controversial climate change mitigation mechanisms deployed in the service of this goal is REDD+, a global programme designed to reduce emissions from deforestation and forest degradation (REDD) by rewarding the protection of forests in developing countries. States party to the United Nations Framework Convention on Climate Change (UNFCCC) and the Paris Agreement commit to supporting policy approaches and positive incentives for activities relating to REDD, including 'conservation, sustainable management of forests and enhancement of forest carbon stocks' (these additional words are encompassed by the plus sign in the acronym REDD+).[2] REDD+ envisages the creation of a financial value for the carbon stored in forests, encouraging developing countries to reduce emissions from forested lands in return for the provision of financial incentives, including 'results-based payments'.[3] Developed countries are called upon to provide financial support[4] to enable such payments and to fund the international organizations and programmes tasked with providing advice and resources to recipient states. Recipient countries agree in turn to constrain the use of forests in their territory so as to reduce rates of deforestation and degradation, which in the context of 'drivers' of deforestation may involve limiting commercial logging, agricultural expansion or other activities.

[1] A sizable proportion of greenhouse gas emissions originates from forest-related activities: K L Denman et al, 'Couplings between Changes in the Climate System and Biogeochemistry' in S Solomon et al (eds), *Climate Change 2007: The Physical Science Basis. Contribution of the Working Group I to the Fourth Assessment Report of the Intergovernmental Panel on Climate Change* (Cambridge University Press, 2007) 499; IPCC, *Climate Change 2013: The Physical Science Basis. Contribution of Working Group I to the Fifth Assessment Report of the Intergovernmental Panel on Climate Change* (Cambridge University Press, 2013).

[2] *Paris Agreement*, Article 5(2). [3] Decision 1/CP.16, para 76. [4] Ibid.

While in its initial conception REDD+ may have seemed a simple and elegant tool to support climate change mitigation efforts,[5] the concept engages highly complex and deeply problematic questions of law and justice. This is especially the case because the rights, property and jurisdictions of indigenous peoples and forest-dependent communities are often directly affected by REDD+ planning and projects. Decisions made in anticipation of REDD+ funding touch on matters very close to the core of national sovereignty and constitutionalism, including the distribution of property rights, the role of the state in regulating markets, and the status and rights of colonised or marginalised peoples. Constitutional arrangements governing property, human rights and indigenous peoples differ widely from country to country and responsibilities for law and policy in these areas is differentially allocated amongst national, regional and local governments. States with tribal and indigenous populations also accommodate, to varying degrees, local customary legal systems and property rights derived from those customary laws. These legal systems have a complicated relationship with state law. As we explain in this book, the claims of indigenous and tribal peoples to property rights sourced in customary law are a major source of tension in many nations that host or plan to host REDD+ programmes. In many instances, indigenous legal systems and embedded rights in respect of forest areas pre-exist the establishment of current nation states and the state's allocation of legal tenure and resource extraction rights to public and private actors. In many nations, the rights of forest-dwelling communities are only partially recognised by state law, if at all, and even where recognised, indigenous tenures are typically overridden by grants of interests to third parties.

A further layer of legal pluralism is represented by the international legal system, which contains a large number of treaties and norms of customary international law relating to forests, biodiversity protection, human rights, the rights of indigenous peoples and food security.[6] These are often grouped within established international legal regimes and supported by discrete intergovernmental organizations and epistemic communities. For the emerging REDD+ regime, questions about whether new rules and practices should integrate or diverge from existing international regimes are unavoidable and promise to impact greatly on indigenous peoples and forest-dwelling communities. In this context, this book presents an analysis

[5] On the influence of 'payment for ecosystems services' models, see Chapter 2.
[6] See further in Chapter 3.

of the circumstances, laws and state-indigenous relationships at play in the emerging international legal regime underpinning REDD+. It considers the institutional dimensions of fragmented laws and arrangements that depend heavily on the policies of the World Bank's Forest Carbon Partnership Facility (FCPF) and the United Nations Collaborative Programme on Reducing Emissions from Deforestation and Forest Degradation in Developing Countries (UN-REDD Programme). In their policy guidance and evaluative standards, these bodies have made significant interventions into legal arrangements laid out by REDD+, by, for instance, promulgating rules dealing with human rights, indigenous claims, free prior and informed consent, and tenure. The book outlines these interventions, analyses their content, and critically examines divergences amongst and within them. At the same time, it is important to keep in mind that the evolution of REDD+ at the international level draws on local activities and projects and is informed by climate change and resource management innovations in the domestic law and policy of states. The relationships between local circumstances and international norms that are structured by REDD+ are a central concern of the authors. The book examines four national settings – Malaysia, Papua New Guinea, Vanuatu and Australia – to illustrate the complexity and diversity of the local situations in which REDD+ or equivalent models will operate. The countries chosen as case studies represent regional Southeast Asian–Oceania nations across the development spectrum, from small nation states (Papua New Guinea and Vanuatu), to a medium-level power with large forest reserves (Malaysia), to a developed country with strong dependency on the land-use sector (Australia). As a group, these nations have not previously been intensively studied in terms of their participation in the expanding REDD+ regime. Moreover, each of these states manages the intersection of human rights, property, and indigenous or forest community rights differently, and each of the developing states considered in this book has a different local response to REDD+ and its international institutions. This diversity poses a serious challenge for the international agencies furthering the REDD+ programme, which are responsible for distributing funding to participating states and ensuring that their national laws and policies meet REDD+'s international standards and safeguards. Our choice of study highlights the need to understand local and national contexts as informing and modifying the articulation of the international REDD+ regime.

Drawing on national and local contextual information and at the urging of civil society groups and indigenous and tribal peoples, the UNFCCC has developed a range of safeguards designed to protect vulnerable communities from the possibly negative impacts of REDD+.[7] As REDD+ has matured, and further information on the progress of its planning at the state level has become available, international bodies such as the UN-REDD Programme and World Bank have become more attentive to the complexity of local circumstances affecting REDD+ operations. This awareness has extended to an appreciation of the difficulty of generating uniform standards at the international level with which to evaluate state performance and condition REDD+ funding. The UNFCCC's safeguards and the derivative standards articulated by the UN-REDD Programme and the FCPF to date have been largely procedural in character. States are invited to have regard to a wide range of rights, interests, international legal instruments and international policies, and to ensure all stakeholders are consulted about proposed reforms and projects. REDD+ standards addressing the distribution of resource rights in land and carbon are far less developed and much more tentative in form. For example, work on the rights of indigenous peoples to free prior and informed consent (FPIC) is well developed in the REDD+ context, but policy on tenure reform is at a very nascent stage, even while both the UN-REDD Programme and the FCPF are of the view that 'secure and clear' tenure is a necessary determinant of REDD+ project success.[8] Further, the UNFCCC safeguards themselves do not directly address state obligations to recognise or protect property rights that might be affected by REDD+ implementation, nor do they require states to address tenure arrangements as a condition of REDD+ funding.[9] In short, the UNFCCC and REDD+ policy bodies have to date been reluctant to propose that a particular set of resource rights and entitlements should attend REDD+ planning, let alone suggest that these can be implemented in a substantially uniform way across all participating REDD+ states. Notwithstanding this reluctance, policy decisions and legal reforms undertaken by states in the early stages of REDD+ Readiness activities have profound impacts on the rights and interests of indigenous peoples and local forest communities, and on the capacity of vulnerable groups to benefit from

[7] Conference of the Parties, United Nations Framework Convention on Climate Change, *Report of the Conference of the Parties on Its Sixteenth Session, Held in Cancun from 29 November to 10 December 2010 – Addendum – Part Two: Action Taken by the Conference of the Parties at Its Sixteenth Session* (UN Doc FCCC/CP/2010/7/Add.1) (15 March 2011) Decision 1/CP.16, Annex; see further in Chapter 2.

[8] See further in Chapters 4 and 5. [9] Ibid.

REDD+ projects once implemented.[10] There is a risk that policies set in train in the REDD Readiness phase, under the guidance of the UNFCCC, the UN-REDD Programme and the FCPF, could further entrench unjust allocations of power and property in states with tribal and indigenous populations. It is this conundrum that animates the analysis we undertake in this book.

As REDD+ has evolved, participating states have become obliged not only to reduce emissions by preventing deforestation and forest degradation but also to ensure that emissions reductions are achieved with due respect for social and economic safeguards and are accompanied by tangible and equitable benefits for affected communities. As the range of safeguards and benefits incorporated within REDD+ grows, states are also required to reconcile conflicts between these goals and standards and to prioritise some over others. In the REDD+ context, as in many others, indigenous peoples and forest-dwelling communities are especially vulnerable to having their interests overridden by those of the state, the public and third-party stakeholders. REDD+ poses a particular challenge in this respect, as most of the targeted forests are inhabited or used by indigenous peoples and forest-dependent communities. On the one hand, the projects supported by REDD+ could provide indigenous and forest-dependent communities with financial, social and environmental benefits. On the other hand, the same projects and incentives could support measures that restrict or deny communities access to and use of forested lands on which they depend for the continuity of their livelihoods, cultural practices and connections to traditional land. For these vulnerable, and socially and economically disadvantaged communities, then, the opportunities offered by REDD+ are attended by significant risks. The interplay of risks and benefits, and the trade-offs between the two, require difficult decisions about how best to reconcile competing interests affected by REDD+, both in the short and in the long term.[11] The variables in play depend largely on law, not only at the national level but also at the international and local ones.

While participating states have agreed to adhere to the UNFCCC safeguards, including those designed to protect indigenous peoples, outcomes for indigenous and forest-dependent communities rely in

[10] See Chapter 6.

[11] UN ECOSOC Permanent Forum on Indigenous Issues, *Indigenous People's Rights and Safeguards in Projects Related to Reducing Emissions from Deforestation and Forest Degradation*, UN Doc E/C.19/2013/7 (5 February 2013).

large part on the interrelationship of human rights and property norms and their application to local circumstances. These norms are variously sourced in international, national, subnational and customary legal systems. The climate change mitigation goals and legal mechanisms that animate REDD+, and the safeguards promulgated by the UNFCCC, operate on and amongst these complex relations and conflict with some of their underlying justifications and assumptions. This book provides a study of REDD+ as an international legal regime alongside the cultural, economic, land, and resource rights of indigenous and local forest-dwelling communities as well as other international and domestic legal regimes relating to forests, the protection of the environment, indigenous rights, property rights and human rights. We address the difficult question of how best to position REDD+ amongst a highly fragmented, contested and crowded area of forest governance and carbon markets, bringing together legal analyses of international, national and local contexts.

For the time being, the body of norms and laws comprising the REDD+ regime is contained in the standards advanced by the UNFCCC and the two major international funding bodies: the UN-REDD Programme; which is a collaborative partnership of the Food and Agriculture Organization (FAO); the UN Development Programme (UNDP) and the UN Environment Programme (UNEP); and the FCPF.[12] The standards applied by the UN-REDD Programme and the FCPF are the major interface between the REDD+ regime and domestic laws governing indigenous rights.[13] By examining the anatomy of REDD+ as an international regime, we consider the ways in which the rights and interests of indigenous and forest-dependent communities are incorporated into REDD+ from external sources and how they are thereafter articulated within it. The book outlines the multiple and contrasting sources of obligation for REDD+ participant states, investigates their resonance with extant human rights standards (such as an obligation on the state to obtain the free prior and informed consent of affected indigenous communities before proceeding with proposed projects) and addresses their internal points of tension. We explore how the global content of REDD+ might shape, and be shaped by, the efforts of states

[12] Aside from the UN-REDD Programme and the FCPF, other international organizations are involved in REDD+ funding; see further in Chapter 2.

[13] These frameworks interface with the active carbon and emissions-offsets markets operating in the voluntary sector, which are not a focus of this book; see further in Chapter 2.

and indigenous peoples to secure indigenous interests in domestic and customary law. We take as cases in point efforts to manage forests, reduce carbon emissions, and accommodate indigenous rights to forests in Papua New Guinea, Vanuatu, Malaysia and Australia.

The complex interrelationship of legal regimes implicated by REDD+, and the influence of these dynamics on the ways REDD+ schemes affect indigenous and local communities, has only recently become the object of sustained scholarly analysis.[14] This book adds to an emerging field of study by examining the challenges posed by the legal complexity of REDD+'s context, including the varied capacities and political will of states involved in REDD+, and assessing the impact of this complexity on the rights and interests of indigenous and forest-dwelling communities affected by REDD+ implementation, particularly their resource and land rights. In this book, we engage important aspects of REDD+ that have not been the major orientation of other studies. Our analysis proceeds by focussing attention upon the issues raised by the need to address: indigenous peoples and local community claims to land and resources in forest areas; FPIC principles; tenure arrangements under REDD+; legal definitions of indigeneity and customary ownership; the intersection of state and customary legal systems; the difficulty of equitable benefit-sharing within and amongst local communities; and the tensions between rights held by indigenous peoples as historic collectives and the human rights of indigenous and non-indigenous individuals. All of these potential points of conflict are implicated by REDD+, but in this area, as in others, the capacity of international organizations and international law to propose ways to resolve conflicts between rights holders is severely limited. Advice so far from the UNFCCC, the UN-REDD Programme and the FCPF emphasises the importance of the matters protected by safeguards and advanced by co-benefits, but contains little in the way of a methodology to rank or prioritise interests where they conflict with one another. Given this complexity and diversity (and the inability for

[14] See Christine Voigt (ed), *Research Handbook on REDD+ and International Law* (Edward Elgar, 2016); Sébastian Jodoin, *Forest Preservation in a Changing Climate: REDD+ and Indigenous and Community Rights in Indonesia and Tanzania* (Cambridge University Press, 2017); Julia Dehm, *Reconsidering REDD+: Authority and Law in the Green Economy* (Cambridge University Press, forthcoming); Handa Abidin, *The Protection of Indigenous Peoples and Reduction of Forest Carbon Emissions: The REDD-Plus Regime and International Law* (Brill Nijhoff, 2015); see also Rosemary Lyster, Catherine MacKenzie and Constance McDermott (eds), *Law, Tropical Forests and Carbon: The Case of REDD+* (Cambridge University Press, 2013).

international organizations to effectively enforce priorities between rights holders), we propose that the institutional framework supporting REDD+ at the global level should move towards a model that facilitates the sharing of local experiences and expertise in REDD+ implementation and encourages the establishment of plausible benchmarks based on the real-world state and indigenous responses to REDD+.

The book's inquiry is timely, given that REDD+ continues to evolve as new understandings about impacts, benefits, and problems arise in different national and local contexts. Within the UNFCCC, parties have agreed upon different phases of REDD+ implementation: first, when financial assistance is given to states to build their institutional and administrative capacity to make them 'ready' to conduct REDD+ projects (known as 'readiness', and involving national strategies or action plans, policies and measures); second, when REDD+ projects are implemented; and third, when results are monitored, reported and verified.[15]

The countries we focus on are generally pursuing the 'readiness' phase, with preparation and 'demonstration' activities commenced in a range of international and transnational contexts with support from the UN-REDD Programme, the FCPF and other bodies. Financing of REDD+ has so far mostly come from bilateral and multilateral aid programmes, but many donor countries expect that REDD+ schemes will transition to market-based mechanisms, thereby accessing private finance that is stimulated by the demand for 'carbon credits' generated by carbon markets operating under offsetting arrangements.[16] The Paris Agreement has signalled that REDD+ will continue to be supported under the UNFCCC. The existing framework for REDD+ as set out in previous conferences of the parties (COPs) (such as Cancun and Warsaw) has been endorsed by states. Given the international community's demonstrated support for REDD+, the legal analysis contained in this book provides useful reflections on the continuing experiment of REDD+ as well as the more general endeavour of achieving climate change

[15] Conference of the Parties, United Nations Framework Convention on Climate Change, *Report of the Conference of the Parties on Its Sixteenth Session, Held in Cancun from 29 November to 10 December 2010 – Addendum – Part Two: Action Taken by the Conference of the Parties at Its Sixteenth Session* (UN Doc FCCC/CP/2010/7/Add.1 (15 March 2011), Decision 1/CP.16 ('The Cancun Agreements: Outcome of the Work of the Ad Hoc Working Group on Long-Term Cooperative Action under the Convention') para 70.

[16] Christina Voigt and Felipe Ferreira, 'The Warsaw Framework for REDD+: Implications for National Implementation and Results-Based Finance' in Voigt, above n 14, 30.

mitigation that justly accommodates the interests of indigenous peoples and forest communities.

The book adopts a thematic approach to examining the interaction between the REDD+ scheme implementation and the interests of indigenous and local communities. First, it explores the primary sources of this complexity, by unpacking the legal regimes and concepts that bear most directly on the intersection of REDD+ with international law regimes (climate change, environmental protection, development and human rights) and with national and international law governing the identification, recognition, and constitution of indigenous and local communities (identity law). Second, it explores the implications of these interactions for national-level recognition of the land tenure and resource-based rights of indigenous and forest-dwelling communities. Third, it examines the ways in which these broadly conceived thematic areas of law (international regimes, identity law, and land and resource tenure) interact in three countries: Malaysia, Papua New Guinea and Vanuatu. We go on to consider the interconnection of indigenous land tenure and carbon-related legislation in Australia in order to demonstrate the in situ implications of REDD+-type project implementation and forest and land-sector governance. The final section offers an assessment of rights and accountability frameworks that underpin REDD+ as it has developed internationally, and considers how REDD+ might fit amongst debates on pluralist global governance. The concluding chapter revisits our arguments on whether REDD+, as currently conceived, can hope to provide the necessary balance between climate change mitigation objectives and the protection of the rights and interests of indigenous and forest-dwelling communities.

# PART I

# Climate Change Mitigation and Forest Carbon Sequestration

# 2

# REDD+ as International Legal Regime

MARGARET YOUNG

The UNFCCC negotiations to establish REDD+ have occurred over the last decade, in the context of earlier failures of the international community to agree on international efforts to address the crisis of forest destruction. The 1992 United Nations Conference on Environment and Development in Rio abandoned the negotiation of a convention for forest conservation and instead agreed upon non-binding principles relating to national sovereignty over forestry resources.[1] Characterising deforestation as a problem of climate provided a new way of seeing the problem[2] and led to the advocacy of solutions within the UNFCCC COPs. This chapter investigates whether REDD+ is emerging as an international legal regime. We invoke a definition of 'regimes' that draws attention to discursive and institutional practices as well as laws, rules and procedures. As we foreshadowed in our introductory chapter, it is not clear that the set of activities surrounding REDD+ in international law can be usefully conceived as a 'scheme', 'programme' or 'action', let alone 'regime'. Relevant constitutive instruments refer variously to 'policy approaches', 'guidance', 'mitigation actions' and 'activities' relating to reducing emissions from deforestation and forest degradation.[3] In addition to invoking the language

[1] Constance McDermott, 'REDD+ and Multi-Level Governance: Governing for What and for Whom?' in Rosemary Lyster, Catherine MacKenzie and Constance McDermott (eds), *Law, Tropical Forests and Carbon: The Case of REDD+* (Cambridge University Press, 2013) 84, 93.

[2] William Boyd, 'Ways of Seeing in Environmental Law: How Deforestation Became an Object of Climate Governance' (2010) 37 *Ecology Law Quarterly* 843.

[3] *Paris Agreement* [2016] ATS 24, entered into force 4 November 2016 (UNTS volume number not yet assigned) art 5; see also Decision on Adoption of the Paris Agreement, Conference of the Parties, United Nations Framework Convention on Climate Change, *Report of the Conference of the Parties on Its Twenty-First Session, Held in Paris from 30 November to 13 December 2015 – Addendum – Part Two: Action Taken by the Conference of the Parties at Its Twenty-First Session* (UN Doc FCCC/CP/2015/10/Add.1 (29 January 2016) Decision 1/CP.21) ('Paris COP Decision'); Conference of the Parties, United Nations Framework Convention on Climate Change, *Report of the Conference of*

of 'regimes',[4] commentators use the language of 'assemblages',[5] 'policies',[6] 'mechanisms'[7] and even 'phenomenon[a]'.[8] We thus draw upon relevant legal sources, our own empirical engagement with participants as well as secondary literature to determine whether the terminology of 'regimes' is appropriate, and to assess the consequences of accepting one nomenclature over another. We show how the UNFCCC, the Kyoto Protocol and the new Paris Agreement have sought to entrench an idea of incentives for forest-based climate mitigation. The concept of 'regime' used in this chapter is useful in structuring our inquiry about key aspects of REDD+, which include treaties, international organizations, and certain key assumptions and behaviours. It also allows for a comparison between REDD+ and already existing legal regimes that relate to forests and indigenous peoples, which we explore in the following chapter.

This chapter begins in Section A with an overview of laws and decision-making procedures relevant to REDD+, as developed under the auspices of the UNFCCC, the Paris Agreement and the Kyoto Protocol. In Section B, we describe international organizations and other key participants, noting in particular the UN-REDD Collaborative Programme and the FCPF. In Section C, we investigate the more ephemeral parts of an international legal regime by pointing to key behaviours, assumptions and biases that are possible to discern from the way in which the concept of REDD+ evolved from earlier, stricter concepts of forest-based climate mitigation, and the impact that this has on indigenous peoples.

*the Parties on Its Sixteenth Session, Held in Cancun from 29 November to 10 December 2010 – Addendum – Part Two: Action Taken by the Conference of the Parties at Its Sixteenth Session* (UN Doc FCCC/CP/2010/7/Add.1 (15 March 2011) Decision 1/CP.16) ('The Cancun Agreements: Outcome of the Work of the Ad Hoc Working Group on Long-Term Cooperative Action under the Convention') paras 69, 70 and para 68 (heading).

4 Esteve Corbera and Heike Schroeder, 'Governing and Implementing REDD+' (2011) 14 *Environmental Science & Policy* 89, 89 (introducing the special issue aiming to 'uncover the nature of this new international regime').

5 William Boyd, 'Climate Change, Fragmentation, and the Challenges of Global Environmental Law: Elements of a Post-Copenhagen Assemblage' (2010) 32 *University of Pennsylvania Journal of International Law* 457.

6 Jan Willem den Besten, Bas Arts and Patrick Verkooijen, 'The Evolution of REDD+: An Analysis of Discursive-Institutional Dynamics' (2014) 35 *Environmental Science & Policy* 40.

7 See, e.g., Annecoos Wiersema, 'Climate Change, Forests, and International Law: REDD's Descent into Irrelevance' (2014) 47 *Vanderbilt Journal of Transnational Law* 1.

8 Christine Voigt, 'Introduction: The Kaleidoscopic World of REDD+' in Christine Voigt (ed), *Research Handbook on REDD+ and International Law* (Edward Elgar, 2016) 1.

Before moving to these areas, it is useful to give some further background to the concept of 'regime', a term popularised in international relations scholarship[9] and invoked by international legal tribunals.[10] A hybrid definition describes 'regimes' as 'sets of norms, decision-making procedures and organizations coalescing around functional issue-areas and dominated by particular modes of behaviour, assumptions and biases'.[11] The notion of 'regimes' has become an important tool for understanding and analysing the fragmentation of public international law. The concept is not restricted to the main sources of public international law (treaties or norms of customary international law) but also incorporates an institutional analysis and recognises that participants of regimes can include a range of state and nonstate actors. Given the loosely coupled arrangements occurring in the climate mitigation context, scholars have advanced a related concept of 'regime complex'.[12]

Insights from systems theory point to the organizational principles of regimes and the inevitability of regime collisions.[13] One factor leading to this inevitability is the way in which discourses *within* regimes have their own biases and preferences.[14] These embedded biases and preferences, especially those involving technical expertise, affect the interaction between regimes in unseen and important ways. Where such collisions implicate the laws of indigenous peoples, the methods to confront and deal with legal pluralism are especially complex.[15]

While REDD+ is increasingly being described as a 'regime' in the secondary literature,[16] the appropriateness of the term is not

[9] Stephen D Krasner, 'Structural Causes and Regime Consequences: Regimes as Intervening Variables' in Stephen D Krasner (ed), *International Regimes* (Cornell University Press, 1983) 1.

[10] *United States Diplomatic and Consular Staff in Tehran* [1980] ICJ Rep 3.

[11] Margaret A Young, 'Introduction: The Productive Friction between Regimes' in Margaret A Young (ed), *Regime Interaction in International Law: Facing Fragmentation* (Cambridge University Press, 2012) 1, 11.

[12] Robert O Keohane and David G Victor, 'The Regime Complex for Climate Change' (2011) 9 *Perspectives on Politics* 7.

[13] Andreas Fischer-Lescano and Gunther Teubner, 'Regime-Collisions: The Vain Search for Legal Unity in the Fragmentation of Global Law' (2004) 25 *Michigan Journal of International Law* 999.

[14] Oren Perez, *Ecological Sensitivity and Global Legal Pluralism: Rethinking the Trade and Environment Conflict* (Hart Publishing, 2004).

[15] Gunther Teubner and Peter Korth, 'Two Kinds of Legal Pluralism: Collision of Transnational Regimes in the Double Fragmentation of World Society' in Margaret A Young (ed), *Regime Interaction in International Law: Facing Fragmentation* (Cambridge University Press, 2012) 23.

[16] Corbera and Schroeder, above n 4.

self-evident. Instead, an assessment of the characteristics of 'regimes', as the term has so far been used, as against the emerging features of REDD +, draws out some useful legal, structural and procedural attributes as well as unstated assumptions and vocabularies. The inquiry that follows is particularly relevant to the question of how REDD+ operates alongside other international regimes relevant to forests and international law, and to our related research question about REDD+'s impact on indigenous peoples.

## A. Laws and Decision-Making Procedures

This section concentrates on the most fundamental and uncontroversial characteristic of a regime: namely, a set of norms and decision-making procedures coalescing around functional issue areas. The functional issue addressed by REDD+ is, broadly, the objective of the UNFCCC (and any related legal instruments adopted by its COP), namely the 'stabilization of greenhouse gas concentrations in the atmosphere at a level that would prevent dangerous anthropogenic interference with the climate system'.[17] This objective clearly draws and depends upon scientific understandings about how greenhouse gas concentrations may be stabilised, both by the reduction of greenhouse gas emissions and by carbon sequestration. The obligations flowing from this objective are, at this point in time, mainly a matter of treaty law and international institutional law (mainly, the newly in-force Paris Agreement, the UNFCCC and the Kyoto Protocol), although customary international law is increasingly relevant, including with respect to indigenous peoples. The Paris Agreement, which entered into force on 4 November 2016 after the requisite number of countries had deposited their instruments of ratification, currently has 159 parties,[18] and now operates in tandem with the UNFCCC, which also enjoys near universal acceptance. The following description thus incorporates relevant aspects of the UNFCCC and the Paris Agreement, before moving to other treaty-based and customary norms.

[17] United Nations Framework Convention on Climate Change (opened for signature 4 June 1992, 1771 UNTS 107, entered into force 21 March 1994) art 2.

[18] See status of ratification at 'Paris Agreement – Status of Ratification', United Nations Framework Convention on Climate Change, http://unfccc.int/paris_agreement/items/9444.php. Entry into force required ratification by at least fifty-five countries that would account for at least 55 per cent of global emissions. As of August 2017, 159 parties have ratified.

### 1. *UNFCCC and the Paris Agreement*

To achieve the objective of 'stabilization of greenhouse gas concentrations', the UNFCCC sets out general obligations for its parties, including aspirations for climate mitigation in the forest sector. All parties (including developed and developing countries) agree to:

> Promote sustainable management, and promote and cooperate in the conservation and enhancement, as appropriate, of sinks and reservoirs of all greenhouse gases . . . including forests.[19]

This objective is given more concrete expression in the Paris Agreement, which is aimed at 'enhancing the implementation of the [UNFCCC]', including its objective, 'in the context of sustainable development and efforts to eradicate poverty',[20] by, for one:

> a) Holding the increase in the global average temperature to well below 2°C above pre-industrial levels and to pursue efforts to limit the temperature increase to 1.5 °C above pre-industrial levels, recognizing that this would significantly reduce the risks and impacts of climate change

The UNFCCC adopted a principle that the participation of all countries ought to be 'in accordance with their common but differentiated responsibilities and respective capabilities'.[21] In accordance with that principle, the UNFCCC sets out commitments for developed countries to provide 'new and additional financial resources'[22] to help developing countries meet their UNFCCC obligations. Also recognised in the convention is the need for 'adequacy and predictability in the flow of funds and the importance of appropriate burden sharing amongst the developed country Parties'.[23] This is to include resources for the transfer of technology. These clauses contain clear indications that developed-country parties will provide funding for developing countries in efforts to address climate change, and provide the general conceptual basis that forest-based climate mitigation will require funding by developed-country parties.[24]

[19] United Nations Framework Convention on Climate Change, above n 17, art 4(1)(d).

[20] *Paris Agreement*, art 2(1).

[21] United Nations Framework Convention on Climate Change, above n 17, art 3(1); see also art 4(d); preamble.

[22] Ibid. art 4(3). [23] Ibid. art 4(3).

[24] UNFCCC art 4(3) cited above is reiterated in the Cancun Agreements: see Appendix I ('Guidance and safeguards for policy approaches and positive incentives on issues relating to reducing emissions from deforestation and forest degradation in developing countries; and the role of conservation, sustainable management of forests and enhancement of forest carbon stocks in developing countries'), Conference of the Parties, United Nations Framework Convention on Climate Change, '*Cancun Agreements*', above n 3, 26 para 1.

The Paris Agreement has a different approach and does not differentiate between developed and developing countries in the same way as the UNFCCC or Kyoto Protocol. Instead of requiring developed countries to assume specific targets, the Paris Agreement takes a 'bottom-up' approach that includes contributions from all parties. Thus, the full range of developed and developing parties – including high-aggregate greenhouse gas emitters such as China, European Union, United States and India, and heavily forested countries such as Australia, Russia, Brazil, Democratic Republic of the Congo, Indonesia and Malaysia – are required to undertake and communicate intended 'nationally determined contributions' (NDCs).[25] The collective effort of all parties is supposed to achieve the aim of holding the increase in the global average temperature to well below 2°C above preindustrial levels, and, although the NDCs submitted so far are unlikely to meet this according to current projections,[26] there is a duty to communicate NDCs every five years, with successive NDCs presenting a progression, or a ratcheting up, of contributions.[27]

The principle of common but differentiated responsibilities has been far from abandoned, however. The Paris Agreement recognises that it will be implemented 'to reflect equity and the principle of common but differentiated responsibilities and respective capabilities, in the light of different national circumstances'.[28] Moreover, the mitigation potential (termed 'ambition') of each country's NDC depends on national circumstances:

> Each Party's successive nationally determined contribution will represent a progression beyond the Party's then current nationally determined contribution and reflect its highest possible ambition, reflecting its common but differentiated responsibilities and respective capabilities, in the light of different national circumstances.[29]

The recognition of differentiation has led to evaluations of whether submitted NDCs can achieve equity across countries,[30] and to criticism

[25] *Paris Agreement*, art 3. As of August 2017, the European Union, China, India, Malaysia, Indonesia and Brazil have deposited instruments of ratification or accession and are therefore bound by the Paris Agreement. Russia and the DRC have signed but not ratified the Paris Agreement, while the United States, which ratified on 4 November 2016, has announced its intention to withdraw.

[26] Conference of the Parties, United Nations Framework Convention on Climate Change, *Aggregate Effect of the Intended Nationally Determined Contributions: An Update – Synthesis Report by the Secretariat* (UN Doc FCCC/CP/2016/2 (2 May 2016)) para 39.

[27] Ibid., art 4(2); 4(9). [28] *Paris Agreement*, art 2(2). [29] Ibid. art 4(3).

[30] See, e.g., work that has assessed NDCs according to the five equity categories presented in the Fifth Assessment Report of the IPCC (IPCCAR5): 'capability', 'equality', 'responsibility-capability-need', 'equal cumulative per capita' and 'staged approaches': Yann Robiou

of particular countries that have not addressed equity issues when producing NDCs, including Australia.[31] The agreement is considered to have utilised the 'common but differentiated responsibilities' concept in divergent ways: on a principled basis, on the basis of the legal content of various provisions of the agreement (such as mitigation, finance and transparency) and on the basis of progressive ambitions.[32]

The NDCs do not themselves lead to binding obligations at the international level, despite calls by some states for a system of responsibility to be entrenched within the treaty.[33] Instead, the Paris Agreement provides for the NDCs to be submitted (and therefore subject to review) and strengthened by states every five years.[34] A global stocktaking exercise will allow for evaluation amongst countries. This emphasis on cooperation, sharing and 'bottom-up' ideas is consistent with an experimentalist framework that promotes mutual learning amongst states in a context where the mitigation actions are diverse and sometimes still to be decided.[35]

The Paris Agreement does not create any new specific rules for REDD+, but it demonstrates that states continue to support the concept of REDD+. Article 5(2) provides:

> Parties are encouraged to take action to implement and support, including through results-based payments, the existing framework as set out in related guidance and decisions already agreed under the Convention for: *policy approaches and positive incentives for activities relating to reducing emissions from deforestation and forest degradation, and the role of conservation, sustainable management of forests and enhancement of forest carbon stocks in developing countries*; and alternative policy approaches, such as joint mitigation and adaptation approaches for the integral and

du Pont et al, 'National Contributions for Decarbonizing the World Economy in Line with the G7 Agreement' (2016) 11 *Environmental Research Letters* 054005.

[31] See, e.g., 'Tracking (I)NDCs', Climate Action Tracker, www.climateactiontracker.org, which evaluates Australia's response as 'inadequate'; 'Australia', Climate Action Tracker, www.climateactiontracker.org/countries/australia.html. See also, 'Are We on Track for 2°C?' Mitigation-Contributions.org, www.mitigation-contributions.org.

[32] Christina Voigt and Felipe Ferreira, 'Differentiation in the Paris Agreement' (2016) 6 *Climate Law* 58.

[33] See, e.g., position advanced by Nicaragua in negotiations, cited by Meinhard Doelle, 'The Paris Agreement: Historic Breakthrough or High Stakes Experiment?' (2016) 6 *Climate Law* 1, 8; see also declarations made upon ratification by a number of developing countries (such as the Marshall Islands, Naura and the Philippines) that state that the Paris Agreement is not a renunciation of any rights under international law concerning State responsibility.

[34] *Paris Agreement*, art 4; see also Paris COP Decision, above n 3, para 27.

[35] Charles F Sabel and David G Victor, 'Governing Global Problems under Uncertainty: Making Bottom-Up Climate Policy Work' (2015) (May) *Climatic Change* 1.

> sustainable management of forests, while reaffirming the importance of incentivizing, as appropriate, non-carbon benefits associated with such approaches.[36]

How REDD+ is to be financed remains inconclusive, but is likely to remain a mix of public and private funding similar to the earlier, heterogeneous funding arrangements supporting the UNFCCC.[37] In the COP decision supporting the Paris Agreement, parties recognise the 'importance of adequate and predictable financial resources, including for results-based payments, as appropriate'.[38] Moreover, the trading of carbon credits (which could include REDD+ 'credits') is permitted, with the agreement recognising the use of 'internationally transferred mitigation outcomes' towards their NDCs as long as double counting is avoided.[39] The Paris Agreement thus implies that REDD+ may be financed in the future through a mix of different scenarios, including: (1) the provision of development assistance from foreign states, transferred through agencies such as FCPF or bilateral arrangements, (2) the purchase of credits for equivalent emissions reductions to offset emissions from foreign countries in an effort to meet their targets, (3) the domestic support for such programmes by (tropical forested) states in order to meet their own NDCs or (4) a mix of all three.[40] It is noteworthy that this mix of scenarios is different from the earlier concept of emissions trading in the Kyoto Protocol, in that there is no need (at least explicitly) for countries to use such emissions schemes in a way that is 'supplemental' to their domestic efforts.[41] This means that the bulk – or even the totality – of one country's emissions reductions could in theory be made through reductions occurring in another country, unless the principle that

[36] *Paris Agreement*, art 5(2). Emphasis added.

[37] See generally Charlotte Streck, 'Financing Mechanisms for Climate Change Mitigation' in Thomas Hale and David Held (eds), *The Handbook of Transnational Governance: Institutions and Innovations* (Polity Press, 2011) 377–383.

[38] Paris COP Decision, above n 3, para 55.

[39] *Paris Agreement*, art 6(2).

[40] The current NDC of Malaysia, for example, does not mention REDD+ but mentions that its mitigation will be advanced, inter alia, by two major initiatives in the forestry sector, the Central Forest Spine and the Heart of Borneo Initiative, 'to ensure sustainable forest management and use of natural resources': see Government of Malaysia, 'Intended Nationally Determined Contribution of the Government of Malaysia', 27 November 2015, available from; *INDC – Submissions*, www4.unfccc.int/submissions/indc/Submission%20Pages/submissions.aspx.

[41] See Kyoto Protocol to the United Nations Framework Convention on Climate Change (opened for signature 16 March 1998, 2303 UNTS 148, entered into force 16 February 2005) ('Kyoto Protocol') art 6. See further in n 44 and accompanying text.

reductions must be 'supplemental' is implied in the Paris Agreement. The implications for REDD+ are manifold, but two are worthy of particular note: first, there may be a higher than anticipated demand for REDD+ credits, and second, there is a risk that existing asymmetrical patterns of industrialisation, economic growth and emissions levels in developed and developing countries will remain unchanged or be exacerbated. These questions remain open for future COPs to address.

In summary, REDD+ is endorsed under the Paris Agreement but an understanding of its form and content requires regard to earlier COP decisions, especially the Cancun Agreements and Warsaw Framework. We turn to those decisions in a moment, but before doing so it is important to consider the relevance of the Kyoto Protocol, which remains a key part of the UNFCCC regime complex.

### 2. *Kyoto Protocol*

Although the legal status of the Kyoto Protocol is now uncertain, it is important to understand some of its key provisions, especially in the context of consistencies and divergences in the current Paris Agreement. The Kyoto Protocol's legal status is uncertain because of the current low number of ratifications of the instrument that sets up the current (second) commitment period, namely the Doha Amendment to the Kyoto Protocol adopted at the Doha COP.[42] The Kyoto Protocol's first commitment period ran from 2008–2012, and set out specific obligations for a subset of UNFCCC parties in pursuit of the same overarching objective of the UNFCCC; namely, the 'stabilization of greenhouse gas concentrations in the atmosphere at a level that would prevent dangerous anthropogenic interference with the climate system'.[43] These specific obligations included a target for emissions reductions for the developed countries that had ratified it.[44] While some countries (principally

[42] The Doha Amendment establishing the second commitment period will enter into force on the ninetieth day after 144 of the 192 parties to the Kyoto Protocol have deposited their instruments of ratification with the UN. The current number of ratifications is well below this number. The Paris COP Decision adopting the Paris Agreement 'Urg[ed] all Parties to the Kyoto Protocol that have not already done so to ratify and implement the Doha Amendment to the Kyoto Protocol': see Paris COP Decision, above n 3, para 106(a).

[43] Kyoto Protocol to the United Nations Framework Convention on Climate Change, above n 41, Preamble ('in pursuit of the ultimate objective of the [UNFCCC] as stated in its Article 2').

[44] Ibid. art 3.

countries within the European Union and Australia)[45] have nominally agreed to extend the commitment period to run from 2013–2020,[46] these are fewer than the original participating states and there is considerable uncertainty about the legal underpinnings of Kyoto targets. Nevertheless, the second commitment period is envisaged as bridging the period between the end of the first commitment period and the start of the Paris Agreement in 2020, and for this and other reasons it is useful to consider the Kyoto Protocol's relevance to emissions reductions in the forest sector.

Pursuant to the Kyoto Protocol obligations, the relevant (and now diminished) set of developed countries are required to undertake emissions reductions within their own countries, but they can also acquire the benefit of reductions in other countries (known as 'carbon off-sets'), provided that that acquisition is 'supplemental to domestic actions'.[47] In the first commitment period, these off-sets were often gained through projects in developing countries grouped under the 'Clean Development Mechanism' (CDM), one of the main market mechanisms established by the Kyoto Protocol.

Forest-based climate mitigation was included in the Kyoto Protocol mechanism, although it was far more limited than the proposed arrangements surrounding REDD+. Importantly, CDM activities did not involve avoided deforestation. Instead, the CDM provided market incentives for afforestation and reforestation activities. Limiting the availability of CDM funds to afforestation and reforestation activities was due to a lack of agreement amongst participating states that avoided deforestation

[45] The United States never ratified the Kyoto Protocol and Canada withdrew from it in 2012. Japan, New Zealand and Russia participated in the first commitment period but have not taken on new targets in the second commitment period.

[46] Conference of the Parties Serving as the Meeting of the Parties to the Kyoto Protocol, United Nations Framework Convention on Climate Change, *Report of the Conference of the Parties Serving as the Meeting of the Parties to the Kyoto Protocol on Its Eighth Session, Held in Doha from 26 November to 8 December 2012 – Addendum – Part Two: Action Taken by the Conference of the Parties Serving as the Meeting of the Parties to the Kyoto Protocol at Its Eighth Session* (UN Doc FCCC/KP/CMP/2012/13/Add.1 (28 February 2013) Decision 1/CMP.8 ('Amendment to the Kyoto Protocol Pursuant to Its Article 3, Paragraph 9 (the Doha Amendment)'), available at http://unfccc.int/files/kyoto_protocol/application/pdf/kp_doha_amendment_english.pdf. See further 'Status of the Doha Amendment', United Nations Framework Convention on Climate Change, http://unfccc.int/kyoto_protocol/doha_amendment/items/7362.php.

[47] Kyoto Protocol to the United Nations Framework Convention on Climate Change, above n 41, art 6.

could be adequately incentivised, monitored and accounted for.[48] With an emerging consensus that these deficiencies could be overcome,[49] REDD+ proposals have been made relating to the inclusion of REDD+ within the second commitment period.[50] It remains uncertain as to whether REDD+ will be used to generate carbon credits and 'off-sets' in the context of the Kyoto Protocol, but given that Kyoto second-commitment targets are now forming country NDCs under the Paris Agreement, this question is more important to answer in the Paris Agreement context. The next subsection thus considers the background decisions that gave content to REDD+, namely the Cancun Agreements and the Warsaw Framework.

### *3. Cancun Agreements and Warsaw Framework*

Although mentioned in the Paris Agreement, REDD+ is given content in a set of earlier UNFCCC COP decisions which seek to agree on its scope and modalities. While the COP decisions lack the status of treaty text (as discussed in subsection 3(b) below), they provide an underpinning set of concepts and norms which have evolved during the annual meetings of the parties to the UNFCCC.

To begin with, a review of COP decisions demonstrates that the idea behind REDD+ has morphed over time.[51] The catalysing 2005 COP proposal of Papua New Guinea and Costa Rica to reduce emissions from deforestation (with the acronym RED)[52] was enlarged to include the reduction of emissions from forest degradation (with the acronym REDD) in 2007.[53] Separately mentioned was 'the role of conservation,

48 McDermott, above n 1, 93. 49 See Section C.

50 Commonwealth of Australia, 'Reducing Emissions from Deforestation and Forest Degradation in Developing Countries (REDD)' (Submission to the AWG-LCA, AWG-KP and SBSTA, 2008), http://climatechange.gov.au/sites/climatechange/files/files/Reducing-emissions-deforestation-forest-degradation-developing-countries-Nov-08.pdf.

51 See further Till Pistorius, 'From RED to REDD+: The Evolution of a Forest-Based Mitigation Approach for Developing Countries' (2012) 4 *Current Opinion in Environmental Sustainability* 638.

52 Conference of the Parties, United Nations Framework Convention on Climate Change, *Reducing Emissions from Deforestation in Developing Countries: Approaches to Stimulate Action – Submission by the Governments of Papua New Guinea and Costa Rica* (UN Doc FCCC/CP/2005/MISC.1, 11 November 2005) (submission of Papua New Guinea and Costa Rica).

53 Conference of the Parties, United Nations Framework Convention on Climate Change, *Report of the Conference of the Parties on Its Thirteenth Session, Held in Bali from 3 to 15 December 2007 – Addendum – Part Two: Action Taken by the Conference of the Parties at*

sustainable management of forests and enhancement of forest carbon stocks',[54] and these additional objectives were added to the whole to become 'REDD-plus' in 2009.[55] These objectives were themselves modified in 2010, so that the broader goal of conservation became limited to conservation of forest carbon stocks,[56] a move that was reversed in 2015 with the Paris Agreement referring to 'REDD+' and the 'role of conservation, sustainable management of forests and enhancement of forest carbon stocks in developing countries'.[57]

These changes in objectives warrant further discussion, particularly because they have generated a new and different set of methodologies, actors and discourses. We discuss these in Section C, but for present purposes it suffices to note that the formal laws surrounding REDD+ are now encapsulated in the 2010 UNFCCC COP16 set of agreements (Cancun Agreements),[58] updated in 2013 by COP19 (Warsaw Framework on REDD+).[59] The most important legal guidance is paragraph 70 of the

*Its Thirteenth Session* (UN Doc FCCC/CP/2007/6/Add.1 (14 March 2008) Decision 2/CP.13) ('Reducing Emissions from Deforestation in Developing Countries: Approaches to Stimulate Action') ('Bali Action Plan') 8.

54 Conference of the Parties, United Nations Framework Convention on Climate Change, 'Bali Action Plan', above n 53, 11. ('Notes the further consideration, under Decision 1/CP.13, of policy approaches and positive incentives on issues relating to reducing emissions from deforestation and forest degradation in developing countries; and the role of conservation, sustainable management of forests and enhancement of forest carbon stocks in developing countries').

55 The Copenhagen Accord refers to 'REDD-plus' but does not define it. A separate decision of the Copenhagen COP refers to 'the importance of reducing emissions from deforestation and forest degradation, and the role of conservation, sustainable management of forests and enhancement of forest carbon stocks in developing countries': see Conference of the Parties, United Nations Framework Convention on Climate Change, *Report of the Conference of the Parties on Its Fifteenth Session, Held in Copenhagen from 7 to 19 December 2009 – Addendum – Part Two: Action Taken by the Conference of the Parties at Its Fifteenth Session* (UN Doc FCCC/CP/2009/11/Add.1 (30 March 2010) Decision 4/CP.15 ('Methodological Guidance for Activities Relating to Reducing Emissions from Deforestation and Forest Degradation and the Role of Conservation, Sustainable Management of Forests and Enhancement of Forest Carbon Stocks in Developing Countries')).

56 Conference of the Parties, United Nations Framework Convention on Climate Change, *Cancun Agreements*, above n 3, 12.

57 *Paris Agreement*, art 5. For extracted text, see above n 36 and surrounding text.

58 Conference of the Parties, United Nations Framework Convention on Climate Change, *Cancun Agreements*, above n 3.

59 The following COP decisions are designated as the 'Warsaw Framework for REDD-plus': Decisions 9/CP.19, 10/CP.19, 11/CP.19, 12/CP.19, 13/CP.19, 14/CP.19, and 15/CP.19. See further Conference of the Parties, United Nations Framework Convention on Climate Change, *Report of the Conference of the Parties on Its Nineteenth Session, Held in Warsaw*

Cancun Agreements.[60] In addition to setting out the scope for REDD+, the Cancun Agreements contained an appendix that sought to provide guidance for how REDD+ activities should be undertaken. These are known as the REDD+ Safeguards, and initiate an important set of procedures relating to other regimes.

### (a) The Cancun Safeguards

The Cancun Safeguards are contained in an appendix (Appendix I) to the Cancun Agreements. Paragraph 1 of that document sets out a list of objectives for REDD+. Paragraph 2 provides (in full):

> 2. When undertaking the activities referred to in paragraph 70 of this decision, the following safeguards should be promoted and supported:
>
> (a) That actions complement or are consistent with the objectives of national forest programmes and relevant international conventions and agreements;
> (b) Transparent and effective national forest governance structures, taking into account national legislation and sovereignty;
> (c) Respect for the knowledge and rights of indigenous peoples and members of local communities, by taking into account relevant international obligations, national circumstances and laws, and noting that the United Nations General Assembly has adopted the United Nations Declaration on the Rights of Indigenous Peoples;
> (d) The full and effective participation of relevant stakeholders, in particular indigenous peoples and local communities, in the actions referred to in paragraphs 70 and 72 of this decision;
> (e) That actions are consistent with the conservation of natural forests and biological diversity, ensuring that the actions referred to in paragraph 70 of this decision are not used for the conversion of natural forests, but are instead used to incentivize the protection and conservation of natural forests and their ecosystem services, and to enhance other social and environmental benefits [footnote: Taking into account the need for sustainable livelihoods of indigenous

*from 11 to 23 November 2013 – Addendum – Part Two: Action Taken by the Conference of the Parties at Its Nineteenth Session* (UN Doc FCCC/CP/2013/10/Add.1 (31 January 2014) Decisions 9–15/CP.19 ('Warsaw Framework for REDD-Plus')) para 44.

[60] Conference of the Parties, United Nations Framework Convention on Climate Change, *Cancun Agreements*, above n 3, para 70. In full, para 70: *Encourages* developing country Parties to contribute to mitigation actions in the forest sector by undertaking the following activities, as deemed appropriate by each Party and in accordance with their respective capabilities and national circumstances: (a) Reducing emissions from deforestation; (b) Reducing emissions from forest degradation; (c) Conservation of forest carbon stocks; (d) Sustainable management of forests; (e) Enhancement of forest carbon stocks.

peoples and local communities and their interdependence on forests in most countries, reflected in the United Nations Declaration on the Rights of Indigenous Peoples, as well as the International Mother Earth Day.];

(f) Actions to address the risks of reversals;

(g) Actions to reduce displacement of emissions.

Rather than making adherence to the safeguards mandatory, the Cancun Agreements require developing countries to create a system for providing information on how the safeguards are being addressed and respected domestically, 'while respecting sovereignty'.[61] The requirement to provide information was watered down after Cancun[62] because the next COP decision at Durban stated that 'national circumstances and respective capabilities' should be taken into account.[63] However, the requirement was later strengthened at the Warsaw COP, when UNFCCC parties agreed that 'results-based payments' would not be paid to developing countries until they have provided 'the most recent summary of information on how all of the safeguards ... have been addressed and respected'.[64]

The transparency of the summaries of information provided by recipient countries was emphasised in a separate Warsaw COP decision, which reiterated that the summary of information 'should be provided periodically and be included in national communications, or communication

61 Ibid. para 71(d).

62 Rosemary Lyster, 'International Legal Frameworks for REDD+: Ensuring Legitimacy' in Rosemary Lyster, Catherine MacKenzie and Constance McDermott (eds), *Law, Tropical Forests and Carbon: The Case of REDD+* (Cambridge University Press, 2013) 3, 13.

63 Conference of the Parties, United Nations Framework Convention on Climate Change, *Report of the Conference of the Parties on Its Seventeenth Session, Held in Durban from 28 November to 11 December 2011 – Addendum – Part Two: Action Taken by the Conference of the Parties at Its Seventeenth Session* (UN Doc FCCC/CP/2011/9/Add.2 (15 March 2012) Decision 12/CP.17 ('Guidance on Systems for Providing Information on How Safeguards are Addressed and Respected and Modalities Relating to Forest Reference Emission Levels and Forest Reference Levels as Referred to in Decision 1/CP.16')) 16.

64 Conference of the Parties, United Nations Framework Convention on Climate Change, *Report of the Conference of the Parties on Its Nineteenth Session, Held in Warsaw from 11 to 23 November 2013 – Addendum – Part Two: Action Taken by the Conference of the Parties at Its Nineteenth Session* (UN Doc FCCC/CP/2013/10/Add.1 (31 January 2014) Decision 9/CP.19 ('Work Programme on Results-Based Finance to Progress the Full Implementation of the Activities Referred to in Decision 1/CP.16, paragraph 70') ('Decision 9/CP.19, Warsaw Framework for REDD-Plus')) 24 [4].

channels agreed by the Conference of the Parties'.[65] Further, the summary could be provided '*on a voluntary basis*, via the web platform on the UNFCCC website', and only after implementation of REDD+ activities has begun.[66]

One implication of these measures promoting information sharing and transparency is that REDD+ funds are conditioned upon the provision of the latest information about adherence to safeguards. However, it is still relatively unclear whether the requirement for summaries to be provided will be subject to the oversight of REDD+ international organizations (such as the UN-REDD Programme or FCPF which we discuss in Section B) or donor states, and if so what criteria will be used to evaluate state compliance with safeguards once the information is provided. Moreover, linking funds to the provision of information only occurs after implementation has begun (i.e. after the REDD+ Readiness funding has been provided).[67] This leads to important questions about which actors will be most responsible for attending to a country's compliance with safeguards for indigenous peoples, which we explore in this book, particularly in the context of asserted guarantees of FPIC. It also leads to questions about the legal status of the COP decisions that set up the Cancun Safeguards.

### (b) Legal Status of the COP Decisions

The legal status of the safeguards and the COP decisions establishing REDD+ is an important feature of the emerging REDD+ regime. The relevant decisions of the Cancun Agreements and the Warsaw Framework were reached by consensus, as is customarily required for all UNFCCC COP decisions.[68] UNFCCC COPs are part of the growing phenomenon within international regimes – particularly environmental regimes – in which a framework treaty is followed by updated

[65] Conference of the Parties, United Nations Framework Convention on Climate Change, *Report of the Conference of the Parties on Its Nineteenth Session, Held in Warsaw from 11 to 23 November 2013 – Addendum – Part Two: Action Taken by the Conference of the Parties at Its Nineteenth Session* (UN Doc FCCC/CP/2013/10/Add.1 (31 January 2014) Decision 12/CP.19 ('The Timing and the Frequency of Presentations of the Summary of Information on How All the Safeguards Referred to in Decision 1/CP.16, Appendix I, are Being Addressed and Respected') ('Decision 12/CP.19,Warsaw Framework for REDD-Plus')) 33 [2].

[66] Ibid. 4 [3]. Emphasis added. [67] On REDD+ Readiness, see below n 83.

[68] Consensus has been said to constitute the absence of formal objections, rather than unanimity: see Robin R Churchill and Geir Ulfstein, 'Autonomous Institutional Arrangements in Multilateral Environmental Agreements: A Little-Noticed Phenomenon in International Law' (2000) 94 *American Journal of International Law* 623, 643.

agreements or decisions adopted in plenary sessions, often on a yearly basis, by constituent member states.[69] Like many treaty-based plenary bodies, the UNFCCC COP is empowered to do more than pass rules of procedure and budgetary matters; it has a general power to '[e]xercise such other functions as are required for the achievement of the objective of the Convention'.[70] Its decision-making procedure is particularly elaborate. All states parties to the UNFCCC are represented at the COP with an equal vote.[71] Amendments to the convention and its annexes require a three-fourths majority of state parties.[72] Yet further voting requirements are applied only by convention, because the relevant rules of procedure remain in draft. Parties have been unable to agree on the voting rule (draft Rule 42),[73] which provides that decisions on matters of substance should be taken by consensus, by two-thirds majority of parties present and voting, or by some other voting majority.[74] In the absence of agreement, UNFCCC COP decisions are taken by consensus.

As such, the Cancun Agreements and the Warsaw Framework can be said to have been agreed by a consensus of UNFCCC parties, unlike the earlier Copenhagen Accord which was only 'take[n] note of' by the UNFCCC COP.[75] However, this does not amount to a treaty obligation. Decisions of COPs are not treaties, although they can agree to amendments to the convention[76] or adopt protocols,[77] as was done by the Paris COP decision attaching the Paris Agreement. There is some debate as to their legal status. On the one hand, the COP decisions have been argued

69 See generally ibid.

70 United Nations Framework Convention on Climate Change, above n 17, art 7(2)(m).

71 Ibid. art 18. 72 Ibid. art 15.

73 Ibid. art 7(3) (requiring Rules of Procedure to be adopted by the first session of the COP).

74 At early UNFCCC COPs, draft Rules of Procedure were presented for approval, but parties failed to agree on Rule 42 (Voting). See Conference of the Parties, United Nations Framework Convention on Climate Change, *Organizational Matters: Adoption of the Rules of Procedure – Note by the Secretariat* (UN Doc FCCC/CP/1996/2 (22 May 1996) 'Draft Rules of Procedure of the Conference of the Parties and Its Subsidiary Bodies' ('Draft Rules of Procedure')).

75 Conference of the Parties, United Nations Framework Convention on Climate Change, *Report of the Conference of the Parties on Its Fifteenth Session, Held in Copenhagen from 7 to 19 December 2009 – Addendum – Part Two: Action Taken by the Conference of the Parties at Its Fifteenth Session* (UN Doc FCCC/CP/2009/11/Add.1 (30 March 2010) Decision 2/CP.15 ('Copenhagen Accord')). See further Lavanya Rajamani, 'The Making and Unmaking of the Copenhagen Accord' (2010) 59 *International and Comparative Law Quarterly* 824.

76 United Nations Framework Convention on Climate Change, above n 17, art 15.

77 Ibid. art 17.

to amount to autonomous institutional arrangements subject to international institutional law.[78] Churchill and Ulfstein argue that the legal status of COPs is subject to the doctrine of 'implied powers' (according to which any decision is allowable if it is functionally necessary: that is, if it is required for the achievement of the objective of the convention). On the other hand, there is recognition of the interpretative influence of COP decisions, especially in interpreting treaty provisions that change over time: COP decisions may even amount to 'subsequent agreement' or 'subsequent practice' reflecting the common intentions of treaty parties as to an evolving concept.[79] A recent report by the International Law Commission (ILC) has noted the important function of COP decisions in 'determining the Parties' common understanding of the meaning of the treaty'.[80] A draft conclusion by the ILC envisages a COP decision as forming a subsequent agreement or practice 'in so far as it expresses agreement in substance between the parties regarding the interpretation of a treaty, regardless of the form and the procedure by which the decision was adopted, including by consensus'.[81] This has important legal implications for REDD+'s interaction with other regimes, a topic to which we return later.[82]

In terms of substance, the COP decisions create aspirations for forest-based climate mitigation which may influence legal obligations and procedures surrounding REDD+ projects. However, more concrete stipulations on rights and responsibilities are developed through agreements between donor states and recipient states, or through agreements between international organizations and recipient states. These obligations emerge in different phases: in assistance for capacity building to make countries 'ready' to conduct REDD+ projects (known as 'readiness')[83] and in the implementation of the REDD+ projects themselves (including the proposed 'result-based payments'). At this stage, the bulk of international activity has been focused on REDD+ Readiness.

78 Churchill and Ulfstein, above n 68, 633.

79 Vienna Convention on the Law of Treaties (opened for signature 23 May 1969, 1155 UNTS 331, entered into force 27 January 1980) art 31(3)(a) and (b).

80 International Law Commission, *Report of the International Law Commission on the Work of Its Sixty-Sixth Session* (UN GAOR, 69th sess, Supp No 10, UN Doc A/69/10 (5 May–6 June and 7 July–8 August 2014) ('Report of the ILC Sixty-Sixth Session')) 208, ch VII.

81 Ibid. 205 (Conclusion 10).

82 See Chapter 10.

83 As introduced in Chapter 1 (p 8), the readiness phase for REDD+ occurs prior to the implementation phase and involves capacity building and the creation of domestic strategies, legal frameworks and institutional bodies.

Alongside these contract-like arrangements, other international actors participate in REDD+ programming and are influential in developing informal or soft-law norms such as those set out in the 'Guidelines on Stakeholder Engagement for REDD+ Readiness with a Focus on the Participation of Indigenous Peoples and Other Forest-Dependent Communities'.[84] Were we to conclude that REDD+ is an international legal regime, such soft-law norms and processes would be considered part of it. Many of these norms and processes are developed outside the formal auspices of the UNFCCC, thus leading to a complicating factor in pointing to an emerging REDD+ regime: the relevant laws and decision-making procedures may not be said to be restricted to the UNFCCC, but have emerged from contributions of other international organizations and participants, albeit acting in accordance with the broad parameters of the REDD+ scheme set out in the UNFCCC's Cancun Agreements and Warsaw Framework. Before turning to those actors, we conclude our overview of laws and decision-making procedures with a special note about customary international law.

As we argue in Chapter 3, there are norms of customary international law that are inextricably linked to the emerging REDD+ regime. These norms pertain to the rights and duties of states, and sometimes the balance between them, as well as the special position of indigenous peoples. It could be said that the functional issue area of reducing emissions from deforestation and forest degradation is inseparable from these norms of customary international law, and therefore any conception of an emerging REDD+ regime must make mention of the norms.

A prominent example is the right of indigenous peoples to be consulted over particular state initiatives, which is more extensive than what would generally be regarded as flowing from obligations to consult with the citizenry. Scholars have asserted an emerging principle of customary international law requiring government authorities to consult indigenous peoples on government policies or matters affecting them.[85] This is closely related to the practices surrounding REDD+ relating to FPIC, which, as we demonstrate in Chapter 3, give rise to conflicting practice amongst REDD+ organizations. Where such norms of consultation have achieved the status of customary international law, any description of the

[84] See especially Chapter 4.

[85] See discussion by International Law Association, *Sofia Conference (2012): Rights of Indigenous Peoples (Final Report)* (2012) 4–7.

REDD+ regime would be under-inclusive if it did not incorporate international legal protections for indigenous peoples. This issue could be said to be one of regime *definition*, rather than regime interaction.

Yet even as norms of customary international law may permeate a REDD+ regime, it is conceptually useful to consider them as part of a broader corpus of public international law. For that reason, we give separate analysis to norms of customary international law in Chapter 3. In the following section, we conclude our discussion of a putative REDD+ regime by considering its dominant organizational make-up (namely, the UN-REDD Programme and the World Bank's FCPF) and other attributes.

## B. International Organizations and Other Key Participants

The international organizations relevant to REDD+ are notably diverse and diffuse and have developed innovative arrangements with states and one another. Rather than a linear arrangement whereby the UNFCCC secretariat administers funding through a single body, separate international organizations have been developed to receive and dispense funds and assist in capacity building amongst recipient states. The two most important international organizations in the initial stages of REDD+ have been the UN-REDD Programme and the World Bank's FCPF, and we focus on these two in Sections 1 and 2. Of more recent importance is the Green Climate Fund,[86] which is the main entity entrusted by the Paris Agreement with the UNFCCC's financial mechanisms (along with the existing Global Environment Facility),[87] and

[86] The Green Climate Fund was established at COP16 in Cancun in Decision 1/CP.16 – Conference of the Parties, United Nations Framework Convention on Climate Change, *Cancun Agreements*, above n 3.; see further Conference of the Parties, United Nations Framework Convention on Climate Change, *Report of the Conference of the Parties on Its Seventeenth Session, Held in Durban from 28 November to 11 December 2011 – Addendum – Part Two: Action Taken by the Conference of the Parties at Its Seventeenth Session* (UN Doc FCCC/CP/2011/9/Add.1 (15 March 2012) Decision 3/CP.17 ('Launching the Green Climate Fund'), Annex ('Governing Instrument for the Green Climate Fund')). For background, see Richard B Stewart, Bryce Rudyk and Kiri Mattes, 'Governing a Fragmented Climate Finance Regime' in Hassane Cissé, Daniel D Bradlow and Benedict Kingsbury (eds), *The World Bank Legal Review: International Financial Institutions and Global Legal Governance: Volume 3* (World Bank and Martinus Nijhoff, 2012), 363, http://siteresources.worldbank.org/INTLAWJUSTICE/Resources/WBLegalReviewVol3.pdf.

[87] Paris COP Decision, above n 3, para 59.

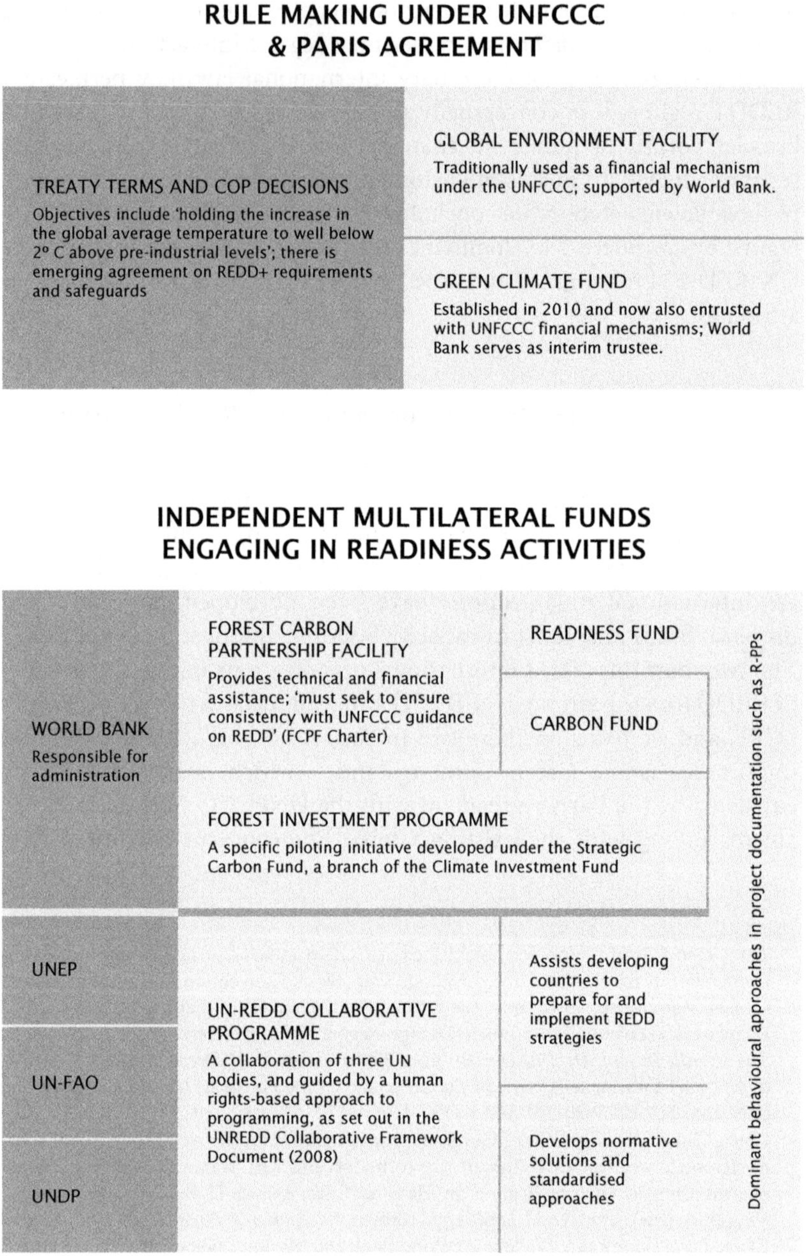


**Figure 2.1.** Rule making and institutional support for REDD+

which is anticipated to play a 'key role'[88] in channeling REDD+ finance in the future.[89]

### 1. *UN-REDD Programme*

The UN-REDD Programme was launched in 2008 after the 2007 UNFCCC COP13 in Bali that first formally endorsed the concept of avoided deforestation. The UN-REDD Programme's aim is expressed as an overall development goal to 'reduce forest emissions and enhance carbon stocks in forests while contributing to national sustainable development'. This is phrased according to a theory of change emphasising political commitments, political will and finance, and a willingness to engage in multi-stakeholder consultative processes at a national level.[90] This goal, expressed in a strategic framework for 2016–2020, updates the previous five-year strategy (2011–2016) to:

> promote the elaboration and implementation of National REDD+ Strategies to achieve REDD+ readiness, including the transformation of land use and sustainable forest management and performance-based payments.[91]

The UN-REDD Programme is a coalition of three UN agencies: UNEP, FAO and UNDP. In carrying out this work, the UN-REDD Programme has drawn on the mandates of these three overseeing agencies, although at times it has been criticised as lacking a clearer and self-standing approach.[92] The UN-REDD Programme adopts

[88] Conference of the Parties, United Nations Framework Convention on Climate Change, Decision 9/CP.19, Warsaw Framework for REDD-Plus, above n 64, para 5.

[89] See further Green Climate Fund, *Decisions of the Board – Eighth Meeting of the Board, 14–17 October 2014* (GCF/B.08/45, 3 December 2014) Annex X: 'Initial Logic Model for REDD+ Results-Based Payments, and Annex XI: Performance Measurement Framework for REDD+ Results-Based Payments.

[90] UN-REDD Programme, *UN-REDD Programme Strategic Framework* 2016–2020 (UNREDD/PB14/2015/III/3, Revised Draft, 7 May 2015).

[91] UN-REDD Programme, *UN-REDD Programme Strategy 2011–2015* (February 2011) 6. See also the FAO, UNDP and UNEP, *UN Collaborative Programme on Reducing Emissions from Deforestation and Forest Degradation in Developing Countries (UN-REDD): Framework Document* (20 June 2008). As mentioned above, REDD+ Readiness refers to the assistance given to countries to build their capacity to host REDD+ projects: see above n 83.

[92] Alain Frechette, Minoli de Bresser and Robert Hofstede, *External Evaluation of the United Nations Collaborative Programme on Reducing Emissions from Deforestation and Forest Degradation in Developing Countries (the UN-REDD Programme): Volume I – Final Report* (July 2014), http://wedocs.unep.org/rest/bitstreams/776/retrieve.

a self-proclaimed 'human rights–based approach' (HRBA) to programming,[93] with particular reference to the UN Development Group (UNDG) guidelines on indigenous peoples' issues, as we discuss further in Chapter 3. In particular, it incorporates an FPIC standard that requires *consent* as opposed to *consultation* as a core principle of stakeholder engagement, in contrast to the World Bank FCPF which we consider in Section 2.

The UN-REDD Programme's governance structures were revised in 2016. Where it was formerly governed by a policy board, a management group and a secretariat, it now comprises an executive board, an assembly, national steering committees and a multi-party trust fund office. While the former policy board had indigenous representatives from three regions (and included the chairperson of the UN Permanent Forum on Indigenous Issues, which has been very active in advocacy on REDD+ issues), the composition of the executive board now includes one permanent observer to represent indigenous peoples.

The influence of the UN-REDD Programme is apparent in major decisions relating to REDD+'s impact on indigenous peoples and forest-dwelling communities, including guidelines on FPIC, but it is not necessarily because of its legally expressed mandate. A 2014 evaluation found that recognition of indigenous rights had been supported by the UN-REDD Programme's implementation of the Cancun Safeguards, although the social and environmental safeguards were difficult to put into practice due in part to 'lack of clear guidance on how to implement, monitor and enforce these'.[94] Indeed, in its most current strategic plan, the UN-REDD Programme has stated that:

> Developing safeguards too early in the readiness phase has led to expansive safeguard systems with literally hundreds of indicator[s], far beyond the scope of the REDD+ interventions. This can lead to high expectations that REDD+ safeguards will be the panacea for all forest-related challenges.[95]

These indicators are found in a plethora of documents including the Readiness Preparation Proposals (R-PPs), which are discussed in Section C.

[93] FAO, UNDP and UNEP, above n 91, 6.
[94] Frechette, de Bresser and Hofstede, above n 92, vi.
[95] UN-REDD Programme, above n 90, 4.

### 2. *Forest Carbon Partnership Facility*

In contrast, the FCPF is based within the World Bank Group. Launched at the Bali COP after consultations between World Bank and The Nature Conservancy, it operates independently of the Bank's main funding programmes and has its own trustee arrangement, and a charter that was most recently revised in 2015.[96] The FCPF Charter states that it 'must seek to ensure consistency with UNFCCC guidance on REDD'.[97] The FCPF's fund-dispensing bodies are the Readiness Fund and the 'Carbon Fund', which seek to build capacity within recipient countries and also pilot performance-based payment systems for emissions reduction.[98] There is a stated ethos of 'learning by doing',[99] and the FCPF's objectives include disseminating knowledge gained in the development and implementation of REDD programmes[100] and 'test[ing] ways to sustain or enhance livelihoods of local communities and to conserve biodiversity'. To date, there are forty-seven country participants, including Papua New Guinea and Vanuatu; Malaysia is not yet a participant.[101]

### 3. *Other Funding Bodies*

In addition to the UN-REDD Programme and the FCPF, a range of other international organizations and coalitions of states aim to advance REDD+ objectives; it is clear that the emerging REDD+ regime will not be limited by the terms of the UNFCCC COP decisions or by the institutions that have been involved in its initial implementation.

Whilst we have concentrated on the UN-REDD Programme and the FCPF in this book, the other prominent multilateral initiatives are the Global Environment Facility (GEF) and the Forest Investment Program (FIP), supported by the World Bank. A separate platform called the

[96] International Bank for Reconstruction and Development, *Charter Establishing the Forest Carbon Partnership Facility* (Revised 23 November 2015), www.forestcarbonpartnership.org/charter-and-governance-documents. See further www.forestcarbonpartnership.org/history (para 1).

[97] Ibid. s 3.1. 'UNFCCC guidance on REDD' is defined in the charter as 'the body of rules, modalities, procedures and guidelines on REDD that is adopted under the auspices of the UNFCCC': Ibid. art 1.1.80.

[98] International Bank for Reconstruction and Development, above n 96, art 2.1.

[99] Ibid. art 3.1(b); art 2.1(d).

[100] 'REDD+ Countries' The Forest Carbon Partnership Facility, www.forestcarbonpartnership.org/redd-countries-1.

[101] International Bank for Reconstruction and Development, above n 96, art 2.1(c).

REDD+ Partnership was established in 2010 to support countries in coordinating efforts to seek funding.[102] Smaller, regional operations include the Pacific Islands Regional Policy Framework for REDD+, endorsed by the Pacific Ministers for Agriculture and Forestry in 2012 and including South Pacific nations such as Papua New Guinea and Vanuatu.[103] The Governors' Climate and Forests Task Force is a subnational collaboration amongst twenty-six states and provinces from Brazil, Indonesia, Mexico, Nigeria, Peru, Spain and the United States.[104] Bilateral arrangements exist, such as the bilateral arrangement between Norway and Indonesia and Norway and Guyana. There are also pilot projects under voluntary market standards. Even within the UN, other approaches have emerged to support REDD+, such as the UN Climate Summit.[105] This plethora of arrangements has led to REDD+ being described as an 'assemblage.'[106]

The diffuse arrangements have caused concern to some, given the risk of overlap, poor coordination, inefficiencies, gaps and uncertainty. At the Warsaw COP in 2013, attempts were made to unify the allocation of REDD+ financing. The result was a decision that provides for the Green Climate Fund (GCF) to have a 'key role' in channeling 'adequate and predictable results-based finance in a fair and balanced manner'.[107] The GCF, which has developed its own models to measure mitigation performance in the land sector,[108] was requested to apply the methodological

[102] See www-cif.climateinvestmentfunds.org/fund/forest-investment-program; www.thegef.org, 'Interim REDD+ Partnership Document' (27 May 2010) 1; Global Witness, 'Understanding REDD+: The Role of Governance, Enforcement and Safeguards in Reducing Emissions from Deforestation and Forest Degradation' (November 2010).

[103] Secretariat of the Pacific Island Community, 'Pacific Islands Regional Policy Framework for REDD+' (Supported by the SPC/GIZ Regional Project, 'Climate Protection through Forest Conservation in Pacific Island Countries', 2013), http://theredddesk.org/countries/policies/pacific-islands-regional-policy-framework-redd. See further in Chapter 7.

[104] 'Governors' Climate and Forests Task Force', www.gcftaskforce.org/about.

[105] The UN Climate Summit is not part of the UNFCCC negotiating process, but was instead convened as a stand-alone process of the United Nations Secretary-General Ban Ki-Moon in 2014. See further 'Climate Summit 2014: Catalyzing Action' (23 September 2014), www.un.org/climatechange/summit/about.

[106] Boyd, above n 5.

[107] Decision 9/CP.19, Warsaw Framework for REDD+, above n 64.

[108] See Green Climate Fund, above n 89. See further Christina Voigt and Felipe Ferreira, 'The Warsaw Framework for REDD+: Implications for National Implementation and Results-Based Finance' in Christina Voigt (ed) *Research Handbook on REDD+ and International Law* (Edward Elgar, 2016) 30, 50–53.

guidance consistent with earlier UNFCCC COP decisions.[109] An enduring challenge will be to ensure that the GCF acts in ways that are compatible with the Cancun Safeguards,[110] as well as more general intentions relating to the implementation of benefit-sharing arrangements at a national level.[111]

### *4. Nonstate Actors*

Aside from these organizations, REDD+ has mobilised a range of other actors, which have participated within UN-REDD Programme and FCPF arrangements as well as more broadly in climate change policy-making and standard setting. As den Besten and others have demonstrated,[112] relevant actors involved in the formation of the REDD+ idea include policy-makers, academics, civil society groups and private-sector representatives.

The most significant group of interested parties for our purposes is indigenous peoples. As we assess in this book, their engagement has been an ongoing challenge in the crafting of REDD+ and has received varied assessment. One study of the UNFCCC negotiations found that, although alliances of indigenous peoples and forest communities have emerged, their ability to directly participate and affect outcomes in the REDD+ design process has been weak.[113] A later study was more positive, finding that indigenous peoples had influenced the REDD+ agenda in a variety of ways.[114] The capacity of these groups to traverse different regimes and institutional contexts is an important issue, which we continue to assess throughout the book.

[109] Decision 9/CP.19, Warsaw Framework for REDD+, above n 64, para 7.

[110] Annalisa Savaresi, 'The Legal Status and Role of Safeguards' in Christina Voigt (ed) *Research Handbook on REDD+ and International Law* (Edward Elgar, 2016) 126, 132.

[111] Sandrine Mathy and Odile Blanchard, 'Proposal for a Poverty-Adaptation-Mitigation Window within the Green Climate Fund' (2016) 16 *Climate Policy* 752.

[112] One of their specific research questions was to ask which actors took part in the negotiation process that culminated in the REDD+ agreements: see den Besten, Arts and Verkooijen, above n 6.

[113] Heike Schroeder, 'Agency in International Climate Negotiations: The Case of Indigenous Peoples and Avoided Deforestation' (2010) 10 *International Environmental Agreements: Politics, Law & Economics* 317.

[114] Linda Wallbott, 'Indigenous Peoples in UN REDD+ Negotiations: "Importing Power" and Lobbying for Rights through Discursive Interplay Management' (2014) 19(1) *Ecology and Society* 166.

Other nonstate actors have been active in observing and reporting on policy developments: the International Institute for Sustainable Development has provided reports about safeguard implementation after capacity-building workshops conducted in Africa in 2011; similarly, the World Resources Institute (WRI) has issued guidelines for instituting REDD+ safeguards at the national level.[115] Groups such as The Forests Dialogue provide innovative arrangements for dialogue and information sharing, while the International Union of Forest Research Organizations (IUFRO) has sought to provide overarching expositions of forest-focused and forest-related regimes.[116] Academics, activists and researchers also engage in advocacy, sometimes issuing 'declarations' writ large, involving informal coalitions.[117] These are just some examples of the ongoing involvement of nonstate actors in the field.

As we demonstrate with our country-specific examples, different groups have shaped national implementation as well as international normative design. This is expressly envisaged by the international REDD+ organizations. The UN-REDD Programme, for example, invited indigenous representatives to take a seat on the UN-REDD Policy Board early in its operations.[118] Meanwhile, the FCPF provides for a 'Participants Assembly' and a 'Participants Committee', and representatives of relevant international organizations, NGOs, forest-dependent indigenous

[115] Gaia Larsen and Florence Daviet, 'Safeguarding Forests and People: A Framework for Designing a National System to Implement REDD+ Safeguards' (World Resources Institute, November 2012), www.wri.org/publication/safeguarding-forests-and-people.

[116] John A Parrotta, Christoph Wildburger and Stephanie Mansourian (eds), *Understanding Relationships between Biodiversity, Carbon, Forests and People: The Key to Achieving REDD+ Objectives – A Global Assessment Report* (International Union of Forest Research Organizations (IUFRO), IUFRO World Series vol. 31, 2012); Jeremy Rayner et al, 'Introduction' in Jeremy Rayner, Alexander Buck and Pia Katila (eds), *Embracing Complexity: Meeting the Challenges of International Forest Governance. A Global Assessment Report* (Global Forest Expert Panel on the International Forest Regime, IUFRO World Series vol. 28, 2010).

[117] For two examples, see Global Environmental Justice Think Tank 'Second Norwich Declaration on Environmental Justice: From Safeguarding to Mainstreaming Local Rights: Democratic Forest Governance as Pre-Requisite for a Just and Effective REDD+' (Declaration of Global Environmental Justice Think Tank, 14–16 May 2014); Forest Peoples Programme, 'The Palangka Raya Declaration on Deforestation and the Rights of Forest Peoples' (Declaration of Looking Ahead: International Workshop on Deforestation Drivers and the Rights of Forest Peoples, Palangka Raya, Central Kalimantan, Indonesia, 18 May 2014), www.forestpeoples.org/sites/fpp/files/news/2014/03/declaration_english.pdf.

[118] See Section B.1 above.

peoples and forest dwellers, and relevant private-sector entities may be invited to observe the annual meetings.[119] Charter provisions establish the definitions of these representatives, which are highly dependent on previous experience and expertise or the representations of affected interests, as follows:[120]

> 34. 'Forest-Dependent Indigenous Peoples and Forest Dwellers' means Indigenous Peoples and local communities living in forests and depending on forest resources for their livelihoods;
>
> . . .
>
> 67. 'Relevant International Organization' means an intergovernmental association of countries, established by and operated according to multilateral treaty, which has experience and expertise in REDD;
>
> 68. 'Relevant Non-Governmental Organization' means a civil society organization or a non-governmental organization that has experience and expertise in REDD or represents interests that are likely to benefit from or be affected by REDD;
>
> 69. 'Relevant Private Sector Entity' means a private company or organization that is not registered as a non-governmental organization that is likely to benefit from or be affected by REDD;

This follows the practice of other regimes such as the Convention on Biological Diversity (CBD) and the Convention on International Trade in Endangered Species of Wild Fauna and Flora (CITES), which make observership contingent on past involvement or expertise or on the representation of key affected interests.

The nonstate participants of the emerging REDD+ regime have been highly influential in shaping the institutional discourses of REDD+. The co-evolution of local and international practices, each drawing on precedents established in the other, has resulted in a form of bench marking, where the experiences of REDD+ pilot schemes and readiness work in particular states have established achievable goals which then inform standards developed by international bodies and organizations. These processes of experimentation and consolidation are explored further in Chapter 10, but also make it difficult to identify a concrete and relatively static set of key participants at this point in time, meaning that it may be premature to conclude that REDD+ is a regime. A similar difficulty arises in trying to identify the key behaviours, assumptions and biases underlying REDD+, as we discuss in the next section.

[119] International Bank for Reconstruction and Development, above n 96, s 10.1.
[120] Ibid. s 1.1.

## C. Key Behaviours, Assumptions and Biases

It is difficult to isolate key discursive practices for REDD+. Even the two main organizations, the UN-REDD Programme and the FCPF, show diverging modes of behaviour, assumptions and biases. A marked division in preferences relates to the differing attitude to the concept of FPIC as between these two sets of organizations, as we discuss in Chapter 4.[121] Moreover, the FCPF has a clearer goal in establishing 'carbon-fund readiness' in its recipient countries than the UN-REDD Programme, which is agnostic about whether REDD+ will develop into a carbon-trading mechanism. The complexity increases when one bears in mind that countries generate their own guidelines and procedures with respect to concepts such as FPIC as part of their national REDD+ strategies.[122] The diverging assumptions and discursive practices therefore make it difficult to essentialise practises with a REDD+ regime. However, there are three areas in which at least some points can be made about the assumptions that are being furthered by an emerging REDD+ regime: in its scope (which evolved from RED to REDD to REDD+), in the evaluative criteria that are being adopted by key institutions and in compliance with safeguards.

### *1. RED to REDD to REDD+*

The original proponents of an enlarged emphasis on forests in UNFCCC climate-mitigation efforts concentrated their attention on deforestation.[123] Flowing from previous economic approaches that supported 'payments for ecosystem services', the idea was to provide appropriate incentives to forested, tropical countries to keep their forests standing. The idea extended previous ecosystem services models because it was

[121] In contexts unrelated to REDD+, empirical studies have suggested that cultural attitudes within the World Bank marginalise human rights concepts: see Galit A Sarfaty, 'Why Culture Matters in International Institutions: The Marginality of Human Rights at the World Bank' (2009) 103 *American Journal of International Law* 647.

[122] See, e.g., Sophia Carodenuto and Kalame Fobissie, 'Operationalizing Free, Prior and Informed Consent (FPIC) for REDD+: Insights from the National FPIC Guidelines of Cameroon' (2015) 9 *Carbon & Climate Law Review* 156.

[123] Submission of Papua New Guinea and Costa Rica, above n 52. See further Antonio GM La Viña, Alaya De Leon and Reginald Rex Barrer, 'History and Future of REDD+ in the UNFCCC: Issues and Challenges' in Christina Voigt (ed), *Research Handbook on REDD+ and International Law* (Edward Elgar, 2016) 11.

based on the assumption that a baseline of carbon could be established in the forest that could be compared to a 'business-as-usual' model, thus establishing the extent of the threat to that stored carbon. As such, the RED proposals extended parallel approaches in other environmental regimes, including the CDM of the Kyoto Protocol. The latter approach sought to incentivise emissions reduction through *positive* emission-reducing projects such as the construction of renewable energy plants (which without CDM funding would not be economical to construct), whereas RED was a negative model that sought to avert threats to existing carbon.

These RED proposals sought to use existing and relatively inexpensive remote-sensing technology to track rates of deforestation, which would be accounted for at the national level. Paying to maintain the 'at-risk' forests was seen as inexpensive relative to alternative sources of emissions reductions, and it was envisioned that the compensation would be sourced from carbon markets established under the UNFCCC.[124] Outside of the UNFCCC, economist Nicholas Stern, reporting to the British Government, stated, "Curbing deforestation is a highly cost-effective way of reducing greenhouse gas emissions and has the potential to offer significant reductions fairly quickly. It also helps preserve biodiversity and protect soil and water quality."[125]

Within the UNFCCC, the RED idea was promoted as a tangible expression of the principle of 'Common but Differentiated Responsibilities', because it would establish a means by which developing countries could contribute to global mitigation without compromising their development needs.[126] Importantly, it was also assumed that a reduction in deforestation would necessarily lead to co-benefits,[127] such as protection of biodiversity and improved ecosystem services. As previously mentioned, Costa Rica and Papua New Guinea, in a joint submission prepared on behalf of the Coalition of Rainforest Nations, formally presented Reduced Emissions from Deforestation (RED) to the UNFCCC at COP11 in Montreal.[128] Seeing the potential of the idea, the COP agreed to establish a two-year review of relevant scientific, methodological and political issues by way of two expert workshops

[124] Submission of Papua New Guinea and Costa Rica, above n 52.
[125] Nicholas Stern, *The Economics of Climate Change: The Stern Review* (Cambridge University Press, 2007) 603.
[126] Pistorius, above n 51, 640. [127] Ibid.
[128] Submission of Papua New Guinea and Costa Rica, above n 52.

under the Subsidiary Body for Scientific and Technological Advice (SBSTA).

The SBSTA-led workshops focused on the scope of RED,[129] apparently to address the fact that not all tropical-forested nations would be eligible to participate under the original formulation. Pistorius, den Besten and others have tracked the emergence, broadly speaking, of two coalitions: countries characterised by low deforestation rates/high degradation that wanted RED expanded to include 'degradation'; and countries with low but stable forest cover or expanding forest cover, that pushed for inclusion of conservation, sustainable forest management and enhancement of forest carbon stocks (the '+'). The latter included India and China,[130] which argued that REDD as then conceived was a reversal of the 'polluter pays' principle, given that it would reward countries that had not taken care to avert deforestation.

The COP officially sanctioned the extension of RED to REDD at COP13 in Bali in 2007 and again at COP14 in Poznan the following year,[131] thus implicitly accepting the argument that had been made by countries that there was scientific capability to account for reduced emissions from forest degradation as well as deforestation. The COP's Bali Action Plan also expanded REDD to the + form, and thus sought to accommodate conservation, sustainable forest management and enhancement of forest carbon stocks within the scope of the programme. There are generally two separately but not necessarily incompatible accounts of why this occurred: first, because it became clear that India's and China's support was contingent on this expanded form being formally adopted by the COP,[132] and second, because countries were so buoyed by the energy surrounding REDD and initial donor commitments that consensus was possible despite the outstanding concerns of those countries opposed to expansion.[133]

[129] Pistorius, above n 51, 640; den Besten, Arts and Verkooijen, above n 6.

[130] Pistorius, above n 51, 640. But note the other possible reasons for the expansion of the scope of RED discussed by Wiersema, above n 7.

[131] Conference of the Parties, United Nations Framework Convention on Climate Change, 'Bali Action Plan', above n 53. den Besten, Arts and Verkooijen, above n 6, 40, 43 (citing various COP decisions at Bali and previous COP reports).

[132] Pistorius, above n 51, 641.

[133] Ingrid J Visseren-Hamakers et al., 'Trade-Offs, Co-Benefits and Safeguards: Current Debates on the Breadth of REDD+' (2012) 4 *Current Opinion in Environmental Sustainability* 646, 649.

Although the shift from RED to REDD+ greatly expanded the number of countries eligible for REDD+ funding, it also had fundamental political and technical implications, as it was a significant move away from core characteristics, such as simplicity and cost-effectiveness, upon which RED's initial widespread appeal was based. As stated by Pistorius, "Although it was well-reasoned, the decision [to expand REDD+'s scope] marked the end of the idea of a simple mechanism ... MRV [measurement, reporting and verification] of carbon stock changes requires much more sophisticated and expensive methods than MRV of land use changes through the application of remote sensing techniques."[134] The expansion would make it harder to implement technical aspects and also to monitor progress.[135] In addition, Visseren-Hamakers et al. have noted that the addition of the '+' meant that the social and biological impacts of REDD would also be greatly expanded, as would the potential trade-offs between competing values.[136] The expanded functional aims of REDD+, together with what den Besten and others describe as a 'discursive-institutional spiral' from 2004 to 2011,[137] would expose the emerging REDD+ regime to the normative ideals and exigencies of other international regimes. Chapter 3 provides an overview of the most relevant of these regimes, in order to foreshadow sites of interaction and opposition as well as key influences. Before turning to those, it is important to mention another set of practices that informs and promotes the key behaviours, assumptions and biases of an emerging REDD+ regime.

### 2. *Evaluative Criteria such as 'Readiness' Proposals or 'R-PPs'*

In assessing the 'readiness' of countries to receive REDD+ funds, the UN-REDD Programme and the FCPF have adopted a common template for readiness preparation proposals or R-PPs.[138] They have also jointly developed various guidelines on technical matters, and sought to share

[134] Pistorius, above n 51, 640; Wiersema, above n 7, 34. [135] Wiersema, above n 7, 6.

[136] Visseren-Hamakers et al, above n 133, 649.

[137] The spiral is said to occur due to the dynamic nature of opening up discourses through enlarged constellations of actors, and then the concentration and refinement of discourses through the subsequent institutionalisation: den Besten, Arts and Verkooijen, above n 6, 46.

[138] See Forest Carbon Partnership Facility (FCPF) and The United Nations Collaborative Programme on Reducing Emissions from Deforestation and Forest Degradation in Developing Countries (UN-REDD), 'Readiness Preparation Proposal (R-PP)' (Template Version 6, Working Draft, 4 April 2012).

information and expertise. The R-PP seeks to follow the Cancun Safeguards as well as be receptive to internal organizational policies or regulations, as we describe further in Chapter 3.[139]

Notwithstanding these efforts at coordination, the evaluative criteria imposed by different agencies at times diverge. This is inevitable given that the organizations are seeking continuous improvement on technical issues.[140] However, it can also leave countries uncertain about the REDD+ Readiness requirements. For example, the UN-REDD Programme has identified seven 'social and environmental principles' attached to the Cancun Safeguards, and twenty-four corollary criteria to give content to those principles. These are identified in the UN-REDD Programme Social and Environmental Principles and Criteria (SEPC) (2012) and the associated UN-REDD Draft Social and Environmental Principles and Criteria – Benefit and Risks Tool (SEPC-BeRT) (2013). These approaches are informed by the human rights-based approach of the UN-REDD Programme. This is contrasted with the FCPF, which, although seeking to ensure consistency with UNFCCC guidance on REDD+, maintains an emphasis on 'risk management' rather than rights. The FCPF has greater experience and interest in managing the implementation of REDD+,[141] given the World Bank's broader work in providing results-based payments. Thus, whether REDD+ is motivated more by 'rights' or by 'risks' is unresolvable and often dependent on which organization is in charge.

The divergences and discrepancies in REDD+ are of course not limited to the international level. Indeed, in much of the evaluations of readiness preparations so far, states are reminded that REDD+ intersects with a whole range of other domestic policies and sectors, and that coordination amongst them is essential.[142] A clear example of this is in the discourse relating to 'co-benefits', which we explore in Chapter 6. A co-benefit in some contexts of REDD+ may well be seen in other contexts as a

[139] For example, the UN-REDD Programme cites its own 'Social and Environmental Principles and Criteria' (SEPC), developed with guidance from the UN-REDD Policy Board and through a public consultation process: Ibid. 3.

[140] Comment of UN-REDD Programme official to author, 27 May 2015.

[141] See Frechette, de Bresser and Hofstede, above n 92, 26.

[142] The R-PP template notes the need for 'some form of cross-sectoral REDD-plus working group ... [to include] key agencies (e.g. agriculture, mining, transportation) and interest forest agency': see Forest Carbon Partnership Facility (FCPF) and The United Nations Collaborative Programme on Reducing Emissions from Deforestation and Forest Degradation in Developing Countries (UN-REDD), 'Readiness Preparation Proposal (R-PP)' (Template Version 6 20 April 2012) 8.

fundamental human right which ought to be provided by the state as a baseline. Different domestic contexts are thus necessary to explore, which we undertake in the book's Part III.

### 3. *Compliance with Safeguards*

The ability of affected communities to seek redress for lack of adherence by states to the Cancun Safeguards, once a project has started to be implemented, is linked to the identity and internal procedures of the relevant implementing international agency. For example, as we previously noted, the FCPF Charter states that it 'must seek to ensure consistency with UNFCCC guidance on REDD'.[143] As such, the FCPF is bound to follow the Cancun Safeguards, at least in the implementation of REDD+ projects (after the readiness stage).[144] Moreover, the FCPF is subject to the World Bank's existing operational policies and procedures, 'taking into account the need for effective participation of Forest-Dependent Indigenous Peoples and Forest Dwellers in decisions that may affect them, respecting their rights under national law and applicable international obligations'.[145] The implication is that affected communities could submit complaints to the World Bank's existing independent Inspection Panel for failures of the FCPF to adhere to the Cancun Safeguards in REDD+ implementation. There are important precedents of such complaints outside of the REDD+ context. For example, members of the Cherangany-Sengwer communities living in the Cherangany Hills in Kenya claimed that they had been forcibly evicted from their lands within the forests as a result of a World Bank–funded conservation project, and the Panel found that the Bank's relevant safeguard policies had not been correctly applied.[146] (The UN-REDD Programme does not contain a similar procedure; at this stage its role has been

143 International Bank for Reconstruction and Development, above n 96, s 3.1(c). For the definition of 'UNFCCC guidance on REDD', see above n 97.

144 On a criticism of the lack of clarity over the application of World Bank policies at the REDD+ Readiness phase or later implementation phase, see Kate Dooley et al, *Smoke and Mirrors: A Critical Assessment of the Forest Carbon Partnership Facility* (FERN and Forest Peoples Programme, February 2011) 7.

145 International Bank for Reconstruction and Development, above n 96, s 3.1(d).

146 The World Bank and The Inspection Panel, 'Joint Statement from World Bank Management and Inspection Panel' (News Release, 2014/143/ECR, 2 October 2014). See further The World Bank and The Inspection Panel, *Kenya: Natural Resource Management Project* (14 January 2013), http://ewebapps.worldbank.org/apps/ip/Pages/ViewCase.aspx?CaseId=89.

limited to 'readiness' activities rather than implementation activities.) It is therefore not sufficient to evaluate key behaviours and assumptions without considering the overarching compliance and grievance mechanisms at play.[147]

## D. Conclusions

We have argued in this chapter that REDD+ may be conceptualised as an emerging international legal regime. Drawing on the definition of regimes outlined in the beginning of the chapter,[148] we consider that there are sets of norms, decision-making procedures, and organizations within and outside of the UNFCCC that are beginning to coalesce around the objectives of reducing emissions from deforestation and forest degradation and promoting activities relating to conservation, sustainable management of forests and enhancement of forest carbon stocks. The objectives have been expressed in treaty law and in COP decisions, but the attendant legal obligations remain open and adaptable to the emerging knowledge and practice of a variety of actors. The actors include UNFCCC parties, organizations such as the UN-REDD Programme and the FCPF, civil society groups and others, whose modes of behaviour, assumptions and biases are combining to reveal dominant trends within an emerging REDD+ regime. Divergences in organizations and behaviours make it difficult to agree on a concrete REDD+ regime, which is instead dynamic and experimental in character. Later chapters explore modes of behaviour, assumptions and biases in more depth, including with respect to tenure, FPIC and benefit-sharing.

This chapter has reviewed the set of decisions and organizations coalescing around the UNFCCC objective of forest-based climate mitigation. These do not, at least at the present time, amount to hard-law obligations for states. Indeed, we have shown that the REDD+ regime has limited hard-law obligations for states, but that soft-law instruments such as the Cancun Safeguards seek to provide substance to the implementation of REDD+ as well as prescriptions for REDD+'s relationship to existing international norms. While these safeguards lack the legal status

[147] Kristen Hite, 'Adjudicating Disputes across Scales: Global Administrative Law Considerations for REDD+' in Christina Voigt (ed), *Research Handbook on REDD+ and International Law* (Edward Elgar, 2016) 408.

[148] See Young, above n 11.

of treaties, they remain relevant for current, past and future relations amongst state and nonstate actors.

In addition to soft-law aspirations, there are bilateral obligations owed by developing countries receiving REDD+ funds to donor states and institutions. Of particular importance have been the activities of key bodies such as the UN-REDD Programme and the FCPF, which have issued guidelines and reviewed country requests for 'readiness' funding. So far, the assumptions have been of an experimentalist kind, seeking to achieve 'learning by doing' through capacity building and pilot projects, and it is important to trace the practical operation through our country-specific examples and through our broader discussion of regime interaction.

Negotiating states within the UNFCCC have sought, through COP decisions, to provide safeguards for participation, indigenous peoples, biodiversity, and complementarity or consistency with relevant international conventions and agreements. International organizations have mediated the relationship with other laws by invoking them in policy documents and programme requirements. The next chapter delves more deeply into the potential for REDD+ to interact with other regimes by setting out forest-focussed and forest-related international laws and custom that have overlaps and possible conflicts with REDD+, including those relating to forestry, human rights (including indigenous rights), and environmental protection.

3

# REDD+'s Broader International Legal Context

MARGARET YOUNG

As highlighted in the previous chapter, the idea that emissions from deforestation can be reduced through economic incentives has emerged amongst an existing suite of international laws and institutions. These are directed not only at forests but also at the people and biological diversity that inhabit them, including indigenous peoples and forest communities. Some of the existing laws and institutions have coalesced around specific legal 'regimes', a shorthand term we utilised in Chapter 2 to denote particular concentrations of normative development and professional specialisation in international law, such as the 'biodiversity protection regime', 'human rights regimes' and 'the climate regime'. Policy approaches and positive incentives to reduce emissions from deforestation and forest degradation (and sustainably manage forests, enhance forest carbon stocks and engage in conservation) (REDD+) exist in tandem with, and sometimes in opposition to, a range of these regimes. For example, while the activities promoted by REDD+ may be effective in sequestering carbon and achieving mitigation goals, they may also cause or allow negative environmental and social impacts, such as the destruction of biological diversity, displacement of indigenous and forest-dependent peoples, loss of traditional knowledge and property, and elimination of subsistence and livelihood means. Legal protections against such impacts are represented in other existing regimes, thus giving rise to normative overlap and even conflict. Chapter 2 demonstrated that REDD+'s formation and implementation has and will continue to be impacted by the highly complex, fragmented and pluralist nature of the international legal system, with especial reference to the safeguards created by COP decisions. In this and later chapters, we delve more deeply into the fragmented legal landscape in order to consider consequences of this pluralism for REDD+, with particular attention on the impact on indigenous peoples. This chapter concentrates on fragmentation at the level of international law.

We begin in Section A with some key international regimes that relate to REDD+. Given that REDD+ is emerging amongst an existing suite of international laws and institutions, it is inevitable that participating states will have existing and sometimes conflicting obligations. We begin our discussion by providing a snapshot – in the form of an infographic – of selected international obligations of countries participating in the UN-REDD Programme and the FCPF. We then provide, in Section B, further details of the selected regimes, namely those relating to forests, human rights and indigenous peoples, environmental protection, and the norms and institutions promoting food security. In Section C, we provide a brief review of relevant norms of customary international law relevant to REDD+. This overview of directly and indirectly relevant international norms and institutions reveals our interests in the boundaries and bridges relevant to REDD+ and its impact on indigenous peoples. Section D provides our preliminary assessment of the types of issues relating to fragmentation and regime interaction which arise in the book.

## A. Fragmentation and REDD+ Obligations

It is trite to observe that much of public international law has developed sporadically and in response to seemingly discrete issues, leading to regimes that are often self-contained and legally isolated from one another, both in terms of their institutional support and their normative application.[1] This has important implications for the making of new laws, such as those structuring REDD+, as well as the coordination, implementation, and interpretation of those laws and decisions and the recognition or resolution of possible conflicting norms. While we do not attempt an exhaustive list of forest-focused and forest-related regimes, the examples discussed in this chapter are those that most closely evoke possibly conflicting norms or practices relating to indigenous peoples, namely the legal regimes that have developed surrounding forestry, human (including indigenous) rights, environmental protection and food security. Others could have been selected; for example, in countries that

[1] See further International Law Commission, *Fragmentation of International Law: Difficulties Arising from the Diversification and Expansion of International Law* (UN GAOR, 58th sess, UN Doc A/CN.4/L.682 (13 April 2006), UN Doc A/CN.4/L.682/Add.1 (2 May 2006), and UN Doc A/CN.4/L.682/Corr.1) (11 August 2006) ('ILC Fragmentation Analytical Study').

have not been our focus of study, REDD+ interacts with energy policy[2] or even the illegal drug trade.

Along with the caveat that our list is not exhaustive, we note further that the degree of potential intersections between REDD+ and other international regimes is often determined by the local conditions surrounding the particular forest targeted by REDD+ schemes – which may be threatened by the demand for logging, agriculture, charcoal fuel, palm oil or other development projects – and which may be subject to existing or future international oversight for a range of reasons quite separate from forest-related climate mitigation. We have been guided in our selection of relevant regimes by the Cancun Safeguards, which as described in Chapter 2 have identified areas of law and governance which need to be supported and promoted when countries undertake REDD+ activities.

This section describes the potential overlaps with respect to selected instruments such as the Convention on Biological Diversity (CBD), the Convention on the International Trade in Endangered Species of Flora and Fauna (CITES), the International Labour Organization (ILO) Convention No 169, the CBD Nagoya Protocol and even sources of law that are not treaties such as the UN Declaration on the Rights of Indigenous Peoples (UNDRIP). The substance and relevance of these laws is assessed in greater detail in Section B. In short, they include REDD+'s possible contradictions with other international objectives such as the protection of biological diversity or indigenous rights.

The resulting complexity is represented graphically by Figure 3.1.[3] It is a static version of an online infographic we developed of REDD+ and related international laws. The infographic enables a user to select a specific country (or sets of countries) participating in the UN-REDD Programme and/or the FCPF (each listed in the left-hand column) so as to identify the relevant, applicable legal regimes (from a preselected and non-exhaustive list in the right-hand column). Figure 3.1 highlights Malaysia and Vanuatu, two countries that are given more in-depth analysis in Part III of the book. Malaysia, a country participating in the

[2] The Forests Dialogue, 'Giving REDD+ Life: Integrating REDD+ with Broader Development Goals' (TFD Review, No 6, 2012), http://theforestsdialogue.org/sites/default/files/tfdreview_givingredd-pluslife_en.pdf.

[3] For background information on ratifications, see Table 4.1 in Margaret A Young, 'REDD+ and Interacting Legal Regimes' in Christina Voigt (ed), *Research Handbook on REDD+ and International Law* (Edward Elgar, 2016) 89, 101–103.

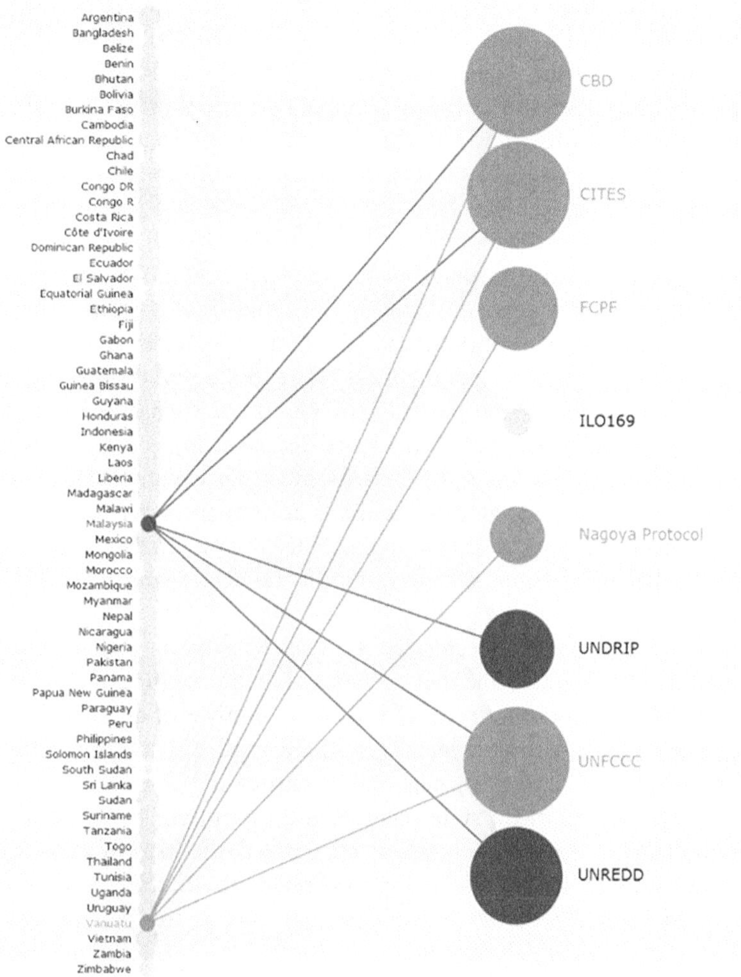


**Figure 3.1.** Infographic of international obligations of countries participating in UN-REDD and/or the FCPF (Malaysia and Vanuatu highlighted)[4]

[4] This infographic lists countries participating in the UN-REDD Programme and/or the FCPF and compares the applicability of selected international treaties and declarations. Malaysia and Vanuatu are highlighted. For the full infographic and dataset, see Margaret A Young and Cobi Smith, *International Laws for Forest Governance* (University of Melbourne, 2016), doi.org/10.4225/49/5567C4E16BB15.

UN-REDD Programme but not the FCPF, has ratified the UNFCCC, CBD and CITES, and voted in favour of UNDRIP. Vanuatu, which participates in the FCPF but not the UN-REDD Programme, shares membership with the CBD and CITES, has ratified the Nagoya Protocol, but was absent for the UNDRIP vote. Figure 3.1 thus emphasises how Malaysia's and Vanuatu's international legal obligations will impact their participation in REDD+.

The size of the balloons in the left- and right-hand columns also imparts important information. For example, the CBD balloon is large because it enjoys near-universal application: *all* of the countries in the left-hand column have ratified it (the United States is of course not on the list as it is not a recipient country for REDD+ funds, but it is perhaps the most infamous non-party to the CBD). As such, CBD norms protecting biological diversity (which are discussed in Section B3) will have immediate applicability to REDD+ for those participating states. The ILO Convention 169, on the other hand, is accompanied by a small balloon, given its much smaller direct applicability – only fifteen of the countries in the left-hand column have ratified it (and less than thirty countries overall).

One might have expected the Cancun Safeguards to have sought to mediate this issue. While they recognised the imperative of ensuring that REDD+ actions do not conflict with the objectives of other international conventions and agreements (see paragraph (a) of the safeguards, excerpted in Chapter 2), negotiating parties did not seek to base this coordinating effort on state membership or consent to the other international conventions and agreements. Indeed, the issue of protecting the sovereignty of states that were *not* members of other regimes (i.e. ensuring that REDD+ parties did not become subject to other international conventions to which they had not consented) arose only obliquely. For example, during the negotiations on the text of the Cancun Safeguards, parties did not show concern about explicitly quoting human rights treaties.[5] However, in the interests of not excluding specific treaties, parties used open-ended language and referred instead to 'relevant international obligations' (paragraph (c)). Some restraint occurred in the safeguard calling for consistency with biological diversity conservation (paragraph (e)); this does not refer to the CBD, which, it will be recalled, has not been ratified by the United States.

[5] Comment of UNFCCC country negotiator to author, 21 October 2014.

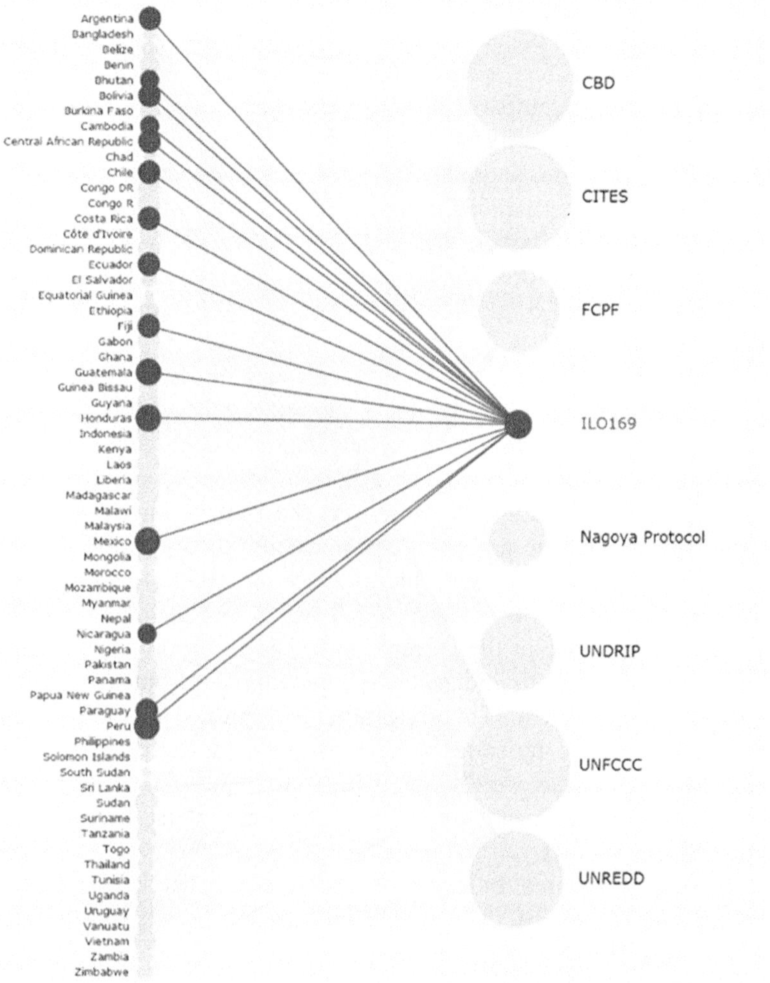


**Figure 3.2.** Direct applicability of ILO Convention 169 for countries participating in UN-REDD and/or the FCPF

As shown in Figure 3.2, ILO Convention 169 – the sole international treaty enshrining rights of indigenous peoples – would appear to have only limited applicability for countries participating in the UN-REDD Programme or the FCPF. For those countries that are parties, there may be uncertainty in the potential for conflicting obligations if their

REDD+ project requirements conform to a different (lesser) standard than what they are obliged to follow. Moreover, the interpretative contexts in which REDD+ itself is understood and applied will be different for these states as compared to nonstate parties.

As we demonstrate in future chapters, however, this rather positivist account of direct international obligations is not necessarily supported by the nuances in international organizational activity which, for example, has seen many concepts of the ILO Convention 169 have relevance to nonstate parties.[6] Moreover, fragmentation gives rise to different relationships of conflict and interpretation, as we explore further in Chapter 10.

A further important caveat about the infographic is related to the fact that the list of legal instruments is non-exhaustive. A range of other treaties and non-binding or even nonstate-based instruments are not included, some of which, like the Forest Principles, are described in greater detail in Section B. Moreover, the infographic does not give expression to specific norms of customary international law, which by definition bind all states and, due to a lack of specific supporting institutions, are often not considered to fall within international regimes. Of particular relevance in the current context is a purported obligation of states to consult with indigenous peoples on government policies or matters affecting them.[7] This obligation finds expression in the UNDRIP, but is also argued by some to have assumed the status of customary international law, as we discuss in Section C. Customary international law protections such as these will impact on the intersecting regimes in the REDD+ context.

Having considered in overview the international laws relating to forests and forest land use, the following section provides greater detail of the most relevant legal regimes.

## B. Overview of Relevant International Regimes

The Cancun Safeguards recognise the imperative of ensuring that REDD+ actions do not conflict with the objectives of other international conventions and agreements. This directive, which is followed in practice by international organizations such as the FCPF and the UN-REDD

[6] See especially Chapters 4 and 10.

[7] International Law Association, *Sofia Conference (2012): Rights of Indigenous Peoples (Final Report)* (2012) 4–7; see further Chapter 2.

Programme, thus elevates the importance of international conventions and agreements to non-parties as well as parties. As a consequence, an understanding of the REDD+ regime is incomplete without awareness of the influence of other international conventions and agreements, and we show their continuing relevance with respect to indigenous rights, tenure arrangements and co-benefits in future chapters. This section provides important details on the content of those other international conventions and agreements.

## 1. *The International Forest Regime*

Legal arrangements surrounding forests and forestry are clearly relevant to policy approaches and positive incentives relating to REDD+, if only because these arrangements give rise to the very need and calibration of 'incentives' to keep forests standing. What becomes clear in the following overview is, as observed by Davenport and others:

> Forest policymaking at the global level does not take place within an engineered structure or a designed architecture.[8]

The following subsection provides a brief overview of the aspects of an international forest 'regime' that are most pertinent to REDD+ and indigenous peoples, including the nature of permanent sovereignty over natural resources, the rise of policies such as 'sustainable forest management' (SFM), and the use of trade and certification frameworks.

### (a) Sovereignty over Natural Resources

Although the failure to agree upon a central forest convention (despite best efforts at the UN Conference on Environment and Development at Rio in 1992, as discussed further in this section) has led some commentators to point to the existence of a 'non-regime' for forests,[9] others argue that a broader set of processes represents part of a 'regime' or even a 'regime-complex'.[10] The absence of a central convention is not fatal to

[8] Deborah Davenport et al, 'Forests and Sustainability' in Jeremy Rayner, Alexander Buck and Pia Katila (eds), *Embracing Complexity: Meeting the Challenges of International Forest Governance. A Global Assessment Report* (Global Forest Expert Panel on the International Forest Regime, IUFRO World Series vol. 28, 2010) 75.

[9] Radoslav S Dimitrov et al, 'International Nonregimes: A Research Agenda' (2007) 9 *International Studies Review* 230, 231.

[10] Jeremy Rayner et al, 'Introduction' in Jeremy Rayner, Alexander Buck and Pia Katila (eds), *Embracing Complexity: Meeting the Challenges of International Forest Governance. A Global*

meeting Krasner's definition of regimes as 'sets of norms, decision-making procedures and organizations coalescing around functional issue-areas'.[11] Moreover, the identification of 'particular modes of behaviour, assumptions and biases', characteristics that form part of Young's definition, assists in showing that international forest governance comprises a body of norms that contains dominant themes and approaches.

For example, instead of a hard-law system, international forest governance promotes a voluntary, 'soft-law' approach that conceives of forests as part of 'national sovereignty' rather than a more global, common-pool good. This translates to an acknowledgment that countries have primary jurisdiction over their forests,[12] and regulation is therefore to proceed at the domestic level. The concept of 'permanent sovereignty' was an important one developed in an era of decolonisation, when newly independent countries were recognised as having full control over their natural resources after a history of colonisation.[13] While the endorsement of this concept in forestry may seem inevitable, there were contrary precedents, such as the 1972 World Heritage Convention, which promotes a concept of outstanding universal value (in that case, as imbued in cultural and natural heritage).[14] Even the Rio Declaration had some elements towards less state-centric modes of governance, as well as reference to the vital role of indigenous peoples and their communities in environmental management.[15] Yet, given the perception that a

*Assessment Report* (Global Forest Expert Panel on the International Forest Regime, IUFRO World Series vol. 28, 2010).

11 Ibid. 12–13.

12 *Report of the United Nations Conference on Environment and Development* (UN Doc A/CONF.151/26 (vol. III) (14 August 1992)) annex III ('Non-Legally Binding Authoritative Statement of Principles for a Global Consensus on the Management, Conservation and Sustainable Development of All Types of Forests') art 1(a).

13 See, e.g., *Permanent Sovereignty over Natural Resources* (GA Res 1803 (XVII), UN GAOR, 17th sess, 1194th plen mtg, UN Doc A/RES/1803 (XVII) (14 December 1962)); *Permanent Sovereignty over Natural Resources* (GA Res 3171 (XXVIII), UN GAOR, 28th sess, 2203rd plen mtg, UN Doc A/RES/3171 (XXVIII) (17 December 1973)).

14 We note that the World Heritage Convention relates closely to the forest regime, given that parties are obliged to act to preserve cultural heritage, which can include natural heritage such as forested areas. For reasons of space, these duties are not considered further in the present chapter.

15 *Report of the United Nations Conference on Environment and Development* (UN Doc A/CONF.151/26 (vol. I) (12 August 1992) annex I ('Rio Declaration on Environment and Development')) principle 22; see further Dinah Shelton, 'Principle 22: Indigenous People and Sustainable Development' in Jorge E Viñuales (ed), *The Rio Declaration on Environment and Development: A Commentary* (Oxford University Press, 2015) 541. See also

conception of 'common heritage' would lead to the appropriation of natural resources by a few wealthy states, a global forestry convention proved politically unacceptable. Indeed, the ongoing tensions between economic development and environmental protection are a long-running theme in forest governance, so it is no surprise that these issues are also a major theme in REDD+, as the next chapters discuss more fully.[16]

Given the dominance of national sovereignty in international forest governance, most forestry laws are dictated domestically, and the national laws of most relevance to forests are often those governing agricultural expansion and forest conversion rather than the forests themselves.[17] This is one of the reasons that a forestry instrument does not appear in Figure 3.1. National laws are often guided, however, by non-legally binding declarations at the international level. These historically emphasised forest utilization, such as the FAO's early view of forests as a process ('forestry') of provision of goods for human consumption.[18] Forests were defined to include natural forests and forest plantations, referring to a certain cover of tree canopy;[19] unsurprisingly, no regard was given to carbon stocks within forested land. As the first global intergovernmental body to address forest sustainability, the FAO's institutional approaches and assumptions were highly influential. The FAO's inability to prevent increasing loss of global forest cover has been attributed by Davenport and others to two main factors: first, its neglect of the ecological requirements of the forests (in part because of its members' focus on maximum timber yield), and second, the absence of a mandate to address the causes of deforestation arising outside the

*Report of the United Nations Conference on Environment and Development*, 'Rio Declaration on Environment and Development' principle 1 and principle 27.

16 See especially Chapter 6 with respect to co-benefits.

17 Constance McDermott, 'REDD+ and Multi-Level Governance: Governing for What and for Whom?' in Rosemary Lyster, Catherine MacKenzie and Constance McDermott (eds), *Law, Tropical Forests and Carbon: The Case of REDD+* (Cambridge University Press, 2013) 84, 88. See McDermott's rich historical account with country examples in that chapter.

18 Food and Agriculture Organization of the United Nations, 'Constitution' in *Basic Texts of the Food and Agriculture Organization of the United Nations (Volumes I and II)* (FAO, 2013), http://fao.org/docrep/meeting/022/k8024e.pdf, art 1 ('In this Constitution, the term 'agriculture' and its derivatives include ... forestry and primary forestry products').

19 FAO has defined forests as land with a tree coverage of more than 10 per cent and an area of more than 0.5 ha. The 10 per cent threshold was first used in FAO's 1963 global assessment and later recommended in 1973 in a UNESCO study: see Food and Agriculture Organization of the United Nations (FAO), 'Forest Resources Assessment 2000: On Definitions of Forest and Forest Change' (Working Paper 33, FAO, 2000).

timber sector, such as agricultural expansion.[20] These insights show that the organizational mandates and technical preoccupations of bodies such as the UN-REDD Programme, the FCPF and the Green Climate Fund will be highly relevant to an assessment of REDD+'s effectiveness. They also justify one of the dominant interests in this book in analysing the interaction between REDD+ and existing regimes, and the need to understand and move beyond the immediate sets of obligations as encapsulated, for example, by Figure 3.1.

### (b) Forest Principles, Traditional Knowledge and SFM

The emphasis on forest utilisation was challenged by the push in the Rio Conference for an international forests convention which would safeguard and protect certain forests. However, this push failed to garner sufficient support by states.[21] Instead, countries agreed to a 'Non-Legally Binding Authoritative Statement of Principles for a Global Consensus on the Management, Conservation and Sustainable Development of All Types of Forests' (Forest Principles).[22] The decision to establish a forests convention was postponed, and a forum was created to continue discussions and generate proposals for action (the Intergovernmental Panel on Forests, later renamed the Intergovernmental Forum on Forests). This forum was succeeded in 2000 by the permanent UN Forum on Forests (UNFF), whose mandate is to promote 'the management, conservation and sustainable development of all types of forests and to strengthen long-term political commitment to this end'.[23] The UNFF has universal membership and in 2007 adopted a 'Non-Legally Binding Instrument on All Types of Forests' (NLBI).[24]

While the Forest Principles endorsed the participation of indigenous peoples and the recognition of their traditional knowledge in forests management, the role and power of these ideas dissipated in subsequent meetings of the UNFF, and the NLBI included reference to traditional

20 Davenport et al, above n 8, 76.

21 See, e.g., David Humphreys, *Logjam: Deforestation and the Crisis of Global Governance* (Earthscan, 2006).

22 Countries also agreed the Convention on Biological Diversity, which is discussed further below n 69 and surrounding text.

23 *Report on the Fourth Session of the Intergovernmental Forum on Forests* (UN ESCOR, 46th mtg, UN Doc 2000/35 (18 October 2000)) para 1.

24 *Non-Legally Binding Instrument on All Types of Forests* (GA Res 62/98, UN GAOR, 62nd sess, 74th plen mtg, UN Doc A/RES/62/98 (31 January 2008)).

knowledge only in passing.[25] Newing suggests that the alliance that had been formed between indigenous peoples and environmentalists had broken down after the early 1990s,[26] and the interests of indigenous peoples were concentrated on the biodiversity regime as well as UN efforts to achieve a declaration on indigenous peoples' rights.

A more long-lasting idea in the forest regime was the NLBI's concept of SFM, recognised as 'a dynamic and evolving concept [that] is intended to maintain and enhance the economic, social and environmental value of all types of forests, for the benefit of present and future generations'.[27] SFM had evolved in a number of places, such as the Ministerial Conference for the Protection of Forests in Europe, which had defined it in a relatively robust way.[28] In contrast, the NLBI entrenched a less robust definition, and further interpretation and implementation of the SFM concept has been undertaken at national levels. A number of different regional and international processes (including from the FAO),[29] have led to the development of criteria and indicators for SFM,[30] some of which acknowledge forest contributions to global carbon cycles.[31] The term SFM has become a slogan whose elaboration remains deeply contextual.[32]

[25] Ibid. para 6(f).

[26] H S Newing, 'Traditional Knowledge in International Forest Policy: Contested Meanings and Divergent Discourses' (2009) 6 *Journal of Integrative Environmental Sciences* 175.

[27] *Non-Legally Binding Instrument on All Types of Forests*, above n 24 annex, preamble.

[28] See Resolution H1: General Guidelines for the Sustainable Management of Forests in Europe (Second Ministerial Conference on the Protection of Forests in Europe, Helsinki, Finland, 16–17 June 1993) para D (see especially, 'for the purposes of this resolution, "sustainable management" means the stewardship and use of forests and forest lands in a way, and at a rate, that maintains their biodiversity, productivity, regeneration capacity, vitality and their potential to fulfil, now and in the future, relevant ecological, economic and social functions, at local, national, and global levels, and that does not cause damage to other ecosystems'.)

[29] See 'Sustainable Forest Management (SFM) Toolbox', Food and Agriculture Organization of the United Nations, www.fao.org/sustainable-forest-management/toolbox/en.

[30] Constance McDermott et al, 'Mapping the Core Actors and Issues Defining International Forest Governance' in Jeremy Rayner, Alexander Buck and Pia Katila (eds), *Embracing Complexity: Meeting the Challenges of International Forest Governance. A Global Assessment Report* (Global Forest Expert Panel on the International Forest Regime, IUFRO World Series vol. 28, 2010) 19, 30.

[31] 'The Montréal Process: Criteria and Indicators for the Conservation and Sustainable Management of Temperate and Boreal Forests' (5th edn, 2015) criterion 5.

[32] Anja Eikermann, *Forests in International Law: Is There Really a Need for an International Forest Convention?* (Springer International Publishing Switzerland, 2015) 139.

Two regional extrapolations of SFM demonstrate that the concept can be responsive to a range of issues affecting indigenous peoples and forest-dependent communities, at least on paper. The Montréal Process, an initiative developed in 1993 for temperate and boreal forests,[33] has developed criteria that recognise the 'maintenance and enhancement of long term multiple socio-economic benefits to meet the needs of society' (criteria 6).[34] This includes the compilation of data, for example, on the area and per cent of forests used for subsistence purposes[35] and the 'extent of forests managed primarily for their cultural, social and spiritual values to people and communities, including indigenous communities and others with strong ties to forests'.[36] A separate initiative for SFM in tropical regions, first developed by the International Tropical Timber Organization (ITTO) in 1990,[37] includes an expectation that countries record information on the '[e]xtent of involvement of indigenous peoples, local communities and other forest dwellers in forest management capacity-building, consultation processes, decision-making and implementation'.[38]

The practice of using criteria and indicators within an SFM framework has generated data about forest practices and sustainability. Yet the framework has been criticised as applying a top-down, industrial approach that insufficiently accommodates community-based forest management.[39] It has also been criticised as being ill suited to address the drivers of

[33] The twelve members are Argentina, Australia, Canada, Chile, China, Japan, Republic of Korea, Mexico, New Zealand, Russian Federation, United States and Uruguay: see 'The Montréal Process: Member Countries', http://montrealprocess.org/The_Montreal_Process/Member_Countries/index.shtml.

[34] 'Technical Notes on Implementation of the Montréal Process Criteria and Indicators: Criteria 1–7' (3rd edn, 2009, rev July 2014), http://montrealprocess.org/documents/publications/techreports/MontrealProcessTechnicalNotes3rdEditionRevisedJuly2014.pdf.

[35] Ibid., indicator 6.3.d. This reflects the understanding that '[t]he practice of forest-based subsistence reflects the dependence of rural communities and individuals on forests for essential resources and may be closely linked to cultural identity and quality of life'. Ibid. 60.

[36] 'Technical Notes on Implementation of the Montréal Process Criteria and Indicators: Criteria 1–7', above n 34, indicators 6.5.a. See further ibid. 65.

[37] The ITTO is mentioned further below, n 44 and accompanying text.

[38] International Tropical Timber Organization, 'Revised ITTO Criteria and Indicators for the Sustainable Management of Tropical Forests Including Reporting Format' (ITTO Policy Development Series No 15, 2005), http://itto.int/direct/topics/topics_pdf_download/topics_id=9630000&no=1&disp=inline. Ibid., indicator 7.1.4.

[39] Reem Hajjar et al, 'Criteria and Indicators for Sustainable Forest Management in the Face of Decentralization: Are They Still Relevant in Their Current Form' (XIII World Forestry Conference, Buenos Aires, Argentina, 18–23 October 2009).

deforestation outside of the timber sector, such as mining and agricultural expansion, which themselves are highly context dependent.

Whether to endorse SFM in the REDD+ context became an important issue for negotiators at the UNFCCC COPs, who had to determine whether an emerging REDD+ regime would replicate or depart from elements of the international forest regime. An early COP decision on demonstration activities used the language of 'sustainable forest management' and expressly noted 'inter alia, the relevant provisions of the United Nations Forum on Forests, the United Nations Convention to Combat Desertification and the Convention on Biological Diversity'.[40] Some commentators who saw SFM as condoning commercial logging were wary that REDD+ could be co-opted into becoming another regime furthering timber extraction. Ultimately (and yet obliquely), negotiators did not use the phrase 'sustainable forest management' in the Paris Agreement or the Cancun Agreements, but instead linked REDD+ approaches to the 'sustainable management of forest'.[41] This may be a barely perceptible difference for some (including, perhaps, negotiators speaking many languages and relying on fast-moving translations), but it is an example of attempts to mould regime interaction. As Davenport notes, while these definitional differences have 'almost no resonance outside convention halls', their different conceptualisations of sustainability will impact the distribution of donor funds, and 'REDD will hinge on resolving these conflicts'.[42]

### (c) The Trade in Timber and Certification Schemes

Aside from the NLBI and its general support for SFM, a range of other forest-focused laws and arrangements exists, especially governing the

[40] Conference of the Parties, United Nations Framework Convention on Climate Change, *Report of the Conference of the Parties on Its Thirteenth Session, Held in Bali from 3 to 15 December 2007 – Addendum – Part Two: Action Taken by the Conference of the Parties at Its Thirteenth Session* (UN Doc FCCC/CP/2007/6/Add.1 (14 March 2008) Decision 2/CP.13) ('Reducing Emissions from Deforestation in Developing Countries: Approaches to Stimulate Action') ('Bali Action Plan') annex ('Indicative Guidance') para 8.

[41] See Conference of the Parties, United Nations Framework Convention on Climate Change, *Report of the Conference of the Parties on Its Sixteenth Session, Held in Cancun from 29 November to 10 December 2010 – Addendum – Part Two: Action Taken by the Conference of the Parties at Its Sixteenth Session* (UN Doc FCCC/CP/2010/7/Add.1 (15 March 2011) Decision 1/CP.16 ('The Cancun Agreements: Outcome of the Work of the Ad Hoc Working Group on Long-Term Cooperative Action under the Convention')) para 70.

[42] Davenport et al, above n 8.

trade in timber. In the face of failed efforts to finalise a forests convention at Rio in 1992, some countries asserted their ability to address environmental concerns through unilateral trade measures and restrictions. The United States, for example, noted that 'in certain situations, trade measures may provide an effective and appropriate means of addressing environmental concerns, including long-term sustainable forest management concerns and environmental concerns outside national jurisdiction'.[43] This parallels approaches taken in environmental regimes, such as the 1975 CITES's ban on trade in certain listed species of endangered timber, which we discuss as part of our overview of environment regimes in Section 3.

Trade is also the focus of the ITTO, founded in 1986,[44] whose focus is on expanding trade rather than curtailing its environmental impacts. ITTO's stated aim is to 'promote the expansion and diversification of international trade in tropical timber from sustainably managed and legally harvested forests and to promote the sustainable management of tropical timber producing forests'.[45] The ITTO supports the 'Forest Law Enforcement Governance and Trade (FLEGT) Action Plan', which sets out the European Union's policy of promoting legal logging and trade in legally licensed timber.[46] It also reinforces an objective of 'sustainable forest management', as mentioned above.[47] The ITTO has a stated recognition of the importance of collaboration 'among members, international organizations, the private sector and civil society, including

[43] *Report of the United Nations Conference on Environment and Development* (UN Doc A/CONF.151/26 (vol. IV) (28 September 1992)) 21 cited in Margaret A Young, 'Principle 12: Trade and the Environment' in Jorge E Viñuales (ed), *The Rio Declaration on Environment and Development: A Commentary* (Oxford University Press, 2015) 325, 330.

[44] *International Tropical Timber Agreement* (opened for signature 18 November 1983, 1393 UNTS 119, provisionally entered into force 1 April 1985); *International Tropical Timber Agreement* (opened for signature 1 April 1994, 33 ILM 1014, entered into force 1 January 1997); United Nations Conference on Trade and Development, *International Tropical Timber Agreement 2006* (UN Doc TD/TIMBER.3/12 (1 February 2006, entered into force 7 December 2011)).

[45] United Nations Conference on Trade and Development, above n 44, art 1.

[46] See 'FLEGT Independent Market Monitoring (IMM)', International Tropical Trade Organization (ITTO), www.itto.int/imm.

[47] Article 2 of the International Tropical Timber Agreement (2006) states that '"Sustainable forest management" will be understood according to the organization's relevant policy documents and technical guidelines': United Nations Conference on Trade and Development, above n 44.

indigenous and local communities, and other stakeholders in promoting sustainable forest management'.[48]

Multi-stakeholder participation is also noted in the 2014 'New York Declaration on Forests' agreed as part of the UN Climate Summit and endorsed by governments, multinational companies, civil society groups and indigenous organizations.[49] The New York Declaration isolates a series of actions for governments, companies and business associations, indigenous peoples, civil society organizations and multilateral institutions. It goes further than commitments undertaken by timber companies: for example, major global traders of palm oil adopted zero-deforestation policies (said to cover 60 per cent of global trade),[50] while a broader set of commodities is covered by a 'Consumer Goods Forum'.[51]

These trade initiatives are at times supported by international financial institutions, including the World Bank. While development cooperation between states could be conceptualised as a regime or set of regimes in its own right,[52] the World Bank funding for forest projects in developing countries – including the financing of commercial logging operations and efforts to promote SFM – can be conceptualised as part of the forestry regime. The financing of some projects has led to criticism and the evolution of internal strategies and policies within the World Bank.[53] This is reminiscent of our earlier overview of the actors within the REDD+ regime, in the sense that the legal obligations of international

[48] Ibid., preamble, para (l). But compare earlier reports of decisions by conservation groups to cease attending ITTC sessions in the mid-1990s due to a perception that conservation issues were not receiving adequate attention: Humphreys, above n 21; cited in McDermott et al, above n 30.

[49] 'New York Declaration on Forests and Action Agenda (Section 1)' in *Forests: Action Statements and Action Plans* (Climate Summit 2014, New York, 23 September 2014).

[50] *Forests: Action Statements and Action Plans* (Climate Summit 2014, New York, 23 September 2014) Section 3A, 'Zero Deforestation Commitments from Commodity Producers and Traders'.

[51] Ibid., Section 3A, 'Zero Deforestation Commitments from Consumer Goods Companies'.

[52] Siobhán McInerney-Lankford, 'Human Rights and Development Regime Interaction and the Fragmentation of International Law' in Hassane Cissé et al (eds), *The World Bank Legal Review: Legal Innovation and Empowerment for Development, Volume 4* (World Bank and Martinus Nijhoff, 2013) 123, 127, http://openknowledge.worldbank.org/bitstream/handle/10986/12229/NonAsciiFileName0.pdf.

[53] See, e.g., OP 4.36 – the World Bank's Operational Policy on Forests – as described by Annie Petsonk, 'Legal Obligations and Institutions of Developing Countries: Rethinking Approaches to Forest Governance' in Hassane Cissé, Daniel D Bradlow and Benedict Kingsbury (eds), *The World Bank Legal Review: International Financial Institutions and Global Legal Governance, Volume 3* (World Bank and Martinus Nijhoff, 2012) 293, http://siteresources.worldbank.org/INTLAWJUSTICE/Resources/WBLegalReviewVol3.pdf.

financial institutions do not flow from treaty sources, but arise from the constitutive instruments or at the transactional level of particular legal agreements underpinning specific projects.

In addition to trade initiatives are private timber certification schemes. The Forest Stewardship Council (FSC) was established in 1993 without state membership, and sought to provide consumers with more information and choice in the purchase of labelled timber. This initiative was reportedly first attempted within the ITTO, but tropical-country negotiators objected to labels as constituting barriers to trade;[54] the subsequent FSC was restrictive in the membership it accorded actors with direct economic interests. The ITTO now encourages information sharing for a better understanding of voluntary mechanisms such as certification.[55] The pattern of exclusion and inclusion (especially of indigenous and local community rights) in the forest certification process has important lessons for REDD+.[56]

### (d) 'Less than the sum of its many parts'

The picture of the international forest regime we have painted above is one of a complex and rather loose set of arrangements which focus on voluntary commitments, seek to mobilise private actors, and conceive of forests in terms of timber utilisation and more broadly in terms of agricultural expansion. The move towards SFM attempted to incorporate broader considerations than mere timber yields, but efforts towards incorporating interests of indigenous peoples and achieving sustainability in forests have been hampered. A defining feature, then, of an international forests regime is its diffuse, polycentric, dynamic character. This has led a prominent commentator to reflect that 'the effect of the international forest regime is rather less than the sum of its many parts'.[57] The challenge for REDD+, itself a highly complex and heterogeneous policy approach, will be to interact with the international forest regime in ways that are consistent with REDD+ objectives, including the

[54] As reported in McDermott et al, above n 30, 31.

[55] United Nations Conference on Trade and Development, above n 44, art 1(o).

[56] See, e.g., Jessica Rae, Mahala Gunther and Lee Godden, 'Governing Tropical Forests: REDD+, Certification and Local Forest Outcomes in Malaysia' (2011) 7(2) *Macquarie Journal of International and Comparative Environmental Law* 40.

[57] Jeremy Rayner, 'Preface' in Jeremy Rayner, Alexander Buck and Pia Katila (eds), *Embracing Complexity: Meeting the Challenges of International Forest Governance. A Global Assessment Report* (Global Forest Expert Panel on the International Forest Regime, IUFRO World Series vol. 28, 2010).

recognition of the needs of indigenous peoples, as we discuss in the following chapters.

### 2. *Other Regimes Recognising Indigenous Peoples' 'Rights', Including Human Rights Regimes*

Additional international regimes contain various protections for peoples inhabiting or governing forested lands. The most significant treaty for indigenous peoples is the ILO Convention 169,[58] although with fewer than thirty parties (concentrated in the Latin American region), its membership is far from wide ranging. A broader base underlies the UNDRIP.[59] Although its status is formally limited as a General Assembly resolution rather than an international treaty, some of its provisions may represent customary international law, as we discuss further in Section C. UNDRIP is supported institutionally by UN bodies, including the UN Permanent Forum on Indigenous Issues,[60] an advisory body to the UN Economic and Social Council (ECOSOC) made up of both indigenous experts and state representatives. Consistent with our discussion in Chapter 2, such bodies make up an emerging set of international laws and practices.[61]

We defer our more detailed overview of indigenous protections to the following chapter, which focusses in particular on the FPIC concept. It is important to note at this point, though, that the Permanent Forum has been very active on REDD+ issues. In 2011, it initiated a survey of the

[58] International Labour Organization, *Convention Concerning Indigenous and Tribal Peoples in Independent Countries* (opened for signature 27 June 1989, 1650 UNTS 383, entered into force 5 September 1991) ('ILO Convention 169').

[59] United Nations Declaration on the Rights of Indigenous Peoples, GA Res 61/295, UN GAOR, 61st Sess, 107th Mtg, Agenda Item 68, Supp No 49, UN Doc A/RES/61/295 (2 October 2007, Adopted 13 September 2007).

[60] Ibid. art 42 (requiring bodies and states to 'follow up on the effectiveness of this Declaration'); see also Carsten Smith, Member of the Permanent Forum on Indigenous Issues, *Draft General Comment No 1 (2009): Article 42 of the Declaration on the Rights of Indigenous Peoples* (UN Doc E/C.19/2009/CRP.12 (5 May 2009)).

[61] We note that the assumption of an institutional role by the Permanent Forum is controversial, as states considered it to be a de facto attempt at changing the legal understanding of the declaration and its mandates, and also considering other institutions such as the Expert Mechanism on the Rights of Indigenous Peoples and the Special Rapporteur on the Rights of Indigenous Peoples: see further Claire Charters, 'Multi-Sourced Equivalent Norms and the Legitimacy of Indigenous Peoples' Rights under International Law' in Tomer Broude and Yuval Shany (eds), *Multi-Sourced Equivalent Norms in International Law* (Hart Publishing, 2011) 289.

impact on indigenous peoples of REDD+ projects. Members of the Permanent Forum reported in 2013 and noted that the impact of REDD+ could be positive or negative depending 'on the recognition of their rights, their level of participation and the nature and effectiveness of safeguards to be applied at the international and national levels'.[62]

Aside from indigenous peoples' rights, it is important to mention general human rights flowing from foundational instruments such as the Universal Declaration on Human Rights, the International Covenant on Civil and Political Rights (ICCPR)[63] and the International Covenant on Economic, Social and Cultural Rights (ICESCR),[64] amongst other treaties on the rights of children and the elimination of racial and gender discrimination.[65] Gender equality, for example, is a major imperative included in the institutional guidance on REDD+ provided by the UN-REDD Programme, as we discuss in Chapter 4.

Increasingly, climate change mitigation expressly incorporates these norms. For example, the preamble to the Paris Agreement acknowledges that rights of indigenous peoples are to be 'respect[ed], promot[ed] and consider[ed]' by states in addressing climate change. Along with the

[62] UN ECOSOC Permanent Forum on Indigenous Issues, *Indigenous People's Rights and Safeguards in Projects Related to Reducing Emissions from Deforestation and Forest Degradation*, UN Doc E/C.19/2013/7 (5 February 2013) ('Indigenous People's Rights and Safeguards in Projects Related to Reducing Emissions from Deforestation and Forest Degradation').

[63] *International Covenant on Civil and Political Rights,* (opened for signature 19 December 1966, 999 UNTS 171, entered into force 23 March 1976) (ICCPR); see also *2200 (XXI). International Covenant on Economic, Social and Cultural Rights, International Covenant on Civil and Political Rights and Optional Protocol to the International Covenant on Civil and Political Rights* (UN GAOR, 21st sess, 1496th plen mtg, UN Doc A/RES/21/2200 (16 December 1966)) annex ('International Covenant on Civil and Political Rights') 52.

[64] *International Covenant on Economic, Social and Cultural Rights* (opened for signature 19 December 1966, 993 UNTS 3, entered into force 3 January 1976) (ICESCR); see also *2200 (XXI). International Covenant on Economic, Social and Cultural Rights, International Covenant on Civil and Political Rights and Optional Protocol to the International Covenant on Civil and Political Rights,* above n 63 annex ('International Covenant on Economic, Social and Cultural Rights') 49.

[65] *Convention on the Rights of the Child* (opened for signature 20 November 1989, 1577 UNTS 3, entered into force 2 September 1990) (CRC); *Convention on the Elimination of All Forms of Discrimination against Women* (opened for signature 18 December 1979, 1249 UNTS 13, entered into force 3 September 1981) (CEDAW); *International Convention on the Elimination of All Forms of Racial Discrimination* (opened for signature 7 March 1966, 660 UNTS 195, entered into force 4 January 1969) (ICERD). See further the ICERD Committee's registered concern over the impacts on indigenous peoples in the land-use context in Papua New Guinea, described in Chapter 8, Section C2.

codified laws and principles relevant to the rights of indigenous peoples, this refers to norms of a customary nature, which we mention in Section C. However, not all human rights standards have been listed in the Cancun Safeguards. Moreover, general rights of equality in human rights law may not be consistent with indigenous protections, and this is explored in our investigation of FPIC in the following chapter.

### 3. *Environmental Protection Regimes*

While recognition of indigenous rights as human rights is complex and unresolved, and the international forest regime's incorporation of the needs of indigenous peoples has waxed and waned,[66] international regimes for the protection of the environment have been more consistent in seeking to address issues relating to traditional knowledge and indigenous rights, though with similar difficulties and sensitivities relating to the fact that indigenous peoples' interests were not always in harmony with conservationist approaches. The following overview describes some of the most relevant environmental regimes.[67]. We mention in brief the CBD, its recent Nagoya Protocol and the CITES. In general, these environmental regimes have a different focus on utilisation than the forest regimes, and also take a more nuanced approach to economic development, while entrenching certain rights and benefits for indigenous peoples.

#### (a) Convention on Biological Diversity and the Nagoya Protocol

The CBD 1992, agreed at Rio, seeks to protect forests in a holistic way as part of an ecosystem,[68] along with other sites of biological diversity. It recognises as a fundamental requirement for the conservation of biological diversity 'the *in-situ* conservation of ecosystems and natural habitats and the maintenance and recovery of viable populations of species in their natural surroundings'.[69] To this end, states parties are

[66] See Newing, above n 26.

[67] The UNFCCC, including afforestation projects within the Kyoto Protocol's CDM mechanism, are not discussed here as they are covered in Chapter 2. On co-benefits, see further Chapter 6.

[68] Harro van Asselt, 'Managing the Fragmentation of International Environmental Law: Forests at the Intersection of the Climate and Biodiversity Regimes' (2012) 44 *New York University Journal of International Law and Politics* 1205.

[69] Convention on Biological Diversity (opened for signature 5 June 1992, 1760 UNTS 79, entered into force 29 December 1993) ('Convention on Biological Diversity') preamble.

to establish systems of 'protected areas'.[70] They shall also adopt measures for *ex situ* conservation of components of biological diversity (such as in botanical gardens or herbaria).[71] Combined with the protection of biological resources is the concept of 'sustainable use', which is defined as use that 'does not lead to the long-term decline in biological diversity',[72] and which is furthered as part of the sovereign right of states to exploit their own resources.[73] The balance between use and protection of biological diversity holds promise for indigenous peoples and local communities that are affected by protected areas, and is said to have the potential 'to overcome the deep chasm between the principle of sovereignty over natural resources and the common concern'.[74] The way in which the CBD achieves this is relevant to the similar challenges faced by a system of protected forests as part of REDD+.

Achieving a balance amongst sovereignty, common concern, and the needs of indigenous and local communities is furthered by treaty obligations, a range of plans and targets for forests,[75] and a new protocol. The Nagoya Protocol on Access to Genetic Resources and the Fair and Equitable Sharing of Benefits Arising from their Utilization[76] relates to the rights of indigenous peoples as holders of traditional knowledge pertaining to genetic resources.[77] This conception of benefit-sharing of

[70] Ibid. art 8. 'Protected areas' are defined in Article 2 as 'a geographically defined area which is designated or regulated and managed to achieve specific conservation objectives': Ibid. art 2.

[71] *Convention on Biological Diversity*, above n 69, art 9.

[72] Ibid. art 10, art 2 (definitions). [73] Ibid. art 3. [74] Eikermann, above n 32, 100.

[75] As listed in Conference of the Parties, Convention on Biological Diversity, *XI/19. Biodiversity and Climate Change Related Issues: Advice on the Application of Relevant Safeguards for Biodiversity with Regard to Policy Approaches and Positive Incentives on Issues Relating to Reducing Emissions from Deforestation and Forest Degradation in Developing Countries; and the Role of Conservation, Sustainable Management of Forests and Enhancement of Forest Carbon Stocks in Developing Countries* (UN Doc UNEP/CBD/COP/DEC/XI/19 (5 December 2012) ('CBD Hyderabad Decision XI/19')) 6, and including decisions such as the 'Tkarihwaié:ri Code of Ethical Conduct (decision X/42) [pertaining to] research into, access to, and use, exchange and management of information concerning traditional knowledge, innovations and practices for the conservation and sustainable use of biological diversity': annex, 6 (13(g)).

[76] *Nagoya Protocol on Access to Genetic Resources and the Fair and Equitable Sharing of Benefits Arising from Their Utilization to the Convention on Biological Diversity* (opened for signature 2 February 2011, UN Doc UNEP/CBD/COP/DEC/X/1, annex I, entered into force 12 October 2014) ('Nagoya Protocol').

[77] See generally Elisa Morgera, Elsa Tsioumani and Matthias Buck, *Unraveling the Nagoya Protocol: A Commentary on the Nagoya Protocol on Access and Benefit-Sharing to the Convention on Biological Diversity* (Brill Nijhoff, 2014); Elisa Morgera, Matthias Buck and

forest biodiversity is distinct and relevant to REDD+, as Chapter 6 describes.

There is no immediate conflict between CBD legal obligations relating to protected areas and REDD+ projects, given that an area subject to CBD protection as a protected area would not be likely to attract financial payments as a REDD+ project as it would likely not be at risk from deforestation or forest degradation. However, the CBD's overarching goal of biodiversity protection may be disrupted by REDD+. Without adequate safeguards, REDD+'s conception of the protection of carbon in forests could inadvertently incentivise monocultural plantations.[78] Given that almost all REDD+ recipient and donor countries are subject to general biodiversity obligations under the CBD (the only non-CBD parties being the United States, Andorra and the Holy See), a REDD+ treaty obligation to sequester forest carbon may conflict with a CBD obligation to preserve biodiversity.

In anticipation of this, CBD parties have issued general guidance on climate mitigation strategies, calling on both states and international organizations to consider a range of perspectives when planning and implementing climate mitigation activities, such as 'considering traditional knowledge, including the full involvement of indigenous and local communities'.[79] Parties also sought to influence the development of the REDD+ safeguards. In a dedicated COP decision, CBD parties asked the executive secretary to provide information on how REDD+ may affect the customary practices of indigenous and local communities.[80] This advice was intended to influence policy approaches for safeguards for

Elsa Tsioumani (eds), *The 2010 Nagoya Protocol on Access and Benefit-Sharing in Perspective: Implications for International Law and Implementation Challenges* (Brill Nijhoff, 2012).

78 Risk to biodiversity, which is now addressed in the REDD+ Cancun Safeguards, is identified, for example, by the Global Expert Workshop on Biodiversity, *Outcomes of the Global Expert Workshop on Biodiversity Benefits of Reducing Emissions from Deforestation and Forest Degradation in Developing Countries* (UN Doc UNEP/CBD/WS-REDD/1/3 (23 September 2010) para 6. But note that the workshop identified the biggest risk to biodiversity is that 'a well-designed REDD-plus mechanism is not agreed upon and successfully implemented': Ibid. para 5.

79 Conference of the Parties, Convention on Biological Diversity, *X/33. Biodiversity and Climate Change* (UN Doc UNEP/CBC/COP/DEC/X/33 (29 October 2010) ('CBD Nagoya Decision X/33')).

80 Conference of the Parties, Convention on Biological Diversity, above n 75; see further Annalisa Savaresi, 'REDD+ and Human Rights: Addressing Synergies between International Regimes' (2013) 18(3) *Ecology and Society* 149.

biodiversity, and also to emphasise the concept of adequate benefit-sharing as developed within the CBD context.

### (b) CITES and Other Approaches

While the CITES is not a general environmental protection regime (given that it only covers cross-border activity and not the utilisation of domestic products within national borders), it is relevant to a range of commercially valuable timber products that are traded across borders. It regulates or restricts the trade of species that have been agreed by the parties to be listed as threatened with extinction. The approach is therefore species specific, rather than directed at the forest ecosystem. The CITES has evolved from an ethos of strict preservation to one of sustainable utilisation, demonstrating the need to provide minimal incentives for local communities to protect species.[81] Its increasing involvement in marine and timber species (alongside its original emphasis on the protection of land-based mammals) has led to more calls for collaboration with the FAO, ITTO and regional fisheries management organizations.[82] Meanwhile, coordination between the CITES and CBD secretariats emphasises shared management practices.[83]

Numerous other multilateral environmental agreements (MEAs) and institutional decisions have forest-related provisions. Some adopt an approach that considers ecosystems more broadly: an example is the UN Convention to Combat Desertification (UNCCD), which is particularly focused on African countries experiencing serious drought or desertification.[84]

81 Rosie Cooney, 'CITES and the CBD: Tensions and Synergies' (2001) 10 *Review of European Community & International Environmental Law* 259.

82 See Conference of the Parties, Convention on International Trade in Endangered Species of Wild Fauna and Flora (CITES), 'Cooperation with Other Organizations' (Seventeenth Meeting of the Conference of the Parties, Johannesburg, South Africa, 24 September–5 October 2016). On coordination between the FAO and CITES Secretariat in the fisheries context, see Margaret A Young, *Trading Fish, Saving Fish: The Interaction between Regimes in International Law* (Cambridge University Press, 2011) 134–188.

83 See Memorandum of Cooperation between CITES and CBD secretariats referred to in Conference of the Parties, Convention on International Trade in Endangered Species of Wild Fauna and Flora (CITES), *Cooperation and Synergy with the Convention on Biological Diversity* (Res Conf 10.4 (Rev. CoP14) (June 1997)); see further Cooney, above n 81.

84 UNCCD, art 2 states that the achievement of its objectives 'will involve long-term integrated strategies that focus simultaneously, in affected areas, on improved productivity of land, and the rehabilitation, conservation and sustainable management of land and water resources, leading to improved living conditions, in particular at the community

Environmental protection regimes sometimes contain discourses and approaches that appear vastly different from the dominant approaches of the forest regime. For example, ecocentric themes have been adopted within some UN forums, which go far beyond the orthodox anthropocentric-utilisation concepts of the forests regime. In 2009, the UN General Assembly proclaimed 22 April to be an annual 'International Mother Earth Day', recognising that the expression 'mother earth' is common to many countries and cultures, and 'reflects the interdependence that exists among human beings, other living species and the planet we all inhabit'.[85] This nod to rights to nature was acknowledged in the Cancun Safeguards (paragraph 2(e)) meaning that REDD+ may well be expected to integrate such notions.

### 4. *Food Security*

The final set of normative developments we have selected as relevant to REDD+ are to do with the amorphous goal of 'food security'. Food security has been expressed as an objective to ensure that 'all people, at all times, have physical and economic access to sufficient, safe and nutritious food to meet their dietary needs and food preferences for an active and healthy life'.[86] The effect of food security goals on climate change is complex, and it has been observed that:

> [t]he threat of climate change to hunger must be understood in the context of the evolution, over the next 10 to 50 years (or longer), of resources . . . economies . . . society . . . and political systems.[87]

Although it would be difficult to argue that norms on the protection of food security form an international regime, there are a number of intergovernmental organizations and other institutions constituted to

level': *United Nations Convention to Combat Desertification in Those Countries Experiencing Serious Drought and/or Desertification, Particularly in Africa* (opened for signature 14 October 1994, 1954 UNTS 3, entered into force 26 December 1996) (UNCCD) art 2 (2). For further discussion of the UNCCD in the forests context, see McDermott et al, above n 30, 24.

85 63/278. International Mother Earth Day (GA Res 63/278, UN GAOR, 63rd sess, 80th plen mtg, Agenda Item 49(d), UN Doc A/RES/63/278 (1 May 2009)).

86 Food and Agriculture Organization of the United Nations (FAO), *Rome Declaration on World Food Security and the World Food Summit Plan of Action* (World Food Summit, Rome, Italy, 13–17 November 1996).

87 Thomas E Downing and Martin L Parry, 'Introduction: Climate Change and World Food Security' (1994) 19 *Food Policy* 99, 102.

achieve this goal. For example, it was to ensure 'humanity's freedom from hunger' that the FAO was constituted in 1945.[88] Its partner body, the Committee on World Food Security, was established in 1974.

Forests are a major source of food and sustainable livelihoods, especially for people living in poverty. In 2012, the CFS endorsed the FAO's 'Voluntary Guidelines on Responsible Governance of Tenure of Land, Fisheries and Forests in the Context of National Food Security', which, as we demonstrate in Chapter 5 on tenure, has become highly relevant in implementing REDD+. Amongst other things, these guidelines provide that:

> State and non-state actors should acknowledge that land, fisheries and forests have social, cultural, spiritual, economic, environmental and political value to indigenous peoples and other communities with customary tenure systems.[89]

In addition, there is an overarching obligation to interpret and apply the guidelines consistently with existing obligations under national and international law, and that applications should 'support … national, regional and international initiatives that address human rights'.[90] This emphasis on a human rights–centred implementation is reinforced in the objective contained in Article 1.1, which expressly refers to the Universal Declaration of Human Rights and other international human rights instruments. A specific requirement for consistency with international law relates to indigenous peoples, as follows:

> In the case of indigenous peoples, States should meet their relevant obligations and voluntary commitments to protect, promote and implement human rights, including as appropriate from the International Labour Organization Convention (No 169) concerning Indigenous and Tribal Peoples in Independent Countries, the Convention on Biological Diversity and the United Nations Declaration on the Rights of Indigenous Peoples.[91]

Unsurprisingly, the relationship between food security and the emerging REDD+ regime is difficult, complex and sometimes contradictory. On the one hand, the search for sources of food, especially through

[88] Food and Agriculture Organization of the United Nations, above n 18, preamble.

[89] FAO, *Voluntary Guidelines on the Responsible Governance of Tenure of Land, Fisheries and Forests in the Context of National Food Security* (Food and Agriculture Organization of the United Nations, 2012) s 9.1.

[90] Ibid. s 2.2. [91] Ibid. s 9.3.

agricultural expansion, is a major driver of deforestation. Reducing emissions from deforestation and forest degradation by protecting forests through REDD+ could in those situations restrict access to food. On the other hand, ignoring the need for climate change mitigation might exacerbate the threat of global warming on food insecurity. Calls for a specific negotiating track for agriculture and food security within the climate regime have not been met.[92] Other sectors also come into play: for example, forests that provide energy to households in the form of fuel might be better protected through the provision of alternative energy sources, yet REDD+ is not conceived as an energy issue.[93] For the international and national laws themselves, there is a lack of intersectoral coordination, with agricultural, forests and climate change regulation often deeply segregated.[94] In response to this acknowledged problem, the FAO urges policy-makers to:

> Develop mechanisms for coordination across the agricultural, forestry, live-stock, fisheries, energy, mining and other relevant sectors to ensure stronger coherence of food security and nutrition interventions and better policy alignment.[95]

This potential for conflict and regime interaction is explored further in Section D, but for completeness it is necessary to take a brief sidestep to mention relevant norms of customary international law that impact climate change mitigation and indigenous and forest communities.

## C. Relevant Norms of Customary International Law

Customary international law, as identified by state practice and *opinio juris*,[96] is a source of public international law and as such is relevant to REDD+ and its operation. Like treaties, customary international law

[92] John Ingram, Polly Ericksen and Diana Liverman (eds), *Food Security and Global Environmental Change* (Earthscan, 2010) 19.

[93] The Forests Dialogue, above n 2 (citing an example from Ghana, and observing that the Ministry of Energy had no contact with the REDD+ Readiness process).

[94] FAO, 'Towards Food Security and Improved Nutrition: Increasing the Contribution of Forests and Trees' (Food and Agriculture Organization of the United Nations, 2013) 7, www.fao.org/docrep/018/i2969e/i2969e.pdf.

[95] Ibid. 1.

[96] Pierre-Marie Dupuy, 'Formation of Customary International Law and General Principles' in Daniel Bodansky, Jutta Brunnée and Ellen Hey (eds), *The Oxford Handbook of International Environmental Law* (Oxford University Press, 2008) 449.

binds states,[97] yet it is less readily categorised within the language of 'regimes'. This is due to its largely unwritten character and the fact that it is by definition applicable as general international law and not constrained to particular functional issue areas. Indeed, the analysis of customary international law as part of one regime may be misleading and counter to the relevant law's normative force in other areas. For this reason, we have not included our overview of relevant customary international law in the earlier part of this chapter, but provide a brief mention of it here.

The most important norms of customary international law for our purposes are those relating to indigenous peoples. The UNDRIP expresses a purported obligation of states to consult with indigenous peoples on government policies or matters affecting them. This obligation is argued by some to have assumed the status of customary international law.[98] As such, it may cast a set of obligations over all the countries appearing in Figure 3.1.

Further norms of customary international law warrant mention in this context. Increasingly, country practices relating to environmental protection have reached a level of obligation, although these tend to focus on *conduct* rather than *result*. Although such norms may also find expression in treaties (and can therefore be discussed as part of a separate regime – see Section B), their status as international custom means that they will have general applicability within an emerging REDD+ regime. A good example is the principle that no state has a right to use or permit the use of its territory in such a manner so as to cause injury to the territory of another or property therein.[99] This principle on transboundary harm can be extended to the obligation for states to undertake an environmental impact assessment, where a proposed industrial activity

[97] Subject, of course, to the rule relating to persistent objectors.

[98] A survey of relevant state and international practice conducted by a committee of experts assembled under the auspices of the International Law Association (ILA), and drawing on the work of the UN Special Rapporteur on Indigenous Peoples, found in 2010 the existence of such a norm of customary international law. This was cited by at least one international tribunal in 2011 – see *Grand River Enterprises Six Nations, Ltd v United States of America (Award)* (Arbitral Tribunal constituted under chapter 11 of the North American Free Trade Agreement, 12 January 2011), www.state.gov/documents/organization/156820.pdf, para 210; and reaffirmed in 2013, in International Law Association, above n 7, 29.

[99] *Trail Smelter Arbitration (United States v Canada)* (1941) 3 UN Rep Int'l Arb Awards 1905.

poses a risk of significant environmental harm to another state.[100] Furthermore, the duty of 'due diligence' is gaining increasing importance.[101] This is related to the 'duty to cooperate', which is recognised in the Paris Agreement and the UNFCCC,[102] as well as finding expression in general international law.[103]

A further important principle of relevance to REDD+ is the concept of the fair and equitable sharing of benefits accruing from the use of natural resources, environmental protection and knowledge. As we discuss in Chapter 6, some of the principles and criteria emerging from the UN-REDD Programme have included reference to benefit-sharing.[104] In addition, this concept finds expression in a number of treaties, such as the Nagoya Protocol (discussed in Section B3(a)). As demonstrated by Morgera, the concept is emerging in biodiversity conservation and the law of the sea, but it still lacks normative coherence.[105] While benefit-sharing is integral to how REDD+ incorporates the needs and interests of indigenous peoples, the concept does not have the obligatory force of a customary norm of international law at this point in time. Instead, its influence comes from other sources, as we show in Chapter 6.

To return to the customary international norms that might be said to be relevant to an emerging REDD+ regime, we consider that duties regarding consultation, cooperation and due diligence will have impending importance in matters of regime interaction. As noted by Dupuy, 'general customary rules and general principles may act, in part at least, as compensation for the institutional deficiencies of the [international legal] system … for example, by promoting the coherent

[100] *Pulp Mills on the River Uruguay (Argentina v Uruguay)* (2010) 2010 ICJ Reports 14, para 204.

[101] *Pulp Mills on the River Uruguay (Argentina v Uruguay)* (2010) 2010 ICJ Reports 14.

[102] Conference of the Parties, United Nations Framework Convention on Climate Change, *Report of the Conference of the Parties on Its Twenty-First Session, Held in Paris from 30 November to 13 December 2015 – Addendum – Part Two: Action Taken by the Conference of the Parties at Its Twenty-First Session* (UN Doc FCCC/CP/2015/10/Add.1 (29 January 2016) Decision 1/CP.21, Annex ('Paris Agreement')) art 12 (with respect to public participation, etc).

[103] Margaret A Young and Sebastian R Sullivan, 'Evolution through the Duty to Cooperate: Implications of the *Whaling* Case at the International Court of Justice' (2015) 16 *Melbourne Journal of International Law* 310–342.

[104] UN-REDD Programme, 'UN-REDD Programme Social and Environmental Principles and Criteria (SEPC)' (UNREDD/PB8/2012/V/1, Asunción, Paraguay, 25–26 March 2012) Criterion 12.

[105] Elisa Morgera, 'The Need for an International Concept of Fair and Equitable Benefit Sharing' (2016) 27 *European Journal of International Law* 353.

interpretation of many diverse rules and standards'.[106] In the context of seeking to establish the influence of an emerging REDD+ regime on indigenous peoples, the identification and analysis of norms of customary international law are as essential as the identification and analysis of other related regimes.

## D. Fragmentation and Regime Interaction: A Preliminary Overview

The presented list of forest-related regimes and customary international law hints at the diversification and fragmentation of international laws relating to forest use and protection of indigenous peoples. Such diversification and fragmentation is a common feature of the international legal system,[107] and indeed in legal responses to the need for climate change mitigation.[108] What is clear from the public international law literature is that fragmentation creates real tensions for sovereignty.[109] States wish to ensure that they are bound *only* by the norms or decision-making procedures to which they have formally agreed (or to which they are otherwise obliged to follow as customary international law): for this reason, the ILC sought to find the common intentions of states parties in resolving any conflicting norms in its seminal 2006 study.[110] Yet, as this section demonstrates, the question of how REDD+ interacts with other regimes is not a strictly positivist one. This section assesses the potential that REDD+ leads to relationships of conflict or interpretation, and then builds on our understanding by considering how issues outside of the state-centric conception of REDD+ influence regime interaction.

### 1. *Fragmentation and Conflicting Norms*

The threat of an expanding corpus of international law on consistency and certainty has been observed over several decades.[111] Public international

[106] Dupuy, above n 96, 465. [107] International Law Commission, above n 1.

[108] Margaret A Young, 'Climate Change Law and Regime Interaction' (2011) 5 *Carbon & Climate Law Review* 147.

[109] Margaret A Young, 'Fragmentation, Regime Interaction and Sovereignty' in Christine Chinkin and Freya Baetens (eds), *Sovereignty, Statehood and State Responsibility* (Cambridge University Press, 2015) 71.

[110] International Law Commission, above n 1.

[111] Ibid. See generally Margaret A Young, 'Fragmentation' in Anthony Carty (ed), *Oxford Bibliographies in International Law* (Oxford University Press, 2014), www.oxfordbibliographies.com.

lawyers have focused their attention on the situation ex post, where regimes have been established and subsequently give rise to multiple and perhaps conflicting obligations for states, as we demonstrated pictorially with Figure 3.1. States party to two sets of obligations may be unable to conform to one without violating the other.

In answer to the quandary of conflicting norms, the ILC has recommended several principles of possible application: the priority of particular or special norms (*lex specialis*) over general norms; the priority of a subsequent rule (*lex posterior*) replacing an earlier conflicting rule; and the priority of particular relations of importance, including jus cogens obligations. The drafting of 'conflict clauses' in treaties is also a common technique; for example, the CBD's overriding of another treaty 'where the exercise of [a Contracting Party's] rights and obligations [under an existing treaty] would cause serious damage or threat to biological diversity'.[112]

However, it is potentially misleading, or at the very least premature, to discuss conflicting norms in the REDD+ context. For states, a situation of normative conflict may only develop if they have been obliged by a treaty or norm of customary international law to do something that conflicts with another of their legal obligations. REDD+ has yet to impose concrete legal obligations upon states; its underlying objective, now reinforced in the Paris Agreement, is the broadly worded aspiration of GHG emission reduction, and its specific application is the 'promotion and support' for certain forest-mitigation activities. As we have previously mentioned, the legal obligations so far are emerging through agreements by international organizations to conduct 'readiness' activities in particular countries, and the provision by particular countries to provide information about how they undertake the future implementation of REDD+ projects. For example, in the scenario of CBD treaty obligations conflicting with REDD+ objectives that we previously discussed, the CBD would give rise to the only treaty obligation. Although there may be conceptual conflicts between REDD+ and other regime objectives, an ex post account of conflicting legal obligations and norms has yet to emerge. Moreover, even if REDD+ develops into a hard-law regime, principles of conflict resolution will be extremely difficult to apply, especially with difficulties in ascertaining the timing of particular rules given their periodic updating by COPs.[113]

112 Convention on Biological Diversity, above n 69, art 22. See further van Asselt, above n 68, 1249.

113 As described in other climate law contexts: see Margaret A Young, 'Climate Change Law and Regime Interaction', above n 108, 150–151.

The implications of REDD+ being a soft- rather than hard-law regime are explored throughout this book. They are most apparent in its conflict-avoidance strategies: the Cancun Safeguards are the primary way in which states have sought to mediate REDD+'s relationship with other regimes, rather than any express conflict clause.[114] How these safeguards operate in difficult contexts, and which actors are most important in their implementation, are thus a necessary focus in further chapters.

### 2. *Fragmentation and Relationships of Interpretation*

Aside from the potential for conflicting norms, situations of fragmentation give rise to the possibility that norms will exist in relationships of interpretation. For example, a future REDD+ treaty obligation may be interpreted according to the broader international legal context, as required by principles of treaty interpretation[115] that require treaties to be interpreted in the light of their object and purpose and in the context in which the terms appear.[116] Moreover, treaty interpreters may take into account 'any subsequent practice in the application of the treaty which establishes the agreement of the parties regarding its interpretation'.[117] The Cancun Safeguards may well amount to 'subsequent practice', so that the COP decision's references to existing international laws and conventions have legal consequences.[118]

One example relates to the customary international law obligation to consult with affected communities, as previously mentioned. Rules on treaty interpretation could determine that a REDD+ treaty obligation must be interpreted subject to this duty. Similar arguments have been advanced with respect to the conceptually similar situation of a state seeking to inscribe indigenous properties on the World Heritage List

[114] Annalisa Savaresi, 'The Role of REDD in the Harmonisation of Overlapping International Obligations' in Erkki J Hollo, Kati Kulovesi and Michael Mehling (eds), *Climate Change and the Law* (Springer Netherlands, 2013) 391, 400.

[115] Vienna Convention on the Law of Treaties (opened for signature 23 May 1969, 1155 UNTS 331, entered into force 27 January 1980) arts 31–32.

[116] Ibid. art 31(1). [117] Ibid. art 31(3)(b).

[118] See further International Law Commission, *Report of the International Law Commission on the Work of Its Sixty-Sixth Session* (UN GAOR, 69th sess, Supp No 10, UN Doc A/69/10 (5 May–6 June and 7 July–8 August 2014) ('Report of the ILC Sixty-Sixth Session')) ch VII, 208, as mentioned in Chapter 2, Section A3.

according to the World Heritage Convention.[119] A second example could draw on CBD obligations and find that a future set of REDD+ treaty obligations would need to take into account norms of the CBD due to the principle of systemic integration.

Yet, this scenario once again assumes that REDD+ amounts to a set of hard-law obligations, a scenario that is not borne out by the Paris Agreement. While the ILC was well aware of the growth in 'quasi-autonomous sources' arising at the international level,[120] and of the complexity associated with identifying the obligations of nongovernmental participants and other actors,[121] it did not include these factors in its analysis.

Instead, a broader approach to fragmentation seeks to understand 'regime interaction' in many manifestations: not just in the resolution of conflicting norms at dispute settlement, but at the ex ante involvement of a range of actors in lawmaking, or the ongoing influence of institutional arrangements in the implementation of laws.[122] It calls for attention to issues such as 'path dependency, higher transaction costs, "tunnel vision" and even "solipsistic and imperial" tendencies within regimes'.[123] As such, this book seeks to understand REDD+'s relationship to other regimes by considering not only prospective situations of legal conflict or interpretation (seen through the prism of states' obligations), but also the legal relevance and implications of current, past and future interactions by a range of actors across relevant regimes.

[119] The three relevant scenarios identified by the ILA are when a state (1) proposes inscription of an indigenous property; (2) refuses an indigenous community's request to propose the inscription of a property belonging to it or (3) manages an indigenous property after its inscription: see International Law Association, above n 7, 18–20. The report concludes that there is a customary international law status of a right of indigenous peoples to recognition and preservation of their cultural identity, and this is a 'relevant rule of international law applicable in the relations between the parties to the World Heritage Convention', thus providing interpretative context to the World Heritage Convention pursuant to VCLT article 31(3)(c).

[120] International Law Commission, above n 1, 249. [121] Ibid. 252.

[122] Margaret A Young, 'Regime Interaction in Creating, Implementing and Enforcing International Law' in Margaret A Young (ed), *Regime Interaction in International Law: Facing Fragmentation* (Cambridge University Press, 2012) 85; see also Jeffrey L Dunoff, 'A New Approach to Regime Interaction' in Margaret A Young (ed), *Regime Interaction in International Law: Facing Fragmentation* (Cambridge University Press, 2012) 136.

[123] Margaret A Young, 'Introduction: The Productive Friction between Regimes' in Margaret A Young (ed), *Regime Interaction in International Law: Facing Fragmentation* (Cambridge University Press, 2012) 1, 11 (footnotes omitted).

### 3. *Regime Interaction by States, International Organizations and Nonstate Actors*

Regime interaction on REDD+ involves states, international organizations and nonstate actors, all of whom assert the primacy of particular approaches. Most clearly, states have sought to mediate the boundaries between REDD+ and other regimes through the COP decisions on the scope of REDD+ and through the Cancun Agreements' incorporation of safeguards, and encouraging full and effective participation of relevant stakeholders, including indigenous peoples and local communities.[124] This most immediately involves the countries that develop and implement their own national strategies and implementation of REDD+ – i.e. the recipient countries. But a broader conception of regime interaction draws attention to the legal obligations not only of recipient countries but also donor countries (and private entities). As the COPs were relatively late to recognise, addressing the drivers of deforestation and degradation is an obligation not only of developing countries but *all* parties to the UNFCCC, as well as organizations and the private sector.[125] Countries have been reminded that developed countries must assume responsibilities in fostering responsible business conduct of their multinational companies and in procuring deforestation-free products,[126] which relate to a range of regimes outside of REDD+. Such a perspective further extends the infographic presented in Figure 3.1, and warrants greater attention in our analysis of REDD+'s impact on indigenous peoples.

Issues of regime interaction call for attention to the role and influence of international organizations and other implementing agencies. We pointed in Chapter 2 to the diverging institutional factors affecting REDD+ Readiness and implementation activities. We saw, for example, how the Cancun Safeguards are to be augmented by the guidelines and

[124] Conference of the Parties, United Nations Framework Convention on Climate Change, '*Cancun Agreements*', above n 41, para 72; see also Cancun Safeguards in Appendix I, para 2.

[125] Conference of the Parties, United Nations Framework Convention on Climate Change, *Report of the Conference of the Parties on Its Nineteenth Session, Held in Warsaw from 11 to 23 November 2013 – Addendum – Part Two: Action Taken by the Conference of the Parties at Its Nineteenth Session* (UN Doc FCCC/CP/2013/10/Add.1 (31 January 2014) ('Warsaw Framework for REDD-Plus') Decision 15/CP.19 ('Addressing the Drivers of Deforestation and Forest Degradation') para 3.

[126] Charlotte Streck and Michaela Schwedeler, 'Addressing Drivers of Deforestation and Forest Degradation through International Law' in Christina Voigt (ed), *Research Handbook on REDD+ and International Law* (Edward Elgar, 2016) 213, 225.

rules of bodies such as the UN-REDD Programme and the FCPF.[127] Although the FCPF and the UN-REDD Programme have tried to integrate their approaches, strong differences remain. This is apparent in a variety of contexts: in 'institutional friction' between the two agencies,[128] in a certain 'distance' between the FCPF and the CBD Secretariat over biodiversity issues,[129] and in the application of the standard of FPIC and in the development of tenure models, which we consider in the following chapters. Organizations influencing regime interaction are of course not limited to actors within the emerging REDD+ regime: we have referred, for example, to interventions from outside, such as those of the CBD Secretariat and the Permanent Forum.

The influence of these actors calls for attention to their powers, duties and obligations. For example, the express powers to coordinate with other bodies held by a body such as the UN-REDD Programme may be augmented by implied powers. Implied powers are enlarged in circumstances of fragmentation: where regimes are dependent upon other regimes (such as in standard setting), international governmental organizations have the discretion to apply norms and facts from external sources where such collaboration is necessary for their functions.[130]

In addition to international organizations, nonstate actors have been instrumental and will continue to be instrumental in REDD+'s relationship to other international regimes. This is especially relevant for indigenous groups, as we discuss in the context of benefit-sharing in Chapter 6. In particular, we note the role of the indigenous representatives on the UN-REDD Policy Board in Chapter 5. It is useful to explore the theoretical justifications for their involvement, which often move beyond representativeness and towards a deliberative model. Relatedly, the identification of 'affected groups', including indigenous peoples and other local communities, might be achieved through staggered consultations, as we discuss in Chapter 4. Questions such as these do not often arise in the context of literature on public international law's fragmentation, but the REDD+ example shows how central these activities are to regime interaction.

[127] See, for example, the templates for readiness preparation proposals described in Chapter 2 and further discussed in Chapter 4.

[128] Le Groupe-conseil baastel and Nordic Agency for Development and Ecology, 'First Program Evaluation for the Forest Carbon Partnership Facility (FCPF): Evaluation Report' (2011).

[129] Ibid.

[130] Young, 'Fragmentation, Regime Interaction and Sovereignty', above n 109, 82.

## E. Conclusions

This chapter has identified the international law context for REDD+. We have built on Chapter 2's analysis of REDD's emergence as a legal regime, which conceived REDD+ as a set of norms, decision-making procedures and organizations coalescing around a functional issue area and dominated by particular modes of behaviour, assumptions and biases. That chapter emphasised that a REDD+ regime is emerging amongst conditions of legal fragmentation, with existing treaties, institutions, and customary and other bodies of law and jurisprudence influencing the objective of forest-based climate mitigation in different ways. This is complicated further by the laws and approaches that seek to mediate state and indigenous peoples' relations, and in Part II of the book we examine more closely the specific contexts of FPIC, tenure and benefit-sharing in which such regime interplay exists.

This chapter has listed a range of relevant regimes, including the international forest regime, indigenous peoples' rights, environmental protection regimes and emerging principles of food security. This was not intended to be exhaustive; indeed, as our country-specific examples show, these 'sites of interaction' are themselves evolving and are locally specific. Yet while particular evaluations of REDD+ will differ, there remains a need to continue to account for REDD+'s influences on, and by, other forest-focused and forest-related regimes, and indeed other regimes in international law, and this chapter has sought to provide the most immediate aspects of this.

The overview of international law in this chapter is useful in providing detail about the international norms, institutions and practices that interact with the REDD+ regime in forming a backdrop of the domestic contexts for REDD+ that we explore in Part III of the book. It is also helpful in demonstrating that REDD+ has emerged amongst an established and dense set of legal relationships, not all of which are in harmony with REDD+ objectives or practices.

The fragmented international legal context of REDD+ has consequences for law, for effectiveness and for affected people. How the conditions of fragmentation impact upon an emerging REDD+ regime, and particularly upon indigenous peoples, is a key research question of the book. This chapter provided a preliminary overview of issues of fragmentation and regime interaction and reviewed the stance of the ILC in its seminal approach to the resolution of conflicts or the interpretation of treaties. We showed, however, that the pursuit of normative 'coherence' through satisfying the common intention of states parties is only one aspect of the fragmentation of international law. Broader questions of state and nonstate involvement in regime interaction go to more general issues of effectiveness, accountability and even legitimacy. We bring these questions to our analysis throughout the book.

It is our premise that drawing attention to regime interaction is essential for an emerging regime such as REDD+ that is faced with existing pluralism and uncertainty, limited and yet-to-mature epistemic knowledge, and the need for adaptive responses at multiple levels. We further describe the potential for overlap and normative friction between REDD+-focused and REDD+-related regimes throughout the book, with reference to the substantive issues of FPIC, tenure and benefit-sharing, as well as in selected domestic contexts, in an effort to better understand the impact of climate change mitigation on indigenous and forest communities. Ideas of experimentalist governance are of significance here, as will be explored. 'Horizontal' regime interaction at the international level combines with the 'vertical' relationship between international regimes and domestic action. In particular, we suggest that regime interaction may affect indigenous peoples by negating hard-fought protections or strengthening soft-law aspirations. We test this further in the context of the principle of FPIC in the chapter that follows.

# PART II

## State and Indigenous Peoples' Relations in the Context of REDD+

4

# REDD+, Identity Law and 'Free, Prior and Informed Consent'

KIRSTY GOVER

It is hard, as a matter of principle, to take issue with the 'human rights-based approach' adopted by the UN-REDD Programme.[1] States are obliged by the Cancun Agreements to 'fully respect human rights' when carrying out REDD+ activities, and there is no question that states remain bound by their international human rights commitments while designing and implementing REDD+ projects.[2] Indigenous peoples, however, have good reason to look closely at the content of the human rights articulated within REDD+ and to consider the extent to which they support or undermine distinctive indigenous claims and rights. International human rights treaties offer few avenues for the articulation of indigenous property and self-governance rights, nor do they yet provide much to support an obligation to obtain indigenous consent to activities

[1] UN-REDD Programme and Forest Carbon Partnership Facility (FCPF), 'Joint Guidelines on Stakeholder Engagement in REDD+ Readiness with a Focus on Indigenous Peoples and Other Forest-Dependent Communities' (20 April 2012) ('UN-REDD/FCPF Joint Guidelines on Stakeholder Engagement'), 2 and 15, www.undp.org/content/undp/en/home/ourwork/democraticgovernance/focus_areas/focus_human_rights/empowering_indigenous_peoples/UNREDD.html; HRBA Portal, 'The Human Rights Based Approach to Development Cooperation: Towards a Common Understanding Among UN Agencies' (2003), http://hrbaportal.org/the-human-rights-based-approach-to-development-cooperation-towards-a-common-understanding-among-un-agencies; United Nations Development Group, *UNDG Human Rights Mainstreaming Mechanism: Operational Plan 2011–2013* (November 2011) ('UNDG-Human Rights Operational Plan'); United Nations Development Group, *United Nations Development Group (UNDG) Guidelines on Indigenous Peoples' Issues* (February 2008).

[2] Conference of the Parties, United Nations Framework Convention on Climate Change, *Report of the Conference of the Parties on Its Sixteenth Session, Held in Cancun from 29 November to 10 December 2010 – Addendum – Part Two: Action Taken by the Conference of the Parties at Its Sixteenth Session*, UN Doc FCCC/CP/2010/7/Add.1 (15 March 2011) ('*Cancun Agreements*') Decision 1/CP.16, Appendix I para 2(a); UN-REDD Programme and Forest Carbon Partnership Facility, (FCPF), above n 1, 2: 'Both the UN-REDD Programme and the FCPF also recognize as part of their policies and procedures that for REDD+ to be implemented, participating countries should comply with applicable international obligations, treaties and national laws.'

that affect their interests, which is the strong form of the principle of 'free, prior and informed consent' (FPIC). Where these collective rights are the basis of indigenous claims, in the REDD+ context and elsewhere, they are susceptible to being subordinated to the human and property rights of others. This raises especially acute problems for REDD+ recipient states. As is discussed in Chapter 5, the recognition of indigenous rights to land, self-governance and FPIC typically requires the redistribution of property and public authority within the settler polity, an exercise that goes to the heart of legal and popular sovereignty, and which impacts directly on the existing property and human rights of third parties, including other indigenous groups and non-indigenous minorities. This chapter evaluates the efforts of institutions within REDD+ to articulate safeguards protecting the rights of indigenous peoples, with particular attention to the principle of FPIC.

REDD Readiness financing is conditioned on compliance with standards promulgated by the two major REDD+ funding bodies: the UN-REDD Programme and the World Bank's FCPF. The initial framing and ranking of human rights in the advice offered by these bodies directly shapes the allocation of REDD+ benefits amongst local actors in recipient states, and influences the way that disputes between rights holders are resolved in domestic law. It matters, then, that within the safeguards and standards articulated by the UNFCCC, the UN-REDD Programme and the FCPF, there are varied interpretations of the content and scope of the indigenous rights that must be protected by states participating in REDD+. States face a complex set of directions on indigenous and human rights that contains little guidance about how to reconcile or choose amongst the different approaches encompassed within it.[3] In practice, the UNFCCC safeguards are translated into norms in evaluative criteria used in the UN-REDD Programme's Social and Environmental Principles and Criteria (SEPC) and the FCPF's Strategic Environmental and Social Assessment (SESA) guidelines. This collection of principles and criteria – used by FCPF and the UN-REDD Programme to guide states through REDD Readiness preparations – is dense and variegated.

[3] The UN-REDD Programme seems attentive to the difficulty of trading off of various costs and multiple benefits, but is less willing to frame conflicts between rights in these terms, emphasising instead the universality, inalienability, indivisibility, interdependence, and interrelatedness of human rights, in line with the reasoning used by the UNDG; see HRBA Portal, above n 1, 'The Common Understanding' (No 2). See, e.g., UN-REDD Programme, *The UN-REDD Programme Strategy 2011–2015* (February 2011) 7.

Because they shape and drive state applications for REDD Readiness funding, the UN-REDD Programme and FCPF criteria are currently the only 'enforceable' directions applicable to REDD+ recipient states. They are for the time being the major interface between the policies favoured by international REDD+ agencies and the domestic law of participating states. For this reason, we consider them in detail.

In this chapter, we examine the ways that indigenous rights and interests have been articulated within REDD+, and outline divergences and overlaps in the standards used by the UN-REDD Programme and the FCPF. We take as a case in point the different approaches to FPIC enforced by these two institutions, and the impact of these approaches on the obligations of states. In Section A, we examine the formal legal sources of the UN-REDD Programme and FCPF safeguard policies and standards. In Section B, we address the complexity of international law on the identification of indigenous peoples. In Section C, we consider the body of international law on which the UN-REDD Programme's principle of FPIC depends. We examine the ways that FPIC has been articulated by the UN-REDD Programme and the FCPF, respectively, along with the efforts made by these bodies to coordinate and rationalise their guidance on FPIC requirements. We consider the evolution of the FPIC principle in international and regional human rights law, including in the 2007 UNDRIP, and the new direction taken by the World Bank in its 2016 Environmental and Social Framework (ESF).

## A. REDD+, the Human Rights-Based Approach and Indigenous Peoples

As discussed in Chapter 2, the UN-REDD Programme was established in 2008 as an interagency partnership of FAO, the UNDP, and the UNEP.[4] Its central task is to 'assist [ ] developing countries to prepare and implement national REDD strategies and mechanisms'.[5] While it

[4] UN-REDD Programme, *UN-REDD Programme Handbook for National Programmes and Other National-Level Activities* (2012) 9.

[5] FAO, UNDP and UNEP, *UN Collaborative Programme on Reducing Emissions from Deforestation and Forest Degradation in Developing Countries (UN-REDD): Framework Document* (20 June 2008) ('UN-REDD Framework Document') 7. (This document was replaced by *The UN-REDD Programme Strategy 2011–2015*, adopted in November 2010: UN-REDD Programme, above n 3.) See UN-REDD Programme, *UN-REDD Programme Strategic Framework 2016–2020* (UNREDD/PB14/2015/III/3, Revised Draft, 7 May 2015) 2.

administers and disburses grants from its Readiness Fund, it does not have a role in funding the implementation of REDD+ projects or 'emissions reductions activities', nor does it provide 'results-based payments' for those activities.[6] Its role is confined to assistance with REDD Readiness. In this work, it applies a 'human rights–based approach' derived from the UN Charter, which provides the mandates of its constituent 'collaborating agencies',[7] and from the operating principles of the UNDG to which those organizations belong.[8] Chief among these is the UNDG's UN Statement of Common Understanding on Human Rights–Based Approaches to Development Cooperation and Programming ('the Common Understanding'), adopted in 2003, which 'reiterates the UN commitment to further the realisation of human rights as laid down in the Universal Declaration of Human Rights and other international human rights instruments by ensuring that these instruments guide all development cooperation and programming'.[9] The Common Understanding outlines the human rights principles that guide the UNDG's nineteen member organizations in development programming, and emphasises the 'universality and inalienability; indivisibility; interdependence and inter-relatedness' of human rights, along with the principles of 'non-discrimination and equality; participation and inclusion; accountability and the rule of law'.[10] These are principles that have not yet provided a sound justification for the distinctive qualities of indigenous rights, nor do they provide guidance on how to reconcile rights when they conflict.

[6] 'The UN-REDD Programme is not providing the finance for RBPs, but can ensure that REDD+ readiness processes culminate in the establishment of an institutional landscape where national capacities have been developed such that flows of REDD+ RBPs will be managed in an effective, transparent and equitable manner': UN-REDD Programme, *UN-REDD Programme Strategic Framework 2016–2020*, above n 5, iv.

[7] UN-REDD Programme, 'Operational Guidance: Engagement of Indigenous Peoples and Other Forest Dependent Communities' (Working Document, 25 June 2009) ('UN-REDD Operational Guidance') 5.

[8] HRBA Portal, above n 1; United Nations Development Group, *UNDG Guidelines on Indigenous Peoples' Issues*, above n 1. See also Office of the United Nations High Commissioner for Human Rights, *Frequently Asked Questions on a Human Rights–Based Approach To Development Cooperation* (2006), www.ohchr.org/Documents/Publications/FAQen.pdf.

[9] UN-REDD Programme, *Guidelines on Free, Prior and Informed Consent* (January 2013) ('UN-REDD Programme FPIC Guidelines') 12.

[10] HRBA Portal, above n 1, 'The Common Understanding' (No 2).

### 1. *The World Bank's Forest Carbon Partnership Facility (FCPF)*

In the REDD Readiness phase, the roles of the FCPF and UN-REDD Programme are broadly comparable.[11] Both base their advice to states on the Cancun Safeguards set out in the 2010 UNFCCC Cancun Agreements.[12] Both provide funding to 'assist developing countries in their efforts to reduce emissions from deforestation and forest degradation ... by building national capacity for REDD+ activities'.[13] The FCPF, like the UN-REDD Programme, was formally established in 2008, but is a creature of the World Bank and operates entirely under its auspices. Crucially, the World Bank is not a member of the UNDG and does not pursue a 'human rights–based approach' in its lending activities. The FCPF Charter confirms that its work is governed by the World Bank's ten Operational Policies and Procedures (also known as its 'safeguard policies'), including Operational Policy 4.10 (OP 4.10) on Indigenous Peoples. States have been advised that if funding is sought from the FCPF for REDD Readiness activities, 'the relevant World Bank's safeguard policies apply to this work and have to be complied with',[14] but the application of the policies 'will depend upon the stage of readiness and types of projects, activities, or policies/regulations, and related impacts'.[15] According to the World Bank, 'the FCPF Readiness Fund is not intended

[11] The UN-REDD Programme has reported that the perceived functional overlap between itself and the FCPF has made it more difficult to raise money from donors to support the programme's operations. 'Many donors found it difficult to understand the different roles and function of the FCPF and UN-REDD Programme and while these roles are now much better clarified and cooperation significantly increased, some still perceive overlaps': UN-REDD Programme, *UN-REDD Programme Strategy 2011–2015*, above n 3, 28.

[12] According to the FCPF, the World Bank safeguards 'promote and support' the UNFCCC safeguards. 'While the verbatim texts of the World Bank environmental and social safeguard policies and the UNFCCC REDD+ safeguards are not identical, the required application of the World Bank's safeguards (consisting of policies, procedures, and practices) to Emissions Reduction Programs supported by the FCPF Carbon Fund should be sufficient to ensure that the World Bank's safeguards successfully promote and support the UNFCCC safeguards for REDD+': Forest Carbon Partnership Facility (FCPF) Carbon Fund, 'World Bank Safeguard Policies and the UNFCCC REDD+ Safeguards' (FMT Note CF-2013-3, 28 August 2013) 1.

[13] UN-REDD/FCPF Joint Guidelines on Stakeholder Engagement, above n 1, 1.

[14] Forest Carbon Partnership Facility (FCPF) and The United Nations Collaborative Programme on Reducing Emissions from Deforestation and Forest Degradation in Developing Countries (UN-REDD), 'Readiness Preparation Proposal (R-PP)' (Template Version 6, 20 April 2012) ('R-PP Template Version 6') 3 and [3].

[15] Ibid.

to fund projects or investments on the ground',[16] so that the formal process for ensuring compliance with the bank's safeguard policies will ordinarily not apply to REDD Readiness activities.[17] For the purposes of applying its operational policies, then, the bank draws a distinction between the financing of 'preparation' on the one hand, and the financing of 'implementation of policies and projects' on the other.[18] It is only in respect of the latter that 'the safeguard instruments required by the safeguard policies would have to be prepared and applied as they are in standard Bank-financed projects'.[19] This means that OP 4.10 does not usually govern REDD Readiness activities.[20] The policy does not, consequently, provide a basis on which groups affected by REDD Readiness activities could request an inspection by the World Bank Inspection Panel.[21] The FCPF's allocation of funding for Emissions Reduction Programs from the Carbon Fund, however, are conditional on compliance with the bank's safeguard policies.[22]

OP 4.10 will be replaced in 2018 by the World Bank's Environmental and Social Standard 7 (ESS 7) on Indigenous Peoples/Sub-Saharan

[16] Forest Carbon Partnership Facility (FCPF) Readiness Fund, 'Incorporating Environmental and Social Considerations into the Process of Getting Ready for REDD' (Note FMT 2009–6, Draft, 15 October 2009) 3 [6].

[17] 'In World Bank parlance, the activities carried out with support from the Readiness Fund constitute preparation. The implementation of policies and projects are beyond the scope of what is eligible under the Readiness Fund. However, should it become necessary, on an exceptional basis, to finance the implementation of policies and projects under the Readiness Fund, the safeguard instruments required by the safeguard policies would have to be prepared and applied as they are in standard Bank-financed projects': Ibid. 3 [8].

[18] Ibid. 3 [6]. [19] Ibid. 3 [8].

[20] Section 3.1(d) of the FCPF Charter provides that 'the operation of the Facility, including implementation of activities under Grant Agreements and Emission Reductions Programs, shall ... comply with the World Bank's Operational Policies and Procedures, taking into account the need for effective participation of Forest-Dependent Indigenous Peoples and Forest Dwellers in decisions that may affect them, respecting their rights under national law and applicable international obligations': International Bank for Reconstruction and Development, *Charter Establishing the Forest Carbon Partnership Facility* (Revised 23 November 2015), www.forestcarbonpartnership.org/charter-and-governance-documents ('FCPF Charter').

[21] Bank Information Centre (BiC), 'SESA, Safeguards and the FCPF: A Guide for Civil Society' (January 2013), www.bankinformationcenter.org/wp-content/uploads/2013/01/SESA+informational+doc+for+PC6.pdf. See also Benedict Kingsbury, 'Operational Policies of International Institutions as Part of the Law-Making Process: The World Bank and Indigenous Peoples' in Guy S Goodwin-Gill and Stefan Talmon (eds), *The Reality of International Law: Essays in Honour of Ian Brownlie* (Oxford University Press, 1999) 323.

[22] Forest Carbon Partnership Facility, *Carbon Fund Methodological Framework* (22 June 2016) 18, www.forestcarbonpartnership.org/carbon-fund-methodological-framework.

African Historically Underserved Traditional Local Communities. As discussed in more detail in Section 2, ESS 7 contains important new provisions on FPIC. It requires states to acquire indigenous consent in circumstances where the proposed project will: adversely impact on traditionally owned or used land and natural resources; cause the forced relocation of indigenous peoples from such land; or have significant impacts on indigenous cultural heritage.[23] Existing projects funded by the World Bank will be covered by the current Operational Policies, while projects beginnning after 2018 will be covered by the ESF, so that the two standards will run in tandem until existing projects have been concluded.[24]

Broadly speaking, the World Bank safeguards are designed to reduce risk in the implementation of bank-funded investment projects, while the UN development agencies use safeguards primarily to protect the interests of persons affected by development projects in line with their distinctive institutional mandates, unified by a 'human rights–based approach' to programming. This broadly drawn distinction expresses the institutional divides between the ostensibly apolitical, technocratic, investment-based approach adopted by the World Bank (and its correspondingly conservative stance on human rights and political values),[25] and the facilitative, social inclusion–based development agendas of the UNDP, UNEP and FAO, evidenced by their express reliance on human rights norms.[26] As will be seen, some of these tensions are illustrated by

[23] World Bank, *Environmental and Social Framework: Setting Environmental and Social Standards for Investment Project Financing* (4 August 2016) ('World Bank Environmental and Social Framework') 113.

[24] World Bank, 'World Bank Board Approves New Environmental and Social Framework' (Press Release, 4 August 2016), www.worldbank.org/en/news/press-release/2016/08/04/world-bank-board-approves-new-environmental-and-social-framework.

[25] See generally Ibrahim F I Shihata, 'Prohibition of Political Activities in the Bank's Work: Legal Opinion by the Senior Vice President and General Counsel', *Transnational Dispute Management*, 12 July 1995. See also Ingrid J Visseren-Hamakers et al, 'Trade-Offs, Co-Benefits and Safeguards: Current Debates on the Breadth of REDD+' (2012) 4 *Current Opinion in Environmental Sustainability* 646, 648; Annalisa Savaresi, 'The Role of REDD in the Harmonisation of Overlapping International Obligations' in Erkki J Hollo, Kati Kulovesi and Michael Mehling (eds), *Climate Change and the Law*, Ius Gentium: Comparative Perspectives on Law and Justice, vol. 21 (Springer, 2013) 391, 405.

[26] See, e.g., UNDP, 'Human Rights and UNDP' (Fast Facts, United Nations Development Programme, November 2011), www.undp.org/content/dam/undp/library/corporate/fast-facts/english/FF-Human-Rights.pdf; Office of the United Nations High Commissioner for Human Rights (OHCHR) and the United Nations Environment Programme (UNEP), *Human Rights and the Environment: Rio+20A – Joint Report OHCHR and UNEP* (2012);

the differences between the FPIC and indigenous identity standards applied respectively by the UN-REDD Programme and the FCPF.[27] In this section, we begin first by unpacking the sources of the UN-REDD Programme's human rights mandates, since these underpin its commitments to indigenous rights, including FPIC, before moving to consider the FCPF's approach.

### 2. *The Cancun Safeguards and UN-REDD Programme Advice*

Aside from the funding criteria deployed by the UN-REDD Programme and FCPF, participating states are also obliged independently to meet obligations set out in the COP decisions of the UNFCCC. The formal entry point for indigenous issues in REDD+ programming then, is the UNFCCC's 2010 COP decision (Cancun Agreements),[28] which sets out the basic parameters of the REDD+ regime and outlines in an appendix the 'safeguards' that recipient states should 'promote[] and support[]' when undertaking REDD+ activities in their territories (Cancun Safeguards).[29] While participating in REDD+ activities, state parties are obliged to periodically provide 'a summary of information' on how the safeguards are being 'addressed and respected'.[30] Indigenous interests

Food and Agriculture Organization of the United Nations (FAO), *Voluntary Guidelines to Support the Progressive Realization of the Right to Adequate Food in the Context of National Food Security* (2005). Unlike the FAO, UNDP, and UNEP, the World Bank is an observer, but not a member, of the UNDG and not a participant in its Human Rights Mainstreaming agenda.

27 UN-REDD Programme, above n 4, 14. See also Forest Carbon Partnership Facility (FCPF) Readiness Fund, 'Common Approach to Environmental and Social Safeguards for Multiple Delivery Partners' (Revised 9 August 2012) ('FCPF Readiness Fund Common Approach') 6 [15]: 'The safeguard policies of the DPs that are multilateral investments bank were largely written with investment projects in mind and, in the case of all DPs including those DPs that are part of the United Nations, they serve to support the overarching mission of each organization. However, REDD+ readiness activities in the FCPF context entail no investment projects on the ground. They mostly consist of strategic planning and preparation. Nonetheless, these strategic activities have potentially far-reaching impacts – hopefully positive – but, unless properly addressed, possibly negative (e.g., the definition of rights to forest carbon or the design of benefit-sharing mechanisms).'

28 *Cancun Agreements*, UN Doc FCCC/CP/2010/7/Add.1, Decision 1/CP.16, para 72.

29 Ibid. Appendix I para 2.

30 Conference of the Parties, United Nations Framework on Climate Change, *Report of the Conference of the Parties on Its Seventeenth Session, Held in Durban from 28 November to 11 December 2011 – Addendum – Part Two: Action Taken by the Conference of the Parties at Its Seventeenth Session*, UN Doc FCCC/CP/2011/9/Add.2 (15 March 2012) Decision

affected by REDD+ traverse the full range of risks and benefits referenced in the seven Cancun Safeguards,[31] and indigenous peoples are expressly referenced in two of the safeguards' operative provisions. In the conduct of REDD+ activities, states should ensure 'the full and effective participation of relevant stakeholders, inter alia indigenous peoples and local communities',[32] and promote and support 'respect for the knowledge and rights of indigenous peoples and members of local communities, by taking into account relevant international obligations, national circumstances and laws, and noting that the United Nations General Assembly has adopted the [UNDRIP]'.[33]

In the REDD Readiness phase, the normative force of the Cancun Safeguards is brought to bear on states when funding applications submitted by states are evaluated by the UN-REDD Programme's Executive Board or the FCPF Facility Management Team (FMT). The standards applied by the UN-REDD Programme and the FCPF are designed to be *consistent* with the Cancun Safeguards, although, as will be discussed, the degree to which those standards are *coextensive* with the obligations of states under the Cancun Agreements is a matter for debate. This is particularly the case with the UN-REDD Programme's FPIC standards, which purport to 'build on' existing international standards.[34] The UN-REDD Programme reports that it 'focuses on supporting countries to meet (or exceed as desired) their requirements under the UNFCCC, rather than imposing any novel safeguard requirements of its own', and that its advice is 'designed to enable countries to take a flexible country approach in responding to the UNFCCC decisions on safeguards'.[35]

UN-REDD Programme advice and standards on engagement with indigenous peoples, including its advice on FPIC, addresses the 'effective

12/CP.17, paras 3 and 4. '*Agrees also* that developing country Parties undertaking the activities referred to in decision 1/CP.16, paragraph 70, should provide a summary of information on how all of the safeguards referred to in decision 1/CP.16, appendix I, are being addressed and respected throughout the implementation of the activities': at para 3.

[31] *Cancun Agreements*, UN Doc FCCC/CP/2010/7/Add.1, Decision 1/CP.16, para 72 and Appendix I.

[32] Ibid. para 72 and Appendix I para 2(d).

[33] Ibid. Appendix I para 2(c). UN-REDD Programme, 'UN-REDD Programme Social and Environmental Principles and Criteria' (Version 3, Draft for Consultation, September 2011) ('UN-REDD Draft Social and Environmental Principles and Criteria') criterion 9.

[34] UN-REDD Programme, *UN-REDD Programme Strategy 2011–2015*, above n 3.

[35] UN-REDD Programme, *UN-REDD Programme Strategic Framework 2016–2020*, above n 5, 11.

participation' UNFCCC safeguard,[36] while advice and standards on tenure reform and protection of traditional knowledge address the 'respect for knowledge and rights' safeguard.[37] Indigenous peoples' interests are also referenced by the UN-REDD Programme in its advice to states on the safeguard that directs them 'to enhance other social and environmental benefits'.[38] This safeguard is accompanied in the text of the COP decision by a footnote requiring states to 'Take . . . into account the need for sustainable livelihoods of indigenous peoples and local communities and their interdependence on forests in most countries'.[39] This safeguard and its footnote form the basis for the UN-REDD Programme's advice on the necessity of poverty-reduction measures and the provision of alternative livelihoods for indigenous and forest-dependent communities.[40] Outside of the text of the safeguards, the Cancun Agreements also require states to address 'land tenure issues, [and] forest governance issues',[41] so providing the mandate for UN-REDD Programme guidance on indigenous and forest community tenure and property benefits, as discussed in Chapter 5. Notably, the Cancun Safeguards do not expressly mention human rights, but a preambular passage of the COP decision 'emphasizes that Parties should, in all climate change related actions, fully respect human rights'.[42]

### (a) The Sources and Content of the UN-REDD Programme's Human Rights–Based Approach

The 'human rights–based approach'[43] adopted by the UN-REDD Programme is drawn from the mandates of its three partner organizations and the UNDG (to which those agencies belong). In its programmatic

[36] *Cancun Agreements*, UN Doc FCCC/CP/2010/7/Add.1, Decision 1/CP.16, para 72 and Appendix I para 2(d). See UN-REDD Draft Social and Environmental Principles and Criteria, above n 33, 14, criteria 7 and 11.

[37] UN-REDD Draft Social and Environmental Principles and Criteria, above n 33, principle 7; UN-REDD Programme, 'UN-REDD Programme Social and Environmental Principles and Criteria (SEPC)' (UNREDD/PB8/2012/V/1, 25–26 March 2012) 15.

[38] *Cancun Agreements*, UN Doc FCCC/CP/2010/7/Add.1, Decision 1/CP.16, para 72 and Appendix I para 2(e)

[39] *Cancun Agreements*, UN Doc FCCC/CP/2010/7/Add.1, Decision 1/CP.16, para 72 and Appendix I para 2(e) n 1.

[40] UN-REDD Draft Social and Environmental Principles and Criteria, above n 33, on principle 3, criterion 13.

[41] *Cancun Agreements*, UN Doc FCCC/CP/2010/7/Add.1, Decision 1/CP.16, para 72.

[42] Ibid. 4[8].

[43] UN-REDD/FCPF Joint Guidelines on Stakeholder Engagement, above n 1, 15.

activities and advice, the programme makes particular reference to the 2008 UNDG Guidelines on Indigenous Peoples' Issues[44] and to the responsibility of UN agencies to 'promote respect' for the UNDRIP, which is expressed in Article 42 of the declaration.[45] It further emphasises that the internal human rights 'policies and procedures' of the UNDP, UNEP and FAO[46] will 'help ensure the rights of indigenous and forest-dwelling people are protected and the active involvement of local communities and relevant institutions in the design and implementation of REDD plans'.[47] All three agencies have internal policies governing their engagement with indigenous peoples, which directly inform the approach taken by the UN-REDD Secretariat, and have been endorsed by the UN-REDD Programme Executive Board (formerly the 'Policy Board').[48] The UN-REDD Programme thus draws human rights concepts and methods into its work in two ways: first, by conducting itself 'in a manner which is consistent with the UN Charter as well as

[44] UN-REDD Framework Document, above n 5, 7.

[45] Article 42 requires '[t]he United Nations, its bodies, including the Permanent Forum on Indigenous Issues, and specialized agencies, including at the country level, and States shall promote respect for and full application of the provisions of this Declaration and follow up the effectiveness of this Declaration': UN-REDD Operational Guidance, above n 7, 5.

[46] UN-REDD Draft Social and Environmental Principles and Criteria, above n 33, 2 n 2: 'UNDP's Environmental Sustainability and Climate Change prescriptive policy, UNEP's Framework for Ecosystems Management, and FAO's Environment and Social Impact Assessment'.

[47] UN-REDD Framework Document, above n 5, 7. See also UN-REDD Programme, *UN-REDD Programme Handbook*, above n 4, 10.

[48] Maddie West, 'UN-REDD Programme (2016–2020) Establishes Executive Board, First Meeting to Take Place at FAO Headquarters in June', UN-REDD Programme: Collaborative Online Workspace (5 April 2016), www.unredd.net/announcements-and-news/2390-un-redd-programme-2016–2020-establishes-executive-board-first-meeting-to-take-place-at-fao-headquarters-in-june.html; United Nations Environment Programme (UNEP), 'UNEP and Indigenous Peoples: A Partnership in Caring for the Environment' (Policy Guidance, November 2012); UNDP, 'UNDP and Indigenous Peoples: A Policy of Engagement' (June 2001), www.undp.org/content/undp/en/home/librarypage/environment-energy/local_development/undp-and-indigenous-peoples-a-policy-of-engagement.html; Food and Agriculture Organization, *FAO Policy on Indigenous and Tribal Peoples* (2010), www.fao.org/docrep/013/i1857e/i1857e00.htm. See also Food and Agriculture Organization of the United Nations (FAO), *Respecting Free, Prior and Informed Consent: Practical Guidance for Governments, Companies, NGOs, Indigenous Peoples and Local Communities in Relation to Land Acquisition* (2014), www.fao.org/documents/card/en/c/e5e9c08e-ff02-5509–8929-6f6537883f7b; Food and Agriculture Organization of the United Nations (FAO), *Free Prior and Informed Consent: An Indigenous Peoples' Right and a Good Practice for Local Communities – Manual for Project Practitioners* (2016), www.fao.org/documents/card/en/c/5202ca4e-e27e-4afa-84e2-b08f8181e8c9.

with international treaties and conventions, notably the UNFCCC',[49] and second, by assisting states to meet their obligations under 'relevant international instruments',[50] and in so doing 'ensuring the recognition of and respect for rights, including those of [indigenous peoples], under applicable international law'.[51] The UN-REDD Programme assists states to meet their international law obligations to indigenous peoples by holding states to standards used to evaluate applications for funding, as well as by providing advice to participant states in the form of written 'operational guidance' prepared by the UN-REDD Secretariat (primary responsibility for this advice is vested in UNDP as the lead agency on the UN-REDD Programme's work on 'Engagement with Indigenous Peoples').[52] As this advice provides evidence of the human rights and indigenous rights standards to which states are held in their REDD+ activities, it is this chapter's primary object of investigation.[53]

Broadly speaking, the UN-REDD Programme requires participating states to adhere to two indigenous rights commitments. States are to meet 'standards outlined in key relevant international instruments'[54] and to 'uphold the principle of free, prior and informed consent (FPIC) as stated in the [UNDRIP]'.[55] The obligation on states to demonstrate

[49] UN-REDD Programme, *UN-REDD Programme Strategic Framework 2016–2020*, above n 5, 5.

[50] *Cancun Agreements*, UN Doc FCCC/CP/2010/7/Add.1, Decision 1/CP.16, para 2(c); UN-REDD/FCPF Joint Guidelines on Stakeholder Engagement, above n 1, 1 [2]. See UN-REDD Programme, *UN-REDD Programme Strategy 2011–2015*, above n 3, 14, noting that UN-REDD's work 'must be carried out in accordance with the human rights based approach in a manner that fulfils the requirements set forth in the [UNDRIP] and other UN conventions and declarations on rights and the safeguards articulated for specific application to REDD+, such as those under the UNFCCC negotiations'. See also UN-REDD Draft Social and Environmental Principles and Criteria, above n 33, noting that '[t]he Social and Environmental Principles and Criteria draw on, are consistent with, and seek to help countries meet their commitment to [listed] international agreements': at 2.

[51] UN-REDD Programme, *UN-REDD Programme Strategic Framework 2016–2020*, above n 5, 19.

[52] UN-REDD Programme, *UN-REDD Programme Handbook for National Programmes and Other National-Level Activities*, above n 4, 10 and 12; UN-REDD Programme, *UN-REDD Programme Strategy 2011–2015*, above n 3, 14.

[53] 'The collaborative programme will have two components: (i) assisting developing countries prepare and implement national REDD strategies and mechanisms; (ii) supporting the development of normative solutions and standardized approaches based on sound science for a REDD instrument linked with the UNFCCC. The programme will help empower countries to manage their REDD processes and will facilitate access to financial and technical assistance tailored to the specific needs of the countries': UN-REDD Framework Document, above n 5, 7.

[54] UN-REDD/FCPF Joint Guidelines on Stakeholder Engagement, above n 1, 2 [6].

[55] Ibid.

adherence to 'key relevant international instruments', is drawn from the provision of the Cancun Safeguards requiring states to conduct REDD+ activities by taking actions that 'complement or are consistent with ... relevant international conventions and agreements'.[56] The list of 'relevant international instruments' identified and prescribed by the UN-REDD Programme is extensive and eclectic. It includes a range of binding instruments, along with selected nonbinding declarations, decisions of UN treaty bodies and, finally, interagency policy guidelines (which only indirectly address states) including: the UNDRIP; the UN Common Understanding on the Human Rights Based Approach to Development Cooperation;[57] the UN General Assembly Programme of Action for the Second International Decade of the World's Indigenous People;[58] the CERD Committee's General Recommendation 23 on the Rights of Indigenous Peoples;[59] the UNDG Guidelines on Indigenous Peoples' Issues 2008;[60] the ILO Convention 169;[61] and, of course, the UNFCCC and the CBD.[62]

The variegation of these instruments is striking for several reasons. First, it shows the capacity of the UN-REDD Programme to condition financing

[56] *Cancun Agreements*, UN Doc FCCC/CP/2010/7/Add.1, Decision 1/CP.16, Appendix I para 2(a). UN-REDD/FCPF Joint Guidelines on Stakeholder Engagement, above n 1, 2 [6].

[57] HRBA Portal, above n 1.

[58] *Programme of Action for the Second International Decade of the World's Indigenous People*, UN GAOR, 60th sess, Agenda Item 68, UN Doc A/RES/60/142 (7 February 2006).

[59] *Report of the Committee on the Elimination of Racial Discrimination*, UN GAOR, 52nd sess, Supp No 18, UN Doc A/52/18 (26 September 1997) annex V ('*General Recommendation on the Rights of Indigenous Peoples*') ('General Recommendation 23').

[60] United Nations Development Group, *UNDG Guidelines on Indigenous Peoples' Issues*, above n 1.

[61] *Convention (No 169) Concerning Indigenous and Tribal Peoples in Independent Countries*, opened for signature 27 June 1989, 1650 UNTS 383 (entered into force 5 September 1991) ('ILO Convention 169').

[62] UN-REDD Programme, 'UN-REDD Programme Social and Environmental Principles and Criteria', above n 37, 2. The UN-REDD's 2012 elaboration of safeguard principles is designed, further, to 'help countries meet their commitment to, *inter alia*, the following international agreements': *Non-Legally Binding Instrument on All Types of Forests*, GA Res 62/98, UN GAOR, 62nd sess, 74th plen mtg, UN Doc A/RES/62/98 (31 January 2008); *Convention on the Elimination of All Forms of Discrimination Against Women*, opened for signature 18 December 1979, 1249 UNTS 13 (entered into force 3 September 1981); *United Nations Convention Against Corruption*, opened for signature 31 October 2003, 2349 UNTS 41 (entered into force 14 December 2005); *International Convention on the Elimination of All Forms of Racial Discrimination*, opened for signature 7 March 1966, 660 UNTS 195 (entered into force 4 January 1969): at 2.

on the achievement by states of standards that are non-binding in international law, or which are binding norms to which participant states have not expressly consented.[63] As will be made clear, there is an inherent tension between international human rights law as a body of international norms (in which some are regarded as rules of customary international law binding all states) and the specific treaty obligations *undertaken* by particular states. Second, the UN-REDD Programme's selected body of international law relating to indigenous rights contains some important internal tensions that need to be resolved if they are to be applied in particular contexts. Finally, the list of human rights instruments named by the UN-REDD Programme is selective, excluding some important human rights treaties that deal with minority rights and which have generated important jurisprudence on indigenous rights and property claims, notably the Convention on the Elimination of All Forms of Racial Discrimination (CERD) and the ICCPR – although the recommendations of the relevant UN treaty bodies (Human Rights Committee (HRC) and CERD Committee) are referenced.[64] These features raise secondary questions about the demands made of participating states by REDD+ funding bodies: Are states to apply international norms in the REDD+ context that they have not accepted as binding in any other context? How are they to resolve inconsistencies between international norms and regimes, and between international norms and domestic law in the design and implementation of REDD projects? Finally, does the advice of the UN-REDD Programme, informed by a 'human rights–based approach', assist in these tasks?

Together, then, the institutional mandate for the UN-REDD Programme's human rights–based approach to the interests of indigenous peoples and the guidance it provides to states on these issues derive from the following: decisions on safeguards emerging from the UNFCCC COP meetings (particularly the 2010 Cancun Agreements); institutional mandates and policies of the UN-REDD Programme partner organizations (UNDP, UNEP and FAO); interagency agreements and responsibilities of partner organizations (including those articulated in the governing documents and policies of the UNDG);[65] directions to international

[63] For example, ILO Convention 169 has only 22 state signatories, the majority of which are Latin American states. Australia, Malaysia, Vanuatu and PNG have not signed ILO Convention 169.

[64] UN-REDD Programme FPIC Guidelines, above n 9, 13–14.

[65] HRBA Portal, above n 1; UNDG-Human Rights Operational Plan, above n 1; United Nations Development Group, *United Nations Development Group (UNDG) Guidelines on Indigenous Peoples' Issues*, above n 1.

organizations in the UNDRIP; and finally, international human rights treaties, declarations and jurisprudence.[66]

### 3. *Human Rights and Multiple Benefits for Indigenous Peoples*

Even if it is possible to discern a core set of human rights obligations that could apply as a matter of principle to all states participating in REDD+ activities, it is clear that support for human rights will not always contribute to the realisation of climate change goals. There is, as Savaresi has pointed out, 'clear potential for both frictions and synergies between REDD and human rights instruments'.[67] This is illustrated by the complex interrelationship of safeguards and benefit-distribution requirements in the REDD+ model, and the tensions between various social and environmental benefits. Safeguards allow values and norms from other international regimes to be incorporated into the REDD+ policy framework, and so act as an ex ante prophylactic against regime conflict.[68] They can also provide an opportunity to generate momentum and support for REDD+ projects by providing external justifications for policies favoured by participating international agencies. In both modes, the UN-REDD Programme's efforts to advise participant states show how difficult it is to reconcile bodies of international law that have competing teleologies, and to produce rules that comply with these bodies of law. In short, as the REDD+ agenda becomes more complex, so that states are urged to mitigate more risks and enhance more benefits, the capacity of international bodies like the UN-REDD Programme to assist in the ranking of policy goals at the local level is correspondingly diminished. This was acknowledged in a 2014 External Review of the UN-REDD Programme:

> The UN-REDD Programme has been a strong voice in the prioritisation of free, prior and informed consent (FPIC) principles and the recognition of Indigenous Peoples' rights in the implementation of UNFCCC safeguards, especially in decision-making fora. Programme contributions to

[66] See also UN-REDD Programme, 'Indigenous Peoples and the UN-REDD Programme: An Overview' (Paper delivered at the International Expert Group Meeting on Indigenous Peoples and Forests, 12–14 January 2011) 1.

[67] Annalisa Savaresi, 'The Human Rights Dimension of REDD' (2012) 21 *Review of European Community & International Environmental Law* 102, 104. See also Annalisa Savaresi, 'REDD+ and Human Rights: Addressing Synergies between International Regimes' (2013) 18(3) *Ecology and Society*.

[68] Savaresi, above n 25, 399.

> social and environmental safeguards are helping to raise awareness at national and global levels, but it remains a challenge to put such principles into practice, notably due to the high number of safeguards and the lack of clear guidance on how to implement, monitor, and enforce these.[69]

Indigenous peoples and forest communities are more likely than others to be socially and economically vulnerable, and so are more susceptible to negative impacts accompanying REDD+ activities. They are especially affected by REDD+ projects operating in forests on which they depend for their livelihoods and for food, shelter, and cultural or spiritual sustenance.[70] On the other hand, the REDD+ mechanism can yield benefits for those communities, especially if they are able to assert legal rights to carbon stored in forests or in forested land, or demonstrate the sequestration effects of their forest activities and custodianship. REDD+ benefits not directly derived from climate change mitigation are called 'co-benefits' or 'multiple benefits'. Multiple benefits for indigenous peoples and forest communities are variously identified by the UN-REDD Programme and FCPF. They include the following: 'the sustainable management of biodiversity, the provision of alternative livelihoods, equitable sharing of revenues generated from emissions reductions',[71] and 'improved livelihoods (including from carbon payments), clarification of land tenure, and stronger governance'.[72] Less directly, REDD+ is thought to provide a vehicle for the 'recognition of the value of traditional knowledge systems for forest management', 'opportunities to push for the inclusion of indigenous peoples' rights (including UNDRIP) at the international level', and 'new chances for reforms of policies and laws dealing with [indigenous] rights, ownership, access and control of forests'.[73]

As discussed further in Chapter 6, the proliferation of multiple benefits contributes to the impression that the REDD+ mechanism has become overburdened and unwieldy, and that this has undermined much of the appeal of the programme in its original conception, when it was

[69] Alain Frechette, Minoli de Bresser and Robert Hofstede, *External Evaluation of the United Nations Collaborative Programme on Reducing Emissions from Deforestation and Forest Degradation in Developing Countries (the UN-REDD Programme): Volume 1 – Final Report* (July 2014) vi.

[70] UN-REDD/FCPF Joint Guidelines on Stakeholder Engagement, above n 1, 1 [2] and [4].

[71] Ibid. 1 [2].

[72] Barney Dickson et al, 'REDD+ Beyond Carbon: Supporting Decisions on Safeguards and Multiple Benefits' (UN-REDD Programme Policy Brief, Issue No 2, November 2012) 1.

[73] Annelie Fincke, 'Indigenous Peoples and Climate Change/REDD: An Overview of Current Discussions and Main Issues' (IUCN Briefing Document, 21 July 2010) 3.

presented as a simple and cost-effective 'win-win' mechanism to promote carbon sequestration.[74] The layering of REDD-related objectives complicates the teleology of the REDD+ mechanism by confusing the distinction between activities that are 'ends' of REDD+ and those that are the means of achieving those ends. In the case of indigenous peoples, this question arises in an acute form in the characterisation of consultation requirements. It is not clear, for example, whether consultation and FPIC standards are safeguards, properly so-called, or benefits. In other words, are these aspects of REDD+ planning and activity designed to ensure the success of climate change purposes by mitigating the risks of opposition to the durability of REDD+ projects? Or are they more properly understood as 'benefits', in and of themselves, which should be pursued independently of their role in ensuring the stability of those projects? In the face of inevitable trade-offs between social and environmental benefits, and between mitigation and co-benefits, the question is not simply a semantic one, but directly implicates the central questions posed in this book: the interaction of norms and values within and between legal systems at the international, state and local levels, and their effects on the interests of indigenous and forest-dependent communities.[75] Just what is it that REDD+ recipient states are obliged to do, and how will we know when they are doing it? While the UN-REDD Programme has been 'widely praised' for its efforts to ensure that FPIC is respected,[76] ensuring community buy-in for REDD+ projects does not by necessity generate benefits for indigenous communities. An independent evaluation of the UN-REDD Programme in 2014 noted that:

> [t]he Programme's emphasis on safeguards and the need for the free, prior and informed consent of forest-dependent populations has created more sustained support amongst civil society and Indigenous Peoples' organisations, though effective participation in local decision-making fora and contentious issues such as tenure and benefit distribution remain unresolved challenges.[77]

Finally, any implementation of UN-REDD Programme advice that obliges states to adopt particular risk-management and benefit-conferring approaches in dealings with indigenous peoples will require a method for identifying those peoples, and distinguishing them as necessary from

[74] See, for example, Dickson et al, above n 72, 1; Visseren-Hamakers et al, above n 25, 649.
[75] Visseren-Hamakers et al, above n 25.
[76] Frechette, de Bresser and Hofstede, above n 69, 31. [77] Ibid. 4.

other communities within the national population. This threshold question proves to be more complicated than might be thought at first glance, and is only obliquely addressed by the UNREDD Programme in its advice to states. Section B addresses this body of law and practice.

## B. Indigenous Identity Law: Competing Approaches within REDD+

UN-REDD Programme guidance makes a distinction between the categories of 'indigenous peoples' and 'forest-dependent communities', noting that 'forest-dependent communities shall refer to communities that would not satisfy the commonly accepted definitions of indigenous peoples'.[78] The UN-REDD Programme's 2013 Guidelines on Free, Prior and Informed Consent (UN-REDD Programme FPIC Guidelines) contain an annex setting out 'commonly accepted definitions of indigenous peoples' in international law and commentary. Notably, the programme does not defer to state law or policy, by implication reserving to itself the ultimate authority to decide whether a group is indigenous for the purposes of allocating funding to states for REDD Readiness activities:

> As evidenced by Annex I, the partner country's own recognition or identification of the community as 'indigenous peoples' shall not be the dispositive factor. As such, the determination shall not be dependent on whether the national government has recognized the subject community as indigenous peoples.[79]

This is important, because UN-REDD Programme operational guidance documents specify different standards for indigenous peoples and forest-dependent communities respectively. While indigenous peoples enjoy the full scope of FPIC, including the right to withhold consent (discussed further below), the 2013 UN-REDD Programme FPIC Guidelines provide that 'a blanket application of FPIC is not required for all forest-dependent communities' although 'at a minimum States are required to consult forest-dependent communities in good faith regarding matters that affect them with a view to agreement'.[80] In its 2012 SEPC document, however, the programme takes a different approach, by including forest communities within the class of actors entitled to withhold consent:

[78] UN-REDD Programme FPIC Guidelines, above n 9, 53. [79] Ibid. 29. [80] Ibid. 11.

> Directly or indirectly affected indigenous peoples, tribal groups, ethnic minorities and other forest dependent communities have the right to give or withhold their consent, through their own representative institutions and following their own decision-making processes, regarding activities, proposals, legislative and administrative measures, and policies that may affect their land, territories, resources or livelihoods.[81]

The most that can be said, then, of the UN-REDD Programme's advice to date is that consultation is required for both indigenous and forest-dependent groups, but the question of whether non-indigenous forest-dependent communities are entitled to withhold consent is an unresolved issue. In any case, both the UN-REDD Programme and FCPF consultative standards rely on international law defining indigeneity, so an examination of this body of principle is crucial to understanding the scope and meaning of demands made of states participating in REDD+. Section 1 takes a closer look at this interplay, by reference to the complexities of international law on the categories of 'indigenous peoples' and 'forest-dependent communities'.

## 1. *International Law on 'Indigeneity'*

International instruments contain no definition of 'indigenous peoples' or 'indigeneity', and no such definition has been adopted by a UN body. In the intergovernmental fora tasked with elaborating the UNDRIP, some states and all indigenous representatives strongly opposed the inclusion of a definition in the draft instrument, and the final text of the declaration does not contain one. The UN's development agencies have also become increasingly wary of 'defining' indigenous peoples in state-specific projects, arguing, as the UNDP has done, that:

> [t]he practice of providing a 'definition' of IPs is being abandoned in international forums, as it makes better sense to pragmatically identify partners than to risk excluding peoples and groups because they 'do not fit' in the definition or even risk developing discriminatory and/or exclusionary definitions.[82]

[81] UN-REDD Draft Social and Environmental Principles and Criteria, above n 33, 12.

[82] Max Ooft, *UNDP and Indigenous Peoples: Towards Effective Partnerships for Human Rights and Development* (United Nations Development Programme, Oslo Governance Centre, November 2006) 10. See also United Nations Environmental Programme (UNEP), 'UNEP and Indigenous Peoples', above n 48, 3–4.

In policy-making and project-approval processes, however, UN bodies, including the World Bank and the UNDP,[83] often default to the 'Martínez Cobo' working definition,[84] which emphasises the non-dominance, distinctive ethnic and cultural identity, and historic continuity of indigenous peoples.[85] According to the definition, continuity is demonstrated by 'occupation of ancestral lands', 'common ancestry with the original occupants of those lands', the continuity of culture and language, or 'residence in certain parts of the country or regions of the world'.[86] Because of the lack of an agreed international definition, and the number of states that deny that self-identifying indigenous peoples in their territories are in fact 'indigenous', reports of the demographic features of indigenous populations vary widely. The UN Permanent Forum on Indigenous Issues has estimated that there are '370 million indigenous people spread across 70 countries worldwide'.[87] Using World Bank criteria and demographic studies, in 2012, the Forest Peoples Programme estimated that globally there are '500 million forest-dependent people of which 200 million are indigenous peoples',[88] while the

[83] United Nations Development Programme, 'UNDP and Indigenous Peoples: A Practice Note on Engagement' (2001) 1–2: Listing relevant criteria as follows: '(a) indigenous peoples usually live within (or maintain attachments to) geographically distinct ancestral territories; (b) they tend to maintain distinct social, economic, and political institutions within their territories; (c) they typically aspire to remain distinct culturally, geographically and institutionally rather than assimilate fully into national society and (d) they self-identify as indigenous or tribal.'

[84] UN Secretariat of the Permanent Forum on Indigenous Issues, 'The Concept of Indigenous Peoples' (Background Paper to Workshop on Data Collection and Disaggregation for Indigenous Peoples, New York, United States, 19–21 January 2004) UN Doc PFII/2004/WS.1/3, para 1; José R Martínez Cobo, Special Rapporteur, *Study of the Problem of Discrimination against Indigenous Populations – Volume V: Conclusions, Proposals and Recommendations*, UN Doc E/CN.4/Sub.2/1986/7 (March 1987).

[85] Martinez Cobo, above n 84. According to Erica Irene Daes, the Chairperson of the Working Group on Indigenous Populations, the inclusion in Martínez Cobo's definition of the criteria of 'non-dominance', 'impl[ies] that some form of discrimination or marginalization exists, and justifies action by the international community. It would not follow, however, that a group ceases to be "indigenous" if, as a result of measures taken for the full realization of its rights, it were no longer non-dominant': Erica I Daes, *Working Paper by the Chairperson-Rapporteur on the Concept of 'Indigenous People'*, UN Doc E/CN.4/Sub.2/AC.4/1996/2 (10 June 1996) para 26.

[86] Martinez Cobo, above n 84.

[87] United Nations Permanent Forum on Indigenous Peoples, *Who Are Indigenous Peoples?* (Factsheet, n.d.), www.un.org/esa/socdev/unpfii/documents/5session_factsheet1.pdf. See also Cæcilie Mikkelsen (ed), *The Indigenous World 2013* (The International Work Group for Indigenous Affairs (IWGIA), 2013).

[88] Sophie Chao, 'Forest Peoples: Numbers Across the World' (Forest Peoples Programme, 2012) 3.

UNDP suggests that '[a]pproximately 70 million indigenous peoples depend on forests for their livelihoods and another 350 million rural people reside in or near them'.[89]

### (a) Self-Identification and 'Objective Criteria' in International Law on Indigenous Peoples

In the absence of a widely-accepted international law definition of indigenous peoples, human rights norms and jurisprudence lean heavily on self-identification as the basis on which states should extend recognition to indigenous individuals and communities.[90] This principle is articulated in the only human rights treaty dedicated to indigenous rights – ILO Convention 169 – which provides that '[s]elf-identification as indigenous or tribal shall be regarded as a fundamental criterion for determining the groups to which the provisions of this Convention apply'.[91] It is also the approach taken in the American Declaration on the Rights of Indigenous Peoples, adopted in 2016, which provides that individual or collective 'self-identification as indigenous peoples will be a fundamental criterion (sic) for determining to whom this Declaration applies'.[92] The UNDRIP, however, contained a self-identification provision in its first draft, but this was removed from the text because of concerns expressed by states about its breadth:[93]

[89] United Nations Development Programme (UNDP), 'UN REDD and Indigenous Peoples', www.undp.org/content/undp/en/home/ourwork/democraticgovernance/focus_areas/focus_human_rights/empowering_indigenous_peoples/UNREDD.

[90] See UN Secretariat of the Permanent Forum on Indigenous Issues, 'The Concept of Indigenous Peoples', above n 84, for an overview of the indigenous definitional debate.

[91] ILO Convention 169 art 1(2).

[92] Inter-American Commission on Human Rights (IACHR), *American Declaration on the Rights of Indigenous Peoples*, AG/RES 2888 (XLVI–O/16), 46th reg sess, 3rd plen sess (adopted 15 June 2016) ('ADRIP') art I(2). See also the Committee on the Elimination of Racial Discrimination (CERD), *General Recommendation VIII Concerning the Interpretation and Application of Article 1, Paragraphs 1 and 4 of the Convention*, 38th sess, 884th mtg (21 August 1990) contained in *Report of the Committee on the Elimination of Racial Discrimination*, UN GAOR, 45th sess, Supp No 18 (A/45/18) (30 January 1991) ('General Recommendation 8') recording its opinion that, in respect to State reports questioning the 'ways in which individuals are identified as being members of a particular racial or ethnic group or groups ... such identification shall, if no justification exists to the contrary, be based upon self-identification by the individual concerned'. Also contained in *Compilation of General Comments and General Recommendations Adopted by Human Rights Treaty Bodies*, UN Doc HRI/GEN/1/Rev.9 (vol. II) (27 May 2008) 274.

[93] Statement by the Chinese Delegation to the Third Session of the Working Group Established in Accordance with the Commission on Human Rights Resolution 1995/32 (28 October 1997) (on file with author) 2; Statement by Bill Barker on behalf of the

> Indigenous peoples have the collective and individual right to maintain and develop their distinct identities and characteristics, including the right to identify themselves as indigenous and to be recognized as such.[94]

The UNDRIP, then, does not contain either objective or subjective criteria to guide the identification of the indigenous peoples it protects.

Other instruments, however, list characteristics considered indicative of indigeneity or tribalism. The ILO Convention 169 applies to two categories of community: 'tribal peoples', who are communities 'whose social, cultural and economic conditions distinguish them from other sections of the national community, and whose status is regulated wholly or partially by their own customs or traditions or by special laws or regulations',[95] and 'indigenous peoples', who descend from the populations that inhabited the country or region at the time of colonisation or the time of 'the establishment of present state boundaries' and 'retain some or all of their own social, economic, cultural and political institutions'.[96] A consensus amongst international organizations appears to be developing, in which the term 'tribal peoples' is subsumed within the concept of 'indigenous peoples', so that the latter term encompasses groups who do not claim to have continuity with precolonial or pre-state populations.[97] In commentary on the ILO Convention 169, for example, the chairperson of the UN's Working Group on Indigenous Populations (disestablished in 2006 and replaced by the UN Permanent Forum on Indigenous Issues) offered the view that the phrase 'establishment of present state boundaries' used in the ILO Convention 169 has 'the effect of minimizing any logical differences between the concepts of "indigenous" and "tribal", since both concepts are now chiefly defined by the extent to which the group in question constitutes a distinct society'.[98] What do the UN-REDD Programme and the FCPF make of this indeterminacy when they advise states on their obligations to indigenous peoples and forest communities?

Australian Delegation, Working Group on a Draft Declaration on the Rights of Indigenous Peoples (21 November 1995) (on file with author) 5.

94 Commission on Human Rights, Draft Declaration on the Rights of Indigenous Peoples, UN Doc E/CN.4/Sub.2/1994/2/Add.1 (20 April 1994).

95 ILO Convention 169 art 1(a). 96 Ibid. art 1(b).

97 See World Bank Legal Department, 'Legal Note on Indigenous Peoples' (8 April 2005) para 23, http://siteresources.worldbank.org/INTINDPEOPLE/Publications/20571167/Legal%20Note.pdf: 'In summary, the survey above shows that international practice has shifted to the term "indigenous people," or "peoples." Aside from ILO Conventions 107 and 169, the terms "tribal people," and "tribal population" are no longer in use in any other international instrument. Thus, the use of the term "indigenous peoples" in the draft Bank Policy is consistent with international practice.'

98 Daes, above n 85, para 28.

## 2. *Indigeneity and International Law: UN-REDD Programme and FCPF Approaches*

The indeterminacy of international law on 'indigeneity' and the default reliance by international institutions on 'self-identification' or on the Martínez Cobo definition has direct consequences for REDD+ recipient states. Much depends on whether the activities for which funding is sought are governed by the FCPF, which applies the World Bank's OP 4.10, or by the UN-REDD Programme, which applies the UN-REDD Programme FPIC Guidelines. OP 4.10 expressly states that the policy does not define 'indigenous peoples', noting that 'there is no universally accepted definition' and that indigenous peoples may be referred to in different countries as 'indigenous ethnic minorities', 'aboriginals', 'hill tribes', 'minority nationalities', 'scheduled tribes' or 'tribal groups'.[99] The policy does, however, specify that the term is used in a generic sense to refer to a 'distinct, vulnerable, social and cultural group' that possesses, 'in varying degrees', the following characteristics: 'self-identification as members of a distinct indigenous cultural group and recognition of this identity by others'; 'collective attachment to geographically distinct habitats or ancestral territories'; 'customary cultural, economic, social, or political institutions that are separate from those of the dominant society and culture'; and 'an indigenous language'.[100] Importantly, when OP 4.10 was adopted in 2005, replacing the 1991 Operational Directive 4.20, a major change was made to provisions on the identification of indigenous peoples. While the 1991 directive made no reference to the law of borrower states on indigenous peoples, the current policy specifies that, after consulting with the indigenous peoples concerned and with the borrower government, 'the Bank may follow the borrower's framework for identification of Indigenous Peoples during project screening, when that framework is consistent with this policy'.[101] This expresses a degree of deference to borrower-state law that was not contemplated in the earlier iterations of the bank's indigenous peoples policy. Commentators have emphasised the dangers of international lending institutions relying on borrower-state definitions of indigenous peoples, noting that states have an interest in narrowing the range of entities that may qualify for recognition in order to conserve public resources, suppress possible

[99] World Bank, 'Operational Manual: OP 4.10 – Indigenous Peoples' (July 2005) ('World Bank OP 4.10') para 3.

[100] Ibid. para 4. [101] Ibid. para 8.

challenges to state authority, and maintain control of property claimed by indigenous or tribal groups.[102]

The World Bank's new ESS 7, which will come into force in 2018, takes a similar approach. Like its predecessor OP 4.10, ESS 7 recognises that the groups to whom the standard applies may be referred to in different countries by different terms, including the new designation; 'Sub-Saharan African historically underserved traditional local communities'. The standard provides that it encompasses 'all such alternative terminology', and includes 'forest dwellers, hunter-gatherers, pastoralists or other nomadic groups',[103] provided that those groups meet the other criteria set out in the standard.[104] These other criteria require that the group self-identifies as members of a distinct indigenous social and cultural group and is recognised as such by others; has a collective attachment to geographically distinct habitats, ancestral territories, or areas of seasonal use or occupation; has distinct customary cultural, economic, social or political institutions; and has a distinct language or dialect.[105] As with OP 4.10, the bank must determine whether or not such peoples reside in or have a collective attachment to the project area and, in identifying such peoples, may elect to 'follow national processes', where 'these processes meet the requirements of [ESS 7]'.[106]

In contrast, neither the UN-REDD Programme FPIC Guidelines nor the FCPF/UN-REDD Guidelines on Stakeholder Engagement in REDD+ Readiness (the joint FCPF/UN-REDD Stakeholder Guidelines) contains a definition of 'indigenous and forest-dependent communities'. The UN-REDD Programme FPIC Guidelines affirm the 'fundamental criterion of self-identification'[107] and refer to a set of 'commonly accepted and utilized' definitions and descriptions of indigenous peoples. These include the criteria contained in the ILO Convention 169, the Martínez Cobo report and the Working Paper on the Concept of Indigenous People prepared by

[102] Fergus MacKay, 'The Draft World Bank Operational Policy 4.10 on Indigenous Peoples: Progress or More of the Same?' (2005) 22 *Arizona Journal of International & Comparative Law* 65.

[103] 'Environmental and Social Standard 7: Indigenous Peoples/Sub-Saharan African Historically Underserved Traditional Local Communities' in *Environmental and Social Framework: Setting Environmental and Social Standards for Investment Project Financing* (World Bank, 4 August 2016) ('ESS7') 106 [1] and 109 [9].

[104] Ibid. 106 [1]: 'Such terms include "indigenous ethnic minorities", "aboriginals", "hill tribes", "vulnerable and marginalized groups", "minority nationalities", "scheduled tribes", "first nations" or "tribal groups".' See also 108 [6].

[105] Ibid. 108 [8]. [106] Ibid. 109 [10].

[107] UN-REDD Programme FPIC Guidelines, above n 9, 37 and 38.

the Working Group on Indigenous Populations.[108] These descriptions and definitions vary, however, and it is not clear whether they apply independently of one another or whether all must be satisfied in every case. In lieu of a set of definitive criteria, the UN-REDD Programme FPIC Guidelines contain a list of questions that participating states might ask when identifying indigenous peoples (for example, what is their relationship to the lands and resources they inhabit?).[109] Self-identification is not the determinative measure either. Noting that 'in some cases, however, the notion of being indigenous has pejorative connotations and people may choose to refuse or redefine their indigenous origin' and that 'such choices must be respected', the UN-REDD Programme Guidelines go on to specify that:

> The issue is neither what a people is called nor whether the State in question has recognized them as an indigenous people, but whether the collective satisfies the most commonly accepted definitions of indigenous peoples elaborated here – even where the political situation has dissuaded a people or community to identify themselves as indigenous or prevented them from understanding the rights implications of doing so.[110]

In general, then, international norms and jurisprudence frame indigeneity by reference to the overarching subjective principle of 'self-identification',[111] while the UN-REDD Programme and World Bank policies include additional 'objective' criteria with which to identify indigenous peoples. The UN-REDD Programme makes a distinction between indigenous and forest-dependent communities, and has said that FPIC applies only to indigenous peoples, while the World Bank's OP 4.10 is broad enough to include forest-dependent communities within its remit if they can meet the objective criteria of attachment to land and cultural distinctiveness. Finally, while both institutions allow a determination on indigeneity to be made even if the participating state does not recognise those communities as indigenous, or if the community does not self-identify as such, OP 4.10 contains more detailed objective criteria limiting the category of peoples who are governed by the policy. In short, the relation between international law and domestic law on indigeneity remains underspecified in the advice offered by the UN-REDD

[108] Ibid. 36–37. [109] Ibid. 39.

[110] Ibid. 38. Likewise, a group that satisfies the criteria set out in the Guidelines may qualify as a 'forest-dependent community' 'irrespective of whether they themselves choose to identify themselves as such (owing, for example, to a fear of doing so or a lack of awareness of the legal rights that would attach, etc)': at 53 n 2.

[111] ILO Convention 169 art 1(2); General Recommendation 8, above n 92.

Programme, and the application of World Bank identity criteria to FCPF-funded activities remains opaque, given the likely non-enforceability of OP 4.10 to REDD Readiness activities. Since the application of the World Bank's operational policies to FCPF-funded REDD Readiness activities is in question, the task of applying and navigating international identity law in order to address the interests of indigenous peoples in REDD+ falls largely to states. Further, since the identification of indigenous peoples is an essential component of FPIC, it must be addressed by states in their interpretation of UN-REDD Programme and FCPF advice on how to implement that principle. The following section takes a closer look at UN-REDD Programme and FCPF operational guidance on FPIC.

## C. FPIC in the UN-REDD Programme and FCPF REDD Readiness Standards

The body of norms and standards so far generated by the UN-REDD Programme is immensely variegated, and some of the principles it contains are difficult to reconcile with others. This is particularly the case with respect to the selection of international treaties, declarations and decisions that underpin the Programme's advice on indigenous peoples. This section aims to clarify how the UN-REDD Programme uses external sources of law and policy when assisting states with REDD Readiness activities and when deciding whether to make readiness grants to those states.

### *1. Principles and Criteria on Indigenous Peoples Used by the UN-REDD Programme and FCPF*

The standards articulated by the UN-REDD Programme have real consequences for states applying for REDD Readiness funding. The standards, supplemented by guidelines, are used by the UN-REDD Programme policy secretariat[112] to evaluate draft National Programme Documents (NPDs), and REDD Readiness Preparation Proposals (R-PPs), in which a country 'lays out a roadmap of preparation activities'.[113] Endorsed proposals are

[112] The secretariat engages the expertise of independent reviewers as part of this process: UN-REDD Programme, *UN-REDD Programme Handbook for National Programmes and Other National-Level Activities*, above n 4, 28.

[113] Ibid. 17.

submitted to the UN-REDD Executive Board for approval, and for a decision on the release of funding from the REDD Readiness Fund.[114] The UN-REDD Programme has identified seven 'social and environmental principles' attached to the Cancun Safeguards, and twenty-four corollary criteria to give content to those principles. These are identified in the UN-REDD Programme's SEPC (2012) and the associated UN-REDD draft SEPC-BeRT (2013).[115] The programme also provides 'operational guidance' advising states on how to fulfil safeguard requirements, by reference to general international law and the policies of its member organizations. To date, it has provided operational guidance (in the form of published guidelines) on stakeholder engagement, gender mainstreaming and FPIC.[116]

In the first phases of REDD Readiness preparation, and as a condition of financing, countries are obliged to include a 'Consultation and Participation Plan' in their NPD and R-PP documents to show 'how representative participation of forest-dependent indigenous peoples and other forest dwellers and forest-dependent communities and other civil society would be ensured'.[117] These documents must contain minutes of a 'validation meeting of National Stakeholders', that included appropriately selected indigenous peoples' representatives,[118] and must demonstrate that the states have developed a 'consultation and engagement strategy'.[119]

As further discussed below, because multiple international agencies may be involved in the funding and management of a single REDD project, identifying the relevant consultative standard depends on the interplay of four factors: 1. whether the Delivery Partner (DP) is a

[114] UN-REDD Operational Guidance, above n 7, 10. On the role of the Executive Board in funding decisions, see UN-REDD Programme, *UN-REDD Programme Handbook for National Programmes and Other National-Level Activities*, above n 4, 9.

[115] 'The proposed application of the SEPC is for: 1) assisting countries in formulating national REDD-plus programmes and initiatives for which they seek UN-REDD support; 2) reviewing national programmes prior to submission for a UN-REDD Executive Board decision on funding; and 3) assessing national programme delivery': R-PP Template Version 6, above n 14, 3 [3].

[116] UN-REDD Draft Social and Environmental Principles and Criteria, above n 33, 11.

[117] R-PP Template Version 6, above n 14, 10 and 15; UN-REDD Programme, *UN-REDD Programme Handbook for National Programmes and Other National-Level Activities*, above n 4, 19.

[118] UN-REDD Programme, *UN-REDD Programme Handbook for National Programmes and Other National-Level Activities*, above n 4, 27.

[119] R-PP Template Version 6, above n 14, 23 (Standard 1b); UN-REDD Operational Guidance, above n 7, 10–11.

UN-REDD Programme partner organization or, alternatively, the World Bank; 2. whether that DP is the sole partner or operates jointly with another agency; 3. whether the country in question has enacted legislation implementing the UNDRIP principles on FPIC; and, 4. whether the country has ratified ILO Convention 169.[120] The major difference between the two *currently operative* FPIC standards is that the UN-REDD Programme FPIC Guidelines require states to obtain indigenous consent to measures that impact on their interests,[121] while the World Bank's OP 4.10 refers only to 'free, prior and informed consultation',[122] albeit with a view to obtaining 'broad community support' for the planned measures.[123]

Where funding is sought only from the FCPF, states must comply with the World Bank's safeguards. States applying for funding from the FCPF produce a 'Readiness Package' outlining their progress in REDD Readiness preparations, which is assessed by the FCPF's Participants Committee.[124] REDD Readiness activities must be 'consistent with applicable World Bank and/or other Delivery Partner safeguard policies' including the joint FCPF/UN-REDD Stakeholder Guidelines and the 'relevant United Nations Framework Convention on Climate Change (UNFCCC) guidance on safeguards'.[125] The Package must also include a 'standalone' Environmental and Social Management Framework (ESMF)

[120] R-PP Template Version 6, above n 14, 27–28.

[121] UN-REDD Programme FPIC Guidelines, above n 9, 27.

[122] 'Broad community support, in the case of the FCPF, or free, prior and informed consent, in the case of the UN-REDD Programme, can be withheld at the community level, and such a decision should be respected': UN-REDD/FCPF Joint Guidelines on Stakeholder Engagement, above n 1, 10.

[123] But see the World Bank's 2016 'Environmental and Social Standard 7: Indigenous Peoples/Sub-Saharan African Historically Underserved Traditional Local Communities', which will come into force in 2018, and which requires FPIC in certain named circumstances: ESS7, above n 103.

[124] With advice from the Technical Advisory Panel and input from the Delivery Partner (usually UNDP or the World Bank): Forest Carbon Partnership Facility, 'A Guide to the FCPF Readiness Assessment Framework' (June 2013) 4. The production of the Readiness Package is voluntary for REDD Readiness funding, but compulsory if an application is made to the Carbon Fund for Emissions Reductions payments: at 5.

[125] Ibid. 11. The ESMF Guidelines also specify that the guidelines 'may need to be revised to take account of relevant decisions that emerge from the ongoing REDD-Plus negotiations within the UNFCCC regime': Guidelines for the Development of Terms of Reference (ToRs) for an Environmental and Social Management Framework (ESMF) in the Context of the REDD-Plus Readiness Preparation Process Supported by the FCPF, 'Annex C: Guidelines for the Development of ToRs for the ESMF' in R-PP Template Version 6, above n 14, 2.

setting out 'the principles, rules, guidelines, and procedures to assess potential environmental and social impacts and risks', and outlining 'measures to reduce, mitigate, and/or offset adverse environmental and social impacts and enhance positive impacts and opportunities of said projects'.[126] The ESMF must include a component on indigenous peoples, and states are invited to apply the World Bank's Indigenous Peoples Planning Framework to construct this element of the ESMF.[127]

Navigating this complex and overlapping set of standards and policies, and discerning their various relationships to the Cancun Safeguards, has proved difficult for some states. In an independent evaluation of the FCPF, completed in 2015, reviewers noted that:

> A lack of clarity around the compliance and use of different environmental and social safeguard systems constituted a challenge for the FCPF's readiness implementation. The existence of multiple global safeguard policies for REDD+ (e.g. Institutional: FCPF, Forest Investment Programme and Policy: UNFCCC, UN-REDD) has created concerns among the REDD Countries about additional work burdens and overlapping efforts to comply with the multiple reporting requirements.[128]

On this issue, the complexity of standards on consultation and FPIC are a case in point. While the phrase 'free, prior and informed consent' is not used in the Cancun Safeguards, according to the UN-REDD Programme, FPIC 'is addressed indirectly [in the safeguards] because the text "note[s]" that the General Assembly has adopted UNDRIP (which itself sets out the principle of FPIC)'.[129] This has formed the basis of the programme's efforts to articualte FPIC standards and to coordinate their advice with that of the FCPF. In 2009, following a series of consultations with indigenous peoples and forest communities, the UN-REDD Programme produced a document called Operational Guidance: Engagement of Indigenous Peoples and Other Forest Dependent Communities.[130] In 2012, the

[126] Forest Carbon Partnership Facility, above n 124, 14. See 'Attachment 1: FCPF Readiness Fund: Guidelines and Generic Terms of Reference for SESAs and ESMF' in R-PP Template Version 6, above n 14.

[127] R-PP Template Version 6, above n 14, 11.

[128] Indufor, *Second Program Evaluation of the Forest Carbon Partnership Facility (FCPF)* (Inception Report, The World Bank, 13 November 2015) 5.

[129] UN-REDD Programme FPIC Guidelines, above n 9, 15: 'Securing FPIC is a means to meet the Cancun Agreements' requirement of countries to promote and support "respect for the knowledge and rights of indigenous peoples and members of local communities" and to ensure "the full and effective participation of relevant stakeholders, inter alia, indigenous peoples and local communities".'

[130] UN-REDD Operational Guidance, above n 7.

Operational Guidance framework was revisited in a collaboration with the FCPF that culminated in the joint FCPF/UN-REDD Stakeholder Guidelines. These guidelines set out the consultative standards to be applied by states operating under the UN-REDD Programme and the FCPF, and provide advice on how the relevant standard is to be identified when both agencies are involved in REDD Readiness preparation in the same country.[131] The FCPF has previously identified the need for coordination across different agencies where DPs other than the World Bank are involved in REDD Readiness activities, in its Common Approach to Environmental and Social Safeguards for Multiple Delivery Partners (2011).[132] This document specifies in relevant parts that where World Bank funding is used, but the DP is not the World Bank, the DP should achieve 'substantial equivalence' with the World Bank's social and environmental safeguard policies, including the Joint FCPF/UN-REDD Stakeholder Guidelines.[133] Further, if 'the environmental and social safeguard policies and procedures of the DP are more stringent and/or protective than those of the [World Bank], the DP shall apply its policies and procedures to activities'.[134] More recently, the UN-REDD Programme elaborated its *own* standards on FPIC in its 2013 UN-REDD Programme FPIC Guidelines.[135]

As this history shows, FPIC standards have proved to be a major challenge for the coordination of activities by FCPF and the UN-REDD Programme, one which has catalysed their efforts to resolve substantive inconsistencies in the standards they use and to develop protocols for collaboration. Despite these efforts, major conceptual and programmatic differences between the approaches taken by the two organizations remain.[136] In the first independent evaluation of the FCPF programme, completed in 2011, evaluators noted that:

> Differences with regard to operational guidance provided by FCPF and the United Nations Collaborative Programme on Reducing Emissions from Deforestation in Developing Countries (UN-REDD) on the engagement of stakeholders, in particular of Indigenous Peoples, are

[131] UN-REDD/FCPF Joint Guidelines on Stakeholder Engagement, above n 1.

[132] FCPF Readiness Fund Common Approach, above n 27.

[133] R-PP Template Version 6, above n 14, 3 [4(b)].

[134] FCPF Readiness Fund Common Approach, above n 27, 2 and 3. See also R-PP Template Version 6, above n 14, 4 [4].

[135] UN-REDD Programme FPIC Guidelines, above n 9.

[136] FCPF Readiness Fund Common Approach, above n 27, para 16.

creating a degree of confusion in those countries where both programs operate.[137]

The confusion is understandable. States use a single 'joint' template document to apply for funds from either the UN-REDD Programme or the FCPF, but some questions and standards vary within that document depending on which body funding is sought from. States are obliged to attend to the consultative standards of the fund to which they are applying, while taking into account the requirement of substantial equivalence with FCPF standards where the World Bank is providing funding. The substantive differences are not inconsequential. States must acquire '[b]road community support, in the case of the FCPF, or free, prior and informed consent, in the case of the UN-REDD Programme'.[138] Consultative standards are an area, then, where a divergence between the UN-REDD Programme and the FCPF on REDD+ policy is clearly evident. The following section disaggregates the UN-REDD Programme and UN-REDD/FCPF guidance on FPIC with a view to identifying the types of standards REDD+ participant states are obliged to meet, and to locate the sources of those obligations in general international law, UNFCCC COP decisions and the World Bank's Operational Policies.

## 2. *The Content of FPIC in REDD+: The UN-REDD Programme's Advice*

The FPIC principle, like many elements of international law, is a body of positive law and interpretative jurisprudence, rather than a single authoritative set of rules. Whether this body of law will produce a consensus amongst states on a more fine-grained and determinate set of FPIC obligations remains to be seen. What is clear, however, is that, for better or worse, the UN-REDD Programme itself has taken a stand on controversial and disputed aspects of the principle by endorsing the most protective and onerous interpretations of the right offered in international jurisprudence, and holding states to those standards in the disbursement of REDD Readiness funding. The programme favours an interpretation of the international law principle extending beyond any so far expressly accepted

137 Baastel and Nordic Agency for Development and Ecology (NORDECO), *First Program Evaluation for the Forest Carbon Partnership Facility (FCPF)* (Evaluation Report, 13 June 2011) vi, 10 and 15–16.

138 UN-REDD/FCPF Joint Guidelines on Stakeholder Engagement, above n 1, 10.

by states as a binding international norm. As noted, the UN-REDD Programme FPIC Guidelines effectively oblige states to demonstrate adherence to the ILO Convention 169 (a treaty that has twenty-two parties, mostly Latin American states), and the UNDRIP (a non-binding instrument, albeit with near-universal state support). Further, the UN-REDD Programme relies on the concluding observations and general recommendations of the CERD Committee, which are non-binding as a matter of international law (although they may well be regarded as authoritative interpretative statements in the domestic law of states).[139] Finally, the UN-REDD Programme takes an expansive view of the requirement of consent in FPIC, and a wide view of the range of circumstances in which FPIC is required. In its efforts to bolster its legitimacy with civil society and indigenous communities, the UN-REDD Programme effectively holds developing states to standards that they have not otherwise agreed to, and which very few developed or donor states have been willing to accept.

Identifying the substantive content of states' FPIC obligations by reference to the UN-REDD Programme FPIC Guidelines is not straightforward. The guidelines often restate the phrase without elaborating on its meaning, requiring further investigation of the primary sources used, (some of which are collected in the Legal Companion to the UN-REDD Programme Guidelines on Free, Prior and Informed Consent (FPIC): International Law and Jurisprudence Affirming the Requirement of FPIC).[140] Applying the UN-REDD Programme FPIC Guidelines requires states to reconcile these various sources of law to discover the content and limits of the right. In its fullest sense, this requires an assessment of the 'bindingness' of both the positive law in which the phrase appears, and the jurisprudential interpretations offered by human rights treaty bodies. The latter requires attention to the standing of the forum in which the claim is heard and the nature of the advice offered (which may take the form of a finding in a contentious case, a general recommendation, or concluding observations on state reports). Further, unless and until the principle of consent-based FPIC is accepted as one of customary international law, and therefore binding on all states irrespective of their express undertakings, each inquiry should proceed by reference to the obligations of the state in

[139] UN-REDD Programme, 'Legal Companion to the UN-REDD Programme Guidelines on Free, Prior and Informed Consent (FPIC): International Law and Jurisprudence Affirming the Requirement of FPIC' (January 2013) 16–46.

[140] Ibid.

question rather than as a global proposition enforceable by the UN-REDD Programme via its control of REDD Readiness funding.

As noted above, participants in debates about FPIC disagree on whether the principle encompasses a right on the part of indigenous peoples to *withhold consent* to measures that impact on their interests. Debates on this question can be vociferous and polarising. Some states have insisted that an obligation to obtain indigenous consent is undemocratic (because it can only be exercised by indigenous peoples) and so is incompatible with principles of equality and non-discrimination, or cannot be reconciled with fundamental tenets of sovereignty, including the principle of eminent domain. Indigenous peoples (sometimes asserting their own forms of national sovereignty) argue that states should not interfere with or allow interference with indigenous peoples' control of their property and law. A more nuanced middle path can be discerned in some advice offered by international bodies and officials, rejecting the oppositional nature of the debate, and emphasising the mutual interdependence of the principles of consultation and consent in the context of indigenous rights.[141]

The UN-REDD Programme describes FPIC as a 'corollary ... of universally accepted human rights ... contained in numerous international human rights instruments', which is 'clearly shown to arise from the States' corresponding duties and obligations to give effect to a host of underlying substantive rights' for indigenous peoples, including rights to self-determination, political participation, property, cultural integrity and equality.[142] It is described by the programme as the right of 'potentially impacted peoples ... to participate in and consent to or withhold consent from a proposed action'[143] that arises where 'potential changes in resource uses ... could significantly impact the substantive rights of indigenous peoples'.[144] This description implicates the three most controversial questions attending discussion about FPIC: what activities trigger the obligation of states to acquire FPIC, in respect of which

[141] James Anaya, Special Rapporteur, *Report of the Special Rapporteur on the Rights of Indigenous Peoples*, 21st sess, UN Doc A/HRC/21/47 (6 July 2012): 'It is simply misguided to tend to reduce examination of the rights of indigenous peoples in the context of resource development projects to examination of the contours of a right to be consulted or a right to free, prior and informed consent. ... A better approach appreciates, first, that neither consultation nor consent is an end in itself, nor are consultation and consent stand-alone rights': at paras 48–49.

[142] UN-REDD Programme FPIC Guidelines, above n 9, 13. [143] Ibid. 18. [144] Ibid.

communities does FPIC arise, and under what circumstances does FPIC entail a right to withhold consent?

### (a) Which Activities?

With respect to the first question, the UN-REDD Programme FPIC Guidelines include a list of activities in which 'it is likely that FPIC will be required'.[145] These include several very broadly framed activities, for example, those that involve 'any legislative or administrative measures that will affect the rights, lands, territories and/or resources of indigenous peoples/forest-dependent community', or involve 'any decisions that will affect the status of indigenous peoples'/forest-dependent community's rights to their lands/territories or resources'.[146] The right is also said to arise specifically 'in connection to: mining and oil and gas operations (extraction of subsurface resources); logging; the establishment of protected areas; construction of dams; development of agro-industrial plantations; resettlement; compulsory takings; and any other decisions affecting the status of their land rights'; a list reportedly drawn from provisions of the UNDRIP and the jurisprudence of the HRC, ICESCR and CERD.[147] Likewise, in the UN-REDD SEPC document, FPIC is said to 'appl[y] to all activities, projects, legislative or administrative measures and policies that take place in or impact the lands, territories, resources or otherwise affect the livelihoods of indigenous peoples', a definition that is stated to be 'synthesized from UNDRIP articles'.[148] In the Joint FCPF/UN-REDD Stakeholder Guidelines, however, the UN-REDD Programme's position is expressed in even broader terms, as 'the collective right of indigenous peoples to participate in decision-making and to give or withhold their consent to *activities affecting their lands, territories and resources or rights in general*'.[149] The UN-REDD Programme's broad and open-ended approach to FPIC mirrors (and is probably based on) the approach taken by the UN Permanent Forum on Indigenous Issues, the chairperson of which was a member of the UN-REDD Programme's Policy Board (now the UN-REDD Executive Board).[150] In 2005, the

145 Ibid. 28. 146 Ibid. 27. 147 Ibid. 15.

148 UN-REDD Draft Social and Environmental Principles and Criteria, above n 33, 10 n 20.

149 UN-REDD/FCPF Joint Guidelines on Stakeholder Engagement, above n 1, annex 2.

150 West, above n 47: the Executive Board has an indigenous representative, but it was not clear at the time of writing whether the representative is the Chairperson of the PFII, as it has been in previous years.

Permanent Forum endorsed a concept of FPIC that is triggered by (inter alia) 'policies and legislation dealing with or affecting indigenous peoples'[151] and which 'may include the option of withholding consent'.[152]

### (b) Is Consent Required?

On the question of consent, the UN-REDD Programme FPIC Guidelines specify that the right to withhold consent is a crucial aspect of the FPIC principle:

> [T]he core of FPIC is the right of the peoples concerned to choose to engage, negotiate and decide to grant or withhold consent, as well as the acknowledgement that under certain circumstances, *it must be accepted that the project will not proceed and/or that engagement must be ceased if the affected peoples decide that they do not want to commence or continue with negotiations or if they decide to withhold their consent to the project.*[153]

The only formally binding instrument referenced in the UN-REDD Programme FPIC Guidelines is the ILO Convention 169. Notably, however, except for circumstances involving relocation, the phrase 'free, prior and informed consent' is expressly limited in the convention and does not require states to obtain consent (a point noted in the UN-REDD Programme FPIC Guidelines).[154] The ILO Convention is the only legally binding multilateral treaty dealing specially with indigenous and tribal peoples. It has twenty-two parties (a relatively small number in international law terms), and a strong regional bias towards Central and Latin American states (fifteen of the twenty-two parties are states in this region).[155] References to FPIC appear in ILO Convention 169 sections 6(1) and 6(2). These require states to 'consult the peoples concerned, through appropriate procedures and in particular through their representative institutions, whenever consideration is being given to legislative

[151] Permanent Forum on Indigenous Issues, *Report of the International Workshop on Methodologies Regarding Free, Prior and Informed Consent and Indigenous Peoples*, UN Doc E/C.19/2005/3 (17 February 2005) para 45.

[152] Ibid. para 47.

[153] UN-REDD Programme FPIC Guidelines, above n 9, 20 (emphasis added).

[154] Ibid. 13.

[155] International Labour Organization, *Ratifications of C169 – Indigenous and Tribal Peoples Convention, 1989 (No 169)* (n.d.), www.ilo.org/dyn/normlex/en/f?p=NORMLEXPUB:11300:0::NO::P11300_INSTRUMENT_ID:312314.

or administrative measures which may affect them directly',[156] and specifies that these consultations 'shall be undertaken, in good faith and in a form appropriate to the circumstances, with the objective of achieving agreement or consent to the proposed measures'.[157]

The UN-REDD Programme FPIC Guidelines place more reliance on the UNDRIP. The programme's interpretation of the UNDRIP is, however, controversial. It is of the view that: '[t]he UNDRIP recognizes several situations in which the State is under an obligation to not just seek, but secure the consent of the indigenous peoples concerned'.[158] The guidelines list five situations where consent would be required, including: where indigenous peoples are to be relocated from their lands; with respect to activities involving the 'taking [of] cultural, intellectual, religious and spiritual property'; with respect to 'damages, takings, occupation, confiscation and uses of their lands, territories and resources'; when 'adopting and implementing legislative or administrative measures'; and in respect of 'any project affecting their lands or territories and other resources'. The programme draws these from UNDRIP articles 10, 11(2), 19, 28(1) and 32(2).[159]

However, FPIC is expressed in the UNDRIP in an unqualified form only in respect to relocation (Article 10)[160] and storage of toxic waste on indigenous lands (Article 29(2)).[161] Articles 11(2) and 28(1) direct states to provide redress where 'cultural, intellectual, religious and spiritual property' has been taken from indigenous peoples without their free prior and informed consent and where 'territories and resources which they have traditionally owned or otherwise occupied or used' have been 'confiscated, taken, occupied, used or damaged' without FPIC. On their face, these provisions do not require states to seek FPIC, but rather to provide redress for past involuntary takings. Article 32(2) requires states to 'undertake *effective consultations* with the indigenous peoples

[156] ILO Convention art 6(1)(a). [157] ILO Convention art 6(2).

[158] UN-REDD Programme FPIC Guidelines, above n 9, 24. [159] Ibid.

[160] 'Indigenous peoples shall not be forcibly removed from their lands or territories. No relocation shall take place without the free, prior and informed consent of the indigenous peoples concerned and after agreement on just and fair compensation and, where possible, with the option of return': *United Nations Declaration on the Rights of Indigenous Peoples*, GA Res 61/295, UN GAOR, 61st sess, 107th plen mtg, Agenda item 68, Supp No 49, UN Doc A/RES/61/295 (2 October 2007, adopted 13 September 2007) (UNDRIP) art 10.

[161] 'States shall take effective measures to ensure that no storage or disposal of hazardous materials shall take place in the lands or territories of indigenous peoples without their free, prior and informed consent': UNDRIP art 29(2).

concerned . . . prior to using their lands or territories for military activities'. Finally, Article 19, the more generally framed FPIC provision in UNDRIP, obliges states to consult with indigenous peoples '*in order to obtain* their free, prior and informed consent' (emphasis added):

> States shall consult and cooperate in good faith with the indigenous peoples concerned through their own representative institutions in order to obtain their free, prior and informed consent before adopting and implementing legislative or administrative measures that may affect them.

It is important to note that Article 19, in its draft and final form, proved controversial during the negotiation of the UNDRIP, and attracted persistent objections from some states including from the four 'CANZUS' states that were the only four to vote against the adoption of the UNDRIP in 2007: Australia, Canada, New Zealand and the United States. During the twenty-two-year-long negotiation of the UNDRIP, CANZUS states objected strenuously to the formulation of the provision in its draft and final versions, and all four qualified their subsequent endorsement of the UNDRIP with interpretative statements denying that the article permitted indigenous peoples to veto proposed measures.

The original formulation of the draft article, prepared in 1995, seemed to expressly permit the withholding of consent:

> Indigenous peoples have the right to participate fully, if they so choose, through procedures determined by them, in devising legislative or administrative measures that may affect them. *States shall obtain the free and informed consent of the peoples concerned before adopting and implementing such measures.*[162]

CANZUS states sought unsuccessfully to qualify the draft article by reference to principles of equality and non-discrimination, in order to 'reflect the principle that indigenous individuals have the right to participate fully in public affairs, like any other citizen', while providing for 'special measures to allow participation in decisions of the state which directly affect certain areas of particular concern to indigenous people'.[163] In order to accommodate some of these concerns, changes were made to the draft article by the chairperson of the Working Group,

162 DoCIP, United Nations Declaration on the Rights of Indigenous Peoples: Comparative Table (2011) (on file with author). Emphasis added.

163 See, e.g., Canada on Articles 19 and 20, 'Articles 19, 20, 22 & 23' (28 October 1996), Doc 228a (1996) in doCIP (on file with author).

Luis Enrique Chavez. The chairperson's recollections reveal the rationale for the change:

> My assessment was that the [Working Group on the Draft Declaration on the Rights of Indigenous Peoples] could not accept this [draft Article], neither for practical reasons nor reasons of principle. In practical terms, the state could not renounce either its powers or its responsibility when taking decisions on issues of public order. And, in principle, the declaration could not recognise indigenous peoples (sic) preferential or greater rights than those granted to other members of society, as would be the case with a right of veto. The Chairman's proposals therefore established only an obligation regarding the means (consultation and cooperation in good faith with a view to obtaining consent) but not, in any way, an obligation regarding the result, which would mean having to obtain that consent.[164]

The chairperson's comments accord with the findings of the International Law Association in its 2012 report on the UNDRIP, in which it concluded that:

> Generally speaking, in light of the wording of Article 19, of its comparison with other articles included in the UNDRIP (e.g. Articles 28(1) and 29 (2)), as well as of relevant State practice, it is not possible to conclude that Article 19 may be interpreted as establishing a general right of veto in favour of the indigenous communities concerned to block the adoption or implementation of the governmental measures which may affect them per se.[165]

The newly adopted American Declaration on the Rights of Indigenous Peoples includes a provision that is identical to the UNDRIP's Article 19,[166] and also includes an additional article, not found in the UNDRIP, requiring states to 'consult and cooperate in good faith with the indigenous peoples concerned through their own representative institutions in order to obtain their free and informed consent prior to the approval of any project affecting their lands or territories and other resources, particularly in connection with the development, utilization or exploitation of mineral, water or other resources'.[167]

[164] Luis Enrique Chavez, 'The Declaration on the Rights of Indigenous Peoples – Breaking the *Impasse*: The Middle Ground' in Claire Charters and Rodolfo Stavenhagen (eds), *Making the Declaration Work: The United Nations Declaration on the Rights of Indigenous Peoples* (IWGIA, 2009) 96, 103–104.

[165] International Law Association, *Sophia Conference (2012): Rights of Indigenous Peoples (Final Report)* (2012) 4.

[166] ADRIP art XXIII(2). [167] ADRIP art XXIX(4).

### 3. *FPIC and the World Bank*

Crucially then, the UN-REDD Programme's advice on FPIC does not definitively calibrate state obligations in response to varying types of activity on indigenous land, although it suggests circumstances where FPIC will 'likely be required', and specifies that FPIC includes the right to withhold consent. In the absence of indications to the contrary, states preparing REDD Readiness documents for review by the UN-REDD Programme Executive Board must assume that the consent requirement is triggered by the very wide range of activities noted in the UN-REDD Programme's advice. The scope of this direction is in notable contrast to the World Bank's newly adopted ESS 7, which requires borrowers to undertake 'meaningful consultation, in a culturally appropriate and gender and inter-generationally inclusive manner' and specifies three circumstances in which the *consent* of the consulted group must be sought. Indigenous consent is to be obtained where the project will (1) 'have adverse impacts on land and natural resources subject to traditional ownership or under customary use or occupation'; (2) cause the forced relocation of indigenous peoples from such land; or (3) have significant impacts on cultural heritage that is material to the identity or cultural, ceremonial or spiritual aspects of the affected indigenous peoples.[168] The inclusion of consent-based FPIC requirements in specified circumstances is an important advance on the 'free prior and informed consultation' standard contained in OP 4.10. It also seems to extend beyond the more qualified expressions of the principles in the ILO Convention 169 and the UNDRIP. The ESS 7 does, however, contain an important provision on the meaning of consent, which was added in the last phase of consultations on the draft, '[i]n response to Indigenous Peoples' interest in FPIC and the concerns of some shareholders regarding the potential inconsistency between FPIC and national law'.[169] The new paragraph 26 explains that:

> For the purposes of this ESS, consent refers to the collective support of affected Indigenous Peoples communities/Sub-Saharan African Historically Underserved Traditional Local Communities for the project activities that affect them, reached through a culturally appropriate process. It may exist even if some individuals or groups object to such project activities . . .

[168] World Bank Environmental and Social Framework, above n 23, 113.

[169] World Bank, *Environmental and Social Framework (Proposed Third Draft)* (Review and Update of The World Bank's Safeguard Policies, 4 August 2016) paras 53 and 110–112.

In the face of the controversial status of the FPIC principle in international law, what are the obligations of REDD+ participating states? For countries applying for REDD Readiness funding from the FCPF, the relevant standard is the 'free, prior and informed consultation' directive contained in OP 4.10.[170] OP 4.10 is relatively deferential to the legal frameworks and international commitments (or lack thereof) of borrower states.[171] Accordingly, if a state has ratified ILO Convention 169 or implemented national legislation on FPIC (based on the UNDRIP, for example), the bank, via the FCPF, will defer to those (impliedly) higher standards.[172] Where another organization is the delivery partner under an FCPF scheme, that organization will apply its own safeguard policies 'if these are more stringent and/or protective than those of the World Bank'.[173] Broadly speaking, then, the UN-REDD Programme FPIC Guidelines apply where the UN-REDD Programme is the sole funder or delivery partner, whereas OP 4.10 applies in any project where the FCPF is the sole funder and the World Bank is the delivery partner[174] (at least in respect of indigenous peoples who are 'forest-dependent', in the terms used in the FCPF's charter).[175] For all projects that are proposed for World Bank (FCPF) financing that affect indigenous peoples, OP 4.10 requires a 'process of free, prior, and informed consultation with the affected Indigenous Peoples' communities at each stage of the project, and particularly during project preparation, to fully identify their views

[170] UN-REDD/FCPF Joint Guidelines on Stakeholder Engagement, above n 1, 3 [7].
[171] Ibid. [172] Ibid.
[173] FCPF Readiness Fund Common Approach, above n 27: '27. The UN-REDD Programme Team and FCPF Facility Management Team have drafted guidelines applicable to the key elements of effective stakeholder engagement in the context of REDD+. As per Paragraph 3 above, under this Common Approach all DPs shall treat these guidelines as FCPF requirements. Furthermore, if the standard for stakeholder engagement applied by the DP is higher and/or protective than those of the WB, the DP shall apply its own standard to activities undertaken under the FCPF Readiness Fund': at 9 [27] (citation omitted).
[174] UN-REDD/FCPF Joint Guidelines on Stakeholder Engagement, above n 1, para 6.
[175] FCPF Charter, above n 20, art 1(34): '"Forest-Dependent Indigenous Peoples and Forest Dwellers" means Indigenous Peoples and local communities living in forests and depending on forest resources for their livelihoods.' Note that OP 4.10 includes indigenous peoples who have lost collective attachment to geographically distinct habitats or ancestral territories in the project, because of forced severance, 'occurring within the concerned group members' lifetime because of conflict, government resettlement programs, dispossession from their lands, natural calamities, or incorporation of such territories into an urban area': World Bank OP 4.10, above n 99, para 4. See discussion in Mackay, above n 102, 72–73.

and *ascertain their broad community support for the project*',[176] and the bank 'does not proceed further with project processing if it is unable to ascertain that such support exists'.[177] Therefore, according to the Joint FCPF/UN-REDD Stakeholder Guidelines:

> although OP 4.10 does not expressly mandate FPIC, if the country has ratified ILO Convention No 169 or adopted national legislation on FPIC, or if the Bank is working on a project with a development partner that expressly applies the principle of FPIC, the Bank will in turn support adherence to that principle.[178]

In effect then, the World Bank will not hold states to FPIC obligations beyond the bank's own consultation policies if the state has not otherwise accepted those additional obligations. The reason for this is illustrated by a 2004 World Bank legal opinion (endorsed by the World Bank Executive Board):

> Where a country is not one of the few that have incorporated FPIC into their domestic legal framework, requiring FPIC would be inconsistent with the Bank Group's role as a global institution whose members are sovereign governments, possessed of their own rights to determine whether to follow the terms of any international convention.[179]

As noted above, the UN-REDD Programme, controversially, requires applicant states to adhere to standards in instruments that they have not ratified (in the case of the ILO Convention 169) or in instruments that are non-binding (the UNDRIP, and the recommednations of the CERD and ICESCR committees). Whether this will contribute to or undermine the legitimacy of the REDD+ programmes and projects remains to be seen.[180] The difference in approach reflects the World Bank's long-standing reluctance to deploy

[176] World Bank OP 4.10, above n 99, para 6(c) (emphasis added). [177] Ibid. 11.

[178] UN-REDD/FCPF Joint Guidelines on Stakeholder Engagement, above n 1, para 7.

[179] The World Bank Group and Extractive Industries, Legal Note on Free Prior and Informed Consultation, SecM2004-0369/IDA/SecM2004-0559, 3 August 2004, quoted in World Bank Legal Department, 'Legal Note on Indigenous Peoples', above n 97, para 23.

[180] Galit A Sarfaty, 'Why Culture Matters in International Institutions: The Marginality of Human Rights at the World Bank' (2009) 103 *American Journal of International Law* 647, 681–82. But see Mauro Barelli, 'Free, Prior and Informed Consent in the Aftermath of the UN Declaration on the Rights of Indigenous Peoples: Developments and Challenges Ahead' (2012) 16 *The International Journal of Human Rights* 1, 5: 'while one may speculate on the actual level of protection that this policy guarantees to indigenous peoples, it is fairly clear that the World Bank does not recognize that indigenous peoples have the right to decline unwanted projects taking place on their lands'.

human rights principles in the course of its work, a reluctance that it explains by reference to the Articles of Agreement establishing its relationship with the UN, especially Article IV/10, which provides that:

> The Bank and its officers shall not interfere in the political affairs of any member; nor shall they be influenced in their decisions by the political character of the member or members concerned. Only economic considerations shall be relevant to their decisions, and these considerations shall be weighed impartially in order to achieve the purposes stated in Article I.[181]

The bank's legal team has long insisted that this article obliges them to be agnostic in their financing decisions with regard to the civil and political rights records of borrower states (although it has accepted some institutional responsibility for economic, social and cultural rights where these implicate its 'good governance agenda'):

> The indivisibility of human rights and their relevance to human development are not questioned here. By themselves, they cannot mean, however, that the Bank, which is basically a financial, economic development institution, should ignore its specialized mandate and the limitations of its Articles. What is at issue is the specialization of each international organization and the requirements of its respective charter. . . . The Bank cannot . . . use its operations as a reward for the countries which respect political rights or a punishment for those which do not. Such a criterion is not only alien to the Articles of Agreement, but it runs against the above-quoted provisions of these Articles which enjoin the Bank from interference in the political affairs of its members.[182]

## D. Conclusion

This chapter has presented an analysis of the ways that the two major REDD+ funding bodies, the UN-REDD Programme and the FCPF, have sought to develop standards on indigenous rights. We have taken as a case in point the challenges faced by these bodies in their efforts to identify 'indigenous peoples' and 'forest-dependent communities' in order to require REDD+ recipient states to safeguard the interests and rights of such communities. The lack of consensus in international law on the concept of 'indigeneity' and on the content of FPIC means that the REDD+ funding bodies have struggled to develop standards that could

[181] *Articles of Agreement of the International Monetary Fund*, adopted 22 July 1944, 2 UNTS 39 (entered into force 27 December 1945) art IV/10.

[182] Shihata, above n 25.

apply in a useful and equitable way to the variegated circumstances of REDD+ recipient states. Questions about the category of 'indigenous peoples' and FPIC have been the subject of prolonged debates in international negotiation fora and judicial bodies, including as part of the development of the UNDRIP, and in contentious cases heard by the human rights treaty bodies of the UN. Chief amongst the unsettled aspects of the FPIC principle is whether it entails the right to withhold consent to proposed REDD+ projects; whether the right to withhold consent arises only in specified circumstances; and whether, if it forms part of the principle, such a right is also available to tribal and forest-dependent communities who are not recognised as 'indigenous' by the governments of the states in which they live.

This chapter has described the difficulty faced by REDD+ actors and institutions when they seek to enforce human rights principles in the course of allocating REDD+ funding, and the challenges they face when they draw on the fast-moving international law of indigenous rights to guide states in REDD+ planning and implementation. Neither funding body is equipped to resolve the wide-ranging questions about the normative content and force of human rights and indigenous rights, let alone advise states on how best to give effect to these rights and resolve conflicts between them in anticipation of REDD+ projects and results-based programmes. All of this makes it difficult for the REDD+ regime to find a stable and uncontested place amongst the other international regimes on which it depends, and from which it receives much of its normative direction. Conversely, REDD+ recipient states face a complex set of directions devised by the REDD+ funding bodies that contains little guidance about how to reconcile or choose between the different rights and interests it contains. Whether the solutions that states arrive at in order to progress REDD+ serve the interests of indigenous peoples better than approaches that they may have undertaken independently of REDD+ is likely to remain a controversial question. It is clear however, that the decisions that states make about how to conceptualise and recognise indigenous rights in the REDD Readiness phase – including decisions about which communities are indigenous, which of them should be consulted and which can withhold consent to REDD+ projects – will greatly influence the shape of more concrete changes required to implement REDD+ in those states. For many states, amongst the changes to be faced in REDD Readiness is the need to formalise tenure arrangements and decide on the ownership of the land, forests and carbon on which REDD+ depends. This is the subject of Chapter 5.

# 5

## REDD+, Tenure and Indigenous Property Claims

KIRSTY GOVER

In this chapter, we argue that the emphasis in REDD+ policy on the need for 'secure and clear' tenure, coupled with the 'human rights–based approach' adopted by the UN-REDD Programme, suggests a bias in favour of current possessors and users of land. States embarking on tenure reform in anticipation of REDD+ funding need to address competing claims to forested land, and the methodology so far endorsed by the UN-REDD Programme and the FAO could disadvantage communities claiming tenure rights based on prior possession. This is likely to affect indigenous communities claiming rights sourced in customary law, particularly where those rights are not enforceable in state law and are sought by communities that have been dispossessed of the land they claim. Human rights methodologies do not easily accommodate or protect rights of this kind, particularly if the dispossession was historic or not within the lifespan of members of the group. We note that, in contrast, the World Bank's operational policy (OP) 4.10 on indigenous peoples (which applies in principle to the FCPF in its administration of REDD+ funding)[1] makes some provision (albeit limited) for the rights of indigenous peoples to land they no longer occupy, and provides for compensation where access to those lands cannot be secured.[2] However, the bank's operational policies do not ordinarily apply to REDD Readiness activities, and so will not directly influence state tenure arrangements until REDD+ implementation is underway and funding for Emissions Reductions Programs from the FCPF Carbon Fund is

[1] See further in Chapter 4.

[2] World Bank, 'Operational Manual: OP 4.10 – Indigenous Peoples' (July 2005) ('The World Bank Operations Manual'), http://policies.worldbank.org/sites/ppf3/PPFDocuments/090224b0822f89d5.pdf, para 4. The tenure provisions of the bank's new 'Environmental and Social Standard 7: Indigenous Peoples/Sub-Saharan African Historically Underserved Traditional Local Communities in *Environmental and Social Framework: Setting Environmental and Social Standards for Investment Project Financing* (4 August 2016) ('ESS 7') are similarly framed, as discussed in Section A.

disbursed. This lag means that state practices in tenure reform are likely to be shaped primarily by the criteria and guidance developed by the UN-REDD Programme and the FAO.

We argue in this chapter that REDD+ policy on tenure is at a nascent but crucial stage, and should be informed by the ways in which indigenous property claims have been conceptualised and addressed in North America and Australasia, and in the jurisprudence of international human rights bodies. In particular, we note the marked absence in REDD+ policy of guidance and standards on historic claims or advice on how to reconcile competing claims to land and resources where some claims are premised on customary law and prior possession, and others are based on state law or current usage. Since REDD+ may incentivise powerful actors (including states) to reduce their liability and expenditure by excluding indigenous claims, and in light of the time pressures inherent in the REDD+ agenda, as it stands REDD+ policy could encourage the consolidation of tenure arrangements that do not sufficiently account for the interests of indigenous and tribal peoples. Priorities and evaluative criteria used by the UN-REDD Programme and the FCPF to condition funding for REDD Readiness activities could set in train state practices that disadvantage indigenous and tribal peoples relative to other private and public actors.

In this chapter, we examine the 'state of play' of tenure discourse in the REDD+ context. In Section 1, we discuss, with a critical eye, the small body of interventions on tenure in guidance provided by the UN-REDD Programme and the FCPF, and in the standards deployed by those bodies when evaluating R-PPs. These norms and standards relate in various (and sometimes attenuated) ways to UNFCCC decisions on REDD+, including the safeguard requirements identified in the 2010 Cancun Agreements. We identify within this body of policy a set of conceptual challenges attending efforts to address indigenous property claims based on customary law where the dominant normative framework is premised on human rights: the difficulty of conceiving of property rights as human rights, the problem of circularity in property law, and the relation of these two conceptual hurdles to the official 'recognition' of indigenous property rights. We consider in particular the difficulty of reconciling acts of recognition with the principles of equality and non-discrimination that underpin human rights law. In Section 2, we examine the methods used to address these issues by courts in North America and Australasia, by international human rights bodies and by the Inter-American Court of Human Rights (IACtHR). This body of jurisprudence, we argue, contains two conceptually distinct methodologies that govern the recognition of customary indigenous property rights, both of which could usefully inform the development of REDD+ policy on tenure.

## A. What Are the Entry Points for Tenure Issues in REDD+ Discourse?

There is wide agreement amongst civil society actors, states and international bodies that adequate tenure arrangements will be a pivotal determinant of the success of REDD+ projects and of the REDD+ regime as a whole. In 2013, for example, a 'meeting of experts' convened by the UN-REDD Programme concluded that the need to address tenure issues in the REDD+ context 'is becoming increasingly urgent' and that there is a particular need for 'greater clarity over tenure and the recognition of customary rights to land by indigenous peoples and local communities'.[3] Notwithstanding acknowledgment within REDD+ of the importance of adequate tenure arrangements, including in the Cancun Agreements' direction to states to consider 'tenure issues' in the course of undertaking REDD+ activities, tenure policy is a late starter in REDD+ debates. It was not until 2012, for example, that any policy advice on tenure reform was published by the UN-REDD Programme.[4] An external evaluation conducted in 2014 found that, while the programme's emphasis on safeguards and on the principle of FPIC had attracted support for REDD+ from civil society and indigenous peoples, 'contentious issues such as tenure and benefit distribution remain unresolved challenges'.[5] Evaluators noted that 'relatively little has been done so far' in respect of 'land rights, carbon rights, tenure reform, and benefit sharing mechanisms under REDD+ scenarios',[6] and advised that the UN-REDD Programme Policy Board[7] 'should make resource rights a priority area for programme interventions'.[8]

[3] UN-REDD Programme, 'Background Paper for the Expert Meeting: Options for Addressing Tenure under REDD+' (25–27 February 2013) 3.

[4] Barney Dickson et al, 'REDD+ Beyond Carbon: Supporting Decisions on Safeguards and Multiple Benefits' (UN-REDD Programme Policy Brief, Issue No 2, 2012), http://theredddesk.org/resources/un-redd-policy-brief-redd-beyond-carbon-supporting-decisions-safeguards-and-multiple.

[5] Alain Frechette, Minoli de Bresser and Robert Hofstede, *External Evaluation of the United Nations Collaborative Programme on Reducing Emissions from Deforestation and Forest Degradation in Developing Countries (the UN-REDD Programme): Volume I – Final Report* (July 2014) iii, http://wedocs.unep.org/rest/bitstreams/776/retrieve.

[6] Ibid. 25. [7] Renamed the 'Executive Board' in 2016.

[8] Frechette, de Bresser and Hofstede, above n 5, ix.

The UN-REDD Programme has been hesitant so far to offer substantive direction to states on optimal forest tenure arrangements. In its 2013 policy brief on tenure, the programme acknowledged that tenure reform is 'an enormous, endogenous change process and requires a long-term perspective', and noted that 'resources available for REDD+ readiness are not sufficient to carry out national tenure reforms'.[9] Elsewhere, UN-REDD Programme and FAO officials noted the limitations imposed by the nature and scope of the REDD Readiness phase, and suggested that '[t]he work on tenure under REDD+ must match the scope of the REDD+ instrument in terms of financial and human resources and the time frame for readiness'.[10] At its 2014 Policy Board meeting, for example, board members were advised by officials that '[t]enure is too political an issue for a UN-REDD Programme blueprint to be put in place and the topic must instead be addressed at [the] national level'; and further, that the UN-REDD Programme's capacity to assist states in these endeavours is limited by the fact that '[t]enure is also often dealt with by ministries other than the one where REDD+ focal points work'.[11] At the same meeting, however, more than half of the Policy Board members in attendance 'rated addressing tenure issues as extremely important and an overwhelming majority believe[d] that tenure issues must be addressed at the very start of the REDD+ process'.[12] For now, at least, tenure reform is an area within which the UN-REDD Programme is reluctant to tread too heavily.

The evidence to date suggests that states participating in REDD Readiness activities are themselves ill equipped to address tenure issues as part of these preparations, even while many recognise that it is a crucially important prerequisite for successful REDD+ implementation. A 2012 Country Needs Assessment conducted by the UN-REDD Programme

[9] Dickson et al, above n 4, 6.

[10] UN-REDD Programme, 'Background Paper for the Expert Meeting: Options for Addressing Tenure under REDD+', above n 3, 4. 'Currently, the tenure work has not yet been established in most REDD+ countries, and road maps for the tenure work still need to be developed. The legal framework for tenure is quite well developed in many countries, but compliance with the existing laws and regulations remains a major challenge for many governments and communities. REDD+ work on tenure therefore needs to be based on a solid analysis of the existing legal frameworks for tenure, in particular in view of the clarification of tenure systems and the formalization of customary tenure. In a nutshell: What is specifically needed in the improvement of tenure for effective REDD+ implementation?': Ibid. 9.

[11] UN-REDD Programme, 'Report of the Thirteenth Policy Board Meeting' (6–7 November 2014) 9–10.

[12] Ibid.

and the FCPF concluded that 'legal frameworks to support the implementation of REDD+ and to resolve land tenure and carbon rights issues in the REDD+ context are needed in virtually all existing and nascent REDD+ country strategies'.[13] Likewise, civil society and policy bodies have expressed concern about the lack of reference to tenure in states' REDD Readiness documents.[14] A recent study by the WRI surveying thirty-two readiness proposals reported that while most identified the importance of clear and secure tenure for the success of REDD+ projects, only half went on to discuss 'existing tenure challenges' and fewer than one-third outlined 'specific next steps [for] address[ing] land tenure challenges'.[15] This suggests that while some states are contemplating tenure reform to improve the 'security and clarity' of tenure arrangements ahead of REDD+ implementation, the nature of those reforms is not yet a substantive part of REDD+ planning, leaving open serious questions about how tenure reform will impact on the interests of indigenous and local communities. The WRI was sanguine about the lack of tenure planning in state REDD Readiness activities, noting that readiness proposals necessarily 'reflect preliminary thinking on REDD+ systems and strategies',[16] and further observing that tenure reforms 'can be costly and time consuming and that long-term sources of readiness financing have not yet been identified in most REDD+ countries'.[17] The UN-REDD Programme seems similarly accepting of the lack of tenure proposals in state planning, observing that even where tenure policies have been included in REDD Readiness documents, the complexity of competing claims to tenure in forested lands makes it likely that 'a range of challenges might prevent them from committing to policy reforms in relation to forest tenure'.[18]

[13] UN-REDD Programme and Forest Carbon Partnership Facility (FCPF), 'Country Needs Assessment: A Report on REDD+ Readiness among UN-REDD Programme and Forest Carbon Partnership Facility Member Countries' (October 2012) 118.

[14] Sophie Chao, 'Forest Peoples: Numbers Across the World' (Forest Peoples Programme, 2012), www.forestpeoples.org/sites/fpp/files/publication/2012/05/forest-peoples-numbers-across-world-final_0.pdf; Mary C Thompson, Manali Baruah and Edward R Carr, 'Seeing REDD+ as a Project of Environmental Governance' (2011) 14 *Environmental Science & Policy* 100, 103.

[15] Lauren Goers Williams, 'Putting the Pieces Together for Good Governance of REDD+: An Analysis of 32 REDD+ Country Readiness Proposals' (Working Paper, World Resources Institute, March 2013) 14, www.wri.org/sites/default/files/pdf/putting_the_pieces_together_for_good_governance_of_redd.pdf; see also Chao, above n 14, 8.

[16] Williams, above n 15, 5. [17] Ibid. 7. [18] Dickson et al, above n 4, 8.

As discussed in Chapter 4, REDD+ participant states agree to meet the obligations set out in the COP decisions of the UNFCCC. The UNFCCC's 2010 COP decision (the Cancun Agreements)[19] sets out the basic aims and structure of REDD+, and outlines in an appendix the 'safeguards' that recipient states should 'promote[] and support[]' when undertaking REDD+ activities in their territories (the Cancun Safeguards).[20] The safeguards contained in the annex make no express reference to tenure, but paragraph 72 of the Cancun decision itself *requests* states to consider 'land tenure issues' when developing and implementing national REDD+ strategies.[21] As the UN-REDD Programme has acknowledged, attention to tenure security 'is not a requirement for participating states'.[22] To date, the programme has not provided operational guidance on tenure, as it has on stakeholder engagement, gender mainstreaming and FPIC.[23] The programme has, however, included tenure standards in its SEPC, which are intended to provide a 'guiding framework' for the UN-REDD Programme that assists it to 'address ... social and environmental issues in UN-REDD National Programmes and other UN-REDD Programme funded activities' and 'support ... countries in developing their national approaches to REDD+ safeguards in line with the [UNFCCC]'.[24] The principles and criteria developed by the UN-REDD Programme illustrate its autonomy from the treaty-based mechanisms of the UNFCCC, even while its aim is to develop evaluative standards that are 'consistent with' the UNFCCC's Cancun Safeguards.[25] Because the *safeguards* make no express reference to tenure arrangements or property rights, the UN-REDD Programme's advice on tenure is linked instead to the safeguards provision referencing

[19] Conference of the Parties, United Nations Framework Convention on Climate Change, *Report of the Conference of the Parties on Its Sixteenth Session, Held in Cancun from 29 November to 10 December 2010 – Addendum – Part Two: Action Taken by the Conference of the Parties at Its Sixteenth Session* UN Doc FCCC/CP/2010/7/Add.1 (15 March 2011) Decision 1/CP.16 ('The Cancun Agreements: Outcome of the Work of the Ad Hoc Working Group on Long-Term Cooperative Action under the Convention')) para 72.

[20] Ibid., Appendix II. [21] Ibid., para 72.

[22] UN-REDD Programme, *Tenure Security*, www.unredd.net/knowledge/redd-plus-techni calissues/tenure-security.html.

[23] UN-REDD Programme, 'UN-REDD Programme Social and Environmental Principles and Criteria (SEPC)' (UNREDD/PB8/2012/V/1, Asunción, Paraguay, 25–26 March 2012) ('UN-REDD SEPC') 16.

[24] UN-REDD Programme, 'UN-REDD Programme SEPC: Supporting Document' (2012) 2.

[25] Ibid. 3.

indigenous rights. The provision to which the programme's tenure advice is linked, then, requests states to promote and support respect for:

> the knowledge and rights of indigenous peoples and members of local communities, by taking into account relevant international obligations, national circumstances and laws, and noting that the United Nations General Assembly has adopted the United Nations Declaration on the Rights of Indigenous Peoples.[26]

It is this safeguards provision, then, and not the reference to 'tenure issues' in paragraph 72 of the body of the Cancun decision, that underpins the UN-REDD Programme's directions to states on the recognition of indigenous tenure rights.

As discussed in Chapter 4, the UN-REDD Programme's SEPC are used to guide and evaluate state REDD Readiness Proposals and are communicated to states via the prompts and standards contained in the joint FCPF/UN-REDD Programme R-PP template. UN-REDD Programme SEPC principles require states to '[r]espect and protect stakeholder rights in accordance with international obligations', notably, defining 'stakeholder rights' to include 'human rights, statutory and customary rights, and collective rights'.[27] This principle is attended by criteria that require states to (inter alia) '[r]espect and promote the recognition and exercise of the rights of indigenous peoples, local communities and other vulnerable and marginalized groups to land, territories and resources, including carbon'.[28] In the UN-REDD Programme's SEPC-BeRT tool, these criteria are expanded further into three 'primary questions' (and many more 'guiding questions') asking states to consider whether there are 'existing policies and measures' in place or 'specific actions' in the proposed REDD+ Programme that 'respect and promote the recognition and exercise of equitable land tenure and carbon rights by indigenous peoples and local communities'.[29]

[26] UNFCCC, 'Tenth Session of the United Nations Permanent Forum on Indigenous Issues (UNPFII)' (February 2011), www.un.org/esa/socdev/unpfii/documents/session_10_UNFCCC.pdf, Appendix 1, 2(c); UN-REDD Programme, 'Background Paper for the Expert Meeting: Options for Addressing Tenure under REDD+', above n 3, 11: '[T]he decision of the Convention on REDD+ clearly demands that the work on tenure under REDD+ goes well beyond forest boundaries'.

[27] UN-REDD Programme, UN-REDD SEPC, above n 23, 5.

[28] Ibid. (Criterion 7).

[29] UN-REDD Programme, 'Draft Social and Environmental Principles and Criteria – Benefit and Risks Tool (SEPC-BeRT): UN-REDD Programme 2013' (2013) ('SEPC-BeRT'), www.unredd.net/index.php?option=com_docman&Itemid=134&view=document&alias=6352-draft-sepc-bert-un-redd-6352&category_slug=latest-version-of-the-draft-sepc-1266, Principle 2, Criterion 7. See also, UN-REDD Programme, *UN-REDD*

To assist states to answer these questions, the tool also refers governments to guidelines and policy documents on forest tenure, including most centrally the FAO's 'Voluntary Guidelines on the Responsible Governance of Tenure of Land, Fisheries and Forests in the Context of National Food Security' (the FAO Guidelines).[30]

The R-PP template, however, defers substantive reporting on tenure reform to the final stages of the REDD Readiness phase, so that it does not directly condition the initial provision of REDD+ funding. To complete the R-PP, states need only submit an 'assessment of land use, land use change drivers, forest law, policy and governance', which includes information on 'pertinent laws, policies, and issues surrounding land tenure and resource rights, and traditional land use of indigenous people, extent of titled and untitled indigenous lands, indigenous claims for additional land "extensions" and process of land title demarcations'.[31] They are also directed to prepare a preliminary 'REDD+ Implementation Framework' which must address, inter alia, questions such as:

> [w]ho owns or uses the forest under statutory or customary law? Is there regulatory or legal clarity on who owns carbon benefits generated by REDD-plus activities? Is there a relationship between carbon ownership and land tenure? How would any land tenure, or carbon ownership, issues that arise be resolved or mediated?[32]

Significantly, the UN-REDD Programme and the FCPF admit that they have little concrete advice to offer on the preparation of states' implementation frameworks, in which tenure issues are to be addressed. The R-PP template, for example, records the programme's observation that 'good practices are only just becoming evident for this component, as there is little experience to date'.[33] In lieu of concrete standards, 'promising ideas'

*Programme Launches Benefits and Risks Tool (BeRT): Supporting Countries to Address and Respect the Cancun Safeguards* (19 May 2015), www.unredd.net/announcements-and-news/2131-un-redd-programme-launches-benefits-and-risks-tool-bert-supporting-countries-to-address-and-respect-the-cancun-safeguards.html.

30 The BeRT text notes that 'The development of the BeRT has drawn on a rich body of relevant material, including in particular the REDD+ SES, the World Resources Institute's Governance of Forests Indicators Toolkit, One World Trust Accountability matrix and various multilateral institutions' safeguards policies, as well as Epple et al. (2011)': UN-REDD Programme, 'SEPC-BeRT', above n 29, Table 1.

31 Forest Carbon Partnership Facility (FCPF) and the United Nations Collaborative Programme on Reducing Emissions from Deforestation and Forest Degradation in Developing Countries (UN-REDD), 'Readiness Preparation Proposal (R-PP)' (Template Version 6, 20 April 2012) 33, www.forestcarbonpartnership.org/readiness-fund.

32 Ibid. 40. 33 Ibid.

are offered instead.[34] Thus, for the most part, any tenure strategies proposed by applicant states are not evaluated by the UN-REDD Programme as a condition of REDD Readiness funding.

The tenure standards deployed by the FCPF in the evaluation of REDD Readiness funding applications have a different anatomy. The FCPF Charter confirms that its work is governed by the World Bank's ten Operational Policies and Procedures (also known as its 'safeguard policies'), which include OP 4.10 on Indigenous Peoples.[35] OP 4.10 has had a long-standing influence on the work of the bank, and as with the other safeguard policies, non-compliance by bank staff provides a basis for claims to be made to the World Bank Inspection Panel by affected groups.[36] As discussed in Chapter 4, OP 4.10 will be replaced in 2018 by the bank's new ESS 7, adopted on 4 August 2016. The new policy contains significantly augmented provisions on FPIC. It does not, however, substantively alter the tenure provisions of OP 4.10. It remains the case in both OP 4.10 and ESS 7 that evidence of 'collective attachment' to 'geographically distinct habitats, ancestral territories, or areas of seasonal use or occupation, as well as to the natural resources in these areas' is a primary identity marker of indigenous peoples and sub-Saharan local communities.[37] In both OP 4.10 and ESS 7, 'collective attachment' is further defined to mean that 'for generations there has

[34] Ibid.

[35] Section 3.1(d) of the FCPF Charter provides that 'the operation of the Facility, including implementation of activities under Grant Agreements and Emission Reductions Programs, shall … comply with the World Bank's Operational Policies and Procedures, taking into account the need for effective participation of Forest-Dependent Indigenous Peoples and Forest Dwellers in decisions that may affect them, respecting their rights under national law and applicable international obligations': International Bank for Reconstruction and Development, *Charter Establishing the Forest Carbon Partnership Facility* (Revised 23 November 2015), www.forestcarbonpartnership.org/charter-and-governance-documents.

[36] Bank Information Center (BiC), 'SESA, Safeguards and the FCPF: A Guide for Civil Society' (January 2013), www.bankinformationcenter.org/wp-content/uploads/2013/01/SESA+informational+doc+for+PC6.pdf. See also Benedict Kingsbury, 'Operational Policies of International Relations as Part of the Law-Making Process: The World Bank and Indigenous Peoples' in Ian Brownlie, Guy S Goodwin-Gill and Stefan Talmon (eds), *The Reality of International Law: Essays in Honour of Ian Brownlie* (Oxford University Press, 1999) 323.

[37] World Bank, ESS 7, above n 2, para 8(b); The World Bank Operations Manual, above n 2, OP 4.10, para 4(b). ESS 7 includes 'collective attachment' to 'areas of seasonal use or occupation', an aspect of collective attachment that in OP 4.10 was mentioned only in a footnote: Ibid., OP 4.10, n 7.

been a physical presence in and economic ties to land and territories traditionally owned, or customarily used or occupied, by the group concerned, including areas that hold special significance for it, such as sacred sites'.[38] Importantly, both policies also offer protection to some groups that have 'lost "collective attachment to geographically distinct habitats or ancestral territories in the project area" ... because of forced severance'.[39] In order to qualify under this section of the policy, the loss must have occurred 'within the concerned group members' lifetime because of conflict, government resettlement programs, dispossession from their lands, natural calamities, or incorporation of such territories into an urban area'.[40] Thus, while the policy is designed to protect people who occupy or use land on which bank-funded projects are to operate and does not accommodate claims based on *intergenerational* historic loss, the policy *does* extend to groups who involuntarily lost possession of their land within living memory. In this respect, the policy, when and if deployed by the FCPF in REDD+ programming, could extend to at least some non-occupant indigenous communities in a way that is not contemplated by the FAO Guidelines used by the UN-REDD Programme.

Tenure is addressed most directly in OP 4.10 paragraph 17. The paragraph sets out the tenure considerations that must be addressed by the bank if an initial investigation shows that indigenous peoples will be affected by a proposed project:[41]

> If the project involves (a) activities that are contingent on establishing legally recognized rights to lands and territories that Indigenous Peoples have traditionally owned or customarily used or occupied (such as land titling projects), or (b) the acquisition of such lands, the IPP [Indigenous Peoples Plan] sets forth an action plan for the legal recognition of such ownership, occupation, or usage ... Such legal recognition may take the following forms: (a) full legal recognition of existing customary land tenure systems of Indigenous Peoples; or (b) conversion of customary usage rights to communal and/or individual ownership rights. If neither option is possible under domestic law, the IPP includes measures for legal recognition of perpetual or long-term renewable custodial or use rights.[42]

Additionally, OP 4.10 provides that indigenous peoples should 'share equitably the benefits to be derived from ... commercial development' of

[38] The World Bank Operations Manual, above n 2, OP 4.10, n 7; World Bank, ESS 7, above n 2, footnote 6.
[39] The World Bank Operations Manual, above n 2, OP 4.10, para 4.
[40] Ibid., OP 4.10, para 4, n 8. [41] Ibid., OP 4.10, para 17. [42] Ibid., OP 4.10, para 17.

natural resources on lands they have 'traditionally owned or customarily used or occupied',[43] specifying that:

> [a]t a minimum, the IPP arrangements must ensure that the Indigenous Peoples receive, in a culturally appropriate manner, benefits, compensation, and rights to due process at least equivalent to that which any landowner with full legal title to the land would be entitled in the case of commercial development of their land.[44]

ESS 7 reiterates the tenure directions set out in OP 4.10 paragraph 17, but prefaces them with a stronger statement about the necessity of recognising customary forms of tenure, which affirms that legal recognition of customary rights is the 'objective of [IPP] plans', noting also that:

> While Indigenous Peoples/Sub-Saharan African Historically Underserved Traditional Local Communities may not possess legal title to land as defined by national law, their use of the land, including seasonal or cyclical use, for their livelihoods, or for cultural, ceremonial, and spiritual purposes that define their identity and community, can often be substantiated and documented.[45]

The compensatory provisions of ESS 7 are also more elaborate than those that appear in OP 4.10. As in the earlier policy, where it is proposed that commercial development take place on indigenous lands, those peoples are entitled to benefits, compensation and rights that would be afforded to a person holding 'full legal title'. Unlike OP 4.10, ESS 7 provides examples of the form such benefits might take, including 'fair lease arrangements or, where land acquisition is necessary, ... land-based compensation or compensation in kind in lieu of cash compensation';[46] 'continued access to natural resources, ... replacement resources, or, as a last option, providing compensation and identifying alternative livelihoods',[47] and measures securing 'access, usage, and transit on land the Borrower is developing subject to overriding health, safety, and security considerations'.[48] The new ESS 7 is therefore more detailed and prescriptive than OP 4.10, encompasses a greater range of usages and types of indigenous connection to land, but provides the same (limited) degree of protection for indigenous peoples who are no longer in possession of their traditional lands. Both OP 4.10 and ESS 7 are better equipped to

[43] Ibid., OP 4.10, para 18. [44] Ibid., OP 4.10, para 18.
[45] World Bank, ESS 7, above n 2, para 26. [46] Ibid., para f(i). [47] Ibid., para f(ii).
[48] Ibid., para f(iv).

address the particular claims of indigenous peoples to traditional land and resources than the more general and ahistorical FAO Guidelines used by the UN-REDD Programme.

In practice, however, as discussed in the previous chapter, the question of exactly how the World Bank operational polices are brought to bear on the FCPF's evaluation of REDD Readiness funding applications is controversial. World Bank and FCPF advice on this point is somewhat ambivalent.[49] According to the bank, while the FCPF constitutive documents specify that '[a]ny funding from the Bank would trigger the application of its safeguard policies',[50] in fact 'the FCPF Readiness Fund is not intended to fund projects or investments on the ground',[51] so that the formal process for ensuring compliance with the bank's safeguard policies will ordinarily not apply to the provision of funding for REDD Readiness preparations. In short, while the safeguard policies apply in principle to all 'REDD readiness activities funded by the FCPF Readiness Fund', bank staff are not obliged to apply the operational policies at this stage except 'on an exceptional basis'. Accordingly, persons or communities affected by REDD Readiness activities are not able to contest compliance with the operational standards by appealing to the World Bank's Inspection Panel. The tenure standards included in OP 4.10, although offering a high degree of protection when they are triggered, have been deferred beyond the REDD Readiness phase and are not likely to condition the allocation of funding by the FCPF until REDD+ *implementation* funding is sought. For the foreseeable future, then, tenure issues in the REDD+ context will likely develop in line with the advice so far provided by the UN-REDD Programme and the FAO.

49 Bank Information Center (BiC), above n 36.

50 Forest Carbon Partnership Facility (FCPF), 'Information Memorandum' (13 June 2008) 18, www.forestcarbonpartnership.org/readiness-fund. Including the selection of countries into the REDD Readiness Fund. Forest Carbon Partnership Facility (FCPF) Readiness Fund, 'Incorporating Environmental and Social Considerations into the Process of Getting Ready for REDD' (Note FMT 2009-6, Draft, 15 October 2009) ('Incorporating Environmental and Social Considerations Draft 2009'), www.forestcarbonpartnership.org/sites/fcp/files/Documents/tagged/FCPF_en_soc_guidelines_10-15-09.pdf, 1, para 2(i). See also, Forest Carbon Partnership Facility (FCPF) Readiness Fund, 'Incorporating Environmental and Social Considerations into the Process of Getting Ready for REDD+' (Note FMT 2010-16, 23 June 2010)

51 Forest Carbon Partnership Facility (FCPF) Readiness Fund, 'Incorporating Environmental and Social Considerations Draft, 2009', above n 50, 3.

### B. Tenure in REDD+: Debates and Controversies

There is no doubt that REDD+ has the potential to impact adversely on the well-being of indigenous and forest-dependent communities by incentivising others (including the state) to exclude them from forested lands.[52] REDD+ could also yield tenure-related 'co-benefits' for those communities, by encouraging the legal recognition of their customary property rights and by providing REDD+ payments in compensation for their emissions-reducing activities or custodianship. As one might expect, then, the existing UN-REDD Programme advice on tenure references both risk management and benefit-sharing goals. The report of the UN-REDD Programme's 2013 Expert Meeting on tenure, for example, identifies these as the two central instrumental values of tenure reform, observing that '[s]ecure tenure rights are the foundation for minimizing risks for investments into REDD+ and for sharing the benefits from carbon finance'.[53]

In risk-management reasoning, secure tenure is thought to directly discourage deforestation by allowing rights holders to 'take into account future values in current decision-making'; that is, to think prospectively about the long-term use of forested land rather than to exploit or clear forests to prove ownership or in anticipation of losing access to the resource.[54] Secure and clear tenure is also sometimes presented as a

[52] James Mayers et al, 'Forest-Based Climate Strategies: Making REDD and Other Initiatives Work' (International Institute for Environment and Development (IIED), March 2008), www.fern.org/publications/recommended-reading/forest-based-climate-strategies.

[53] UN-REDD Programme, 'Background Paper for the Expert Meeting: Options for Addressing Tenure under REDD+', above n 3, 12.

[54] Ann-Kristin Rothe and Paul Munro-Faure, 'Tenure and REDD+: Developing Enabling Tenure Conditions for REDD+' (Policy Brief, Issue No 6, UN-REDD Programme, 2013) 2, http://theredddesk.org/resources/tenure-and-redd-developing-enabling-tenure-conditions-redd. See Lisa Westholm et al, 'REDD+ and Tenure: A Review of the Latest Developments in Research, Implementation and Public Policy Debates' in Randall A Bluffstone and Elizabeth JZ Robinson (eds), *Forest Tenure Reform in Asia and Africa: Local Control for Improved Livelihoods, Forest Management, and Carbon Sequestration* (Taylor and Francis, 2015) 221, 224: 'if local community members are denied long-term benefits, they are more likely to look for short-term gains'. See also Lorenzo Cotula and James Mayers, 'Tenure in REDD: Start-Point or Afterthought?' (International Institute for Environment and Development (IIED), 2009) 6, http://pubs.iied.org/13554IIED/?k=tenure+in+redd; Matthew Sommerville, 'Land Tenure and REDD+: Risks to Property Rights and Opportunities for Economic Growth' (Issue Brief, USAID, 2013) 3, www.land-links.org/issue-brief/land-tenure-and-redd-risks-to-property-rights-and-opportunities-for-economic-growth: 'The lack of secure tenure for local populations is recognized as a principal driver of deforestation in many developing countries'.

prerequisite of effective participatory rights, because it is thought to provide a basis for identifying the holders of FPIC and to bolster their power to influence states and private-sector actors in preparatory consultations.[55] Participatory rights, in this line of reasoning, also reduce community resistance to REDD+ projects and so ensure the stability and durability of those schemes. Conversely, insecure tenure is thought to increase the risk that a lack of community buy-in could undermine the success of REDD+ projects by increasing conflict over resources affected by (and benefits provided by) REDD+, creating uncertainty for donors and investors.[56]

On the other hand, in addition to providing the basis for ex ante participatory rights and processes, secure and clear tenure has been presented in UN-REDD Programme and FAO advice as a necessary basis for facilitating ex post benefit-sharing. In this logic, tenure arrangements provide a way to identify the people and communities contributing to REDD+ climate change goals and so enable the equitable distribution of financial and co-benefits flowing from the operation of REDD+ projects.[57] In addition, secure and clear tenure is sometimes identified by the UN-REDD Programme and FAO as an unqualified good or 'right', and so as an end in itself that should be pursued by governments independently of REDD+. In this view, REDD+ simply 'provides an opportunity to revise outdated tenure policy and legal frameworks [and] to strengthen formal and customary tenure rights'.[58] This calls into sharp focus the

[55] Cotula and Mayers, above n 54, 3.

[56] Ibid. See also country concerns about REDD+ related tenure conflict in UN-REDD Programme, 'Review of Tenure Issues and Needs Identified by UN-REDD Countries' (9 January 2013). Likewise, the COP of the CBD has noted that a '[l]ack of benefits and/or lack of equitable distribution of benefits for relevant stakeholders, in particular for indigenous and local communities, are possible threats to the success of [REDD] activities': Conference of the Parties, Convention on Biological Diversity, *XI/19. Biodiversity and Climate Change Related Issues: Advice on the Application of Relevant Safeguards for Biodiversity with Regard to Policy Approaches and Positive Incentives on Issues Relating to Reducing Emissions from Deforestation and Forest Degradation in Developing Countries; and the Role of Conservation, Sustainable Management of Forests and Enhancement of Forest Carbon Stocks in Developing Countries* (UN Doc UNEP/CBD/COP/DEC/XI/19 (5 December 2012) ('CBD Hyderabad Decision XI/19')), annex, para 10. Sommerville, above n 54, 3–4.

[57] See Williams, above n 15, 14. See also, Crystal Davis, Ruth Nogueron and Anne-Gaelle Javelle, 'Analysis of Institutional Mechanisms for Sharing REDD+ Benefits: Property Rights and Resource Governance Project (PRRGP)' (United States Agency for International Development (USAID), March 2012), www.land-links.org/research-publication/analysis-of-institutional-mechanisms-for-sharing-redd-benefits.

[58] Dickson et al, above n 4, 1–2. See also, Sommerville, above n 54, 9.

attributes of tenure that are supported by REDD+ advice and funding. Security and clarity need not promote justice or equity in property distribution. As has been noted, the formalisation of property rights can (and has been) pursued at the expense of the legitimate claims of persons and communities who have the most difficulty proving the existence of their rights in ways that are legible to state law and institutions, including indigenous peoples. Nonetheless, in a 2014 presentation to the UN-REDD Programme Policy Board, for example, an FAO tenure specialist observed that:

> It's also important to remember that regardless of how REDD+ evolves, investment in securing tenure is widely recognized to be beneficial to communities and forests. Basically this is a win-win opportunity.[59]

How then, are states advised to address tenure issues in their REDD Readiness activities? Since most of the advice and assistance provided by the UN-REDD Programme is directed to states in the REDD Readiness phase, and consistently with its human rights–based approach and extensive interventions on the participatory rights of indigenous peoples and forest-dwelling communities, to date UN-REDD Programme advice has emphasised the ways secure tenure can empower communities to influence decision-making on REDD+ projects. In its 2013 policy brief, for example, after canvassing the various benefits attached to secure and clear tenure, the UN-REDD Programme concluded that '[m]ost importantly, tenure security provides communities with some control over whether REDD+ will be implemented in their community and, if so, how it will affect their livelihoods'.[60] The FAO Guidelines, however, are not limited in scope to tenure issues impacted by REDD+, and take a broader development-focused approach, emphasising access to land as a means to ensure food security and reduce poverty. The UN-REDD Programme endorses the FAO Guidelines as a source of guidance for REDD+ participating states that 'can inform REDD+ countries that are developing their own strategies, policy and legal frameworks and activities' in relation to tenure reform.[61] Because of the heavy emphasis placed on the FAO Guidelines by the UN-REDD Programme, and the

[59] Amanda Bradley, 'Tenure and REDD+: Update and Key Messages' (Presentation at the 13th Policy Board Meeting, UN-REDD Programme, Arusha, Tanzania, 5 November 2014). For further information, see UN-REDD Programme, '13th Policy Board Meeting, Arusha (Tanzania), 5–7 November 2014', www.unredd.net/documents/policy-board-86/thirteenth-policy-board-meeting-arusha-tanzania-5-7-november-2014–3553.html.

[60] Rothe and Munro-Faure, above n 54, 2–3.

[61] Ibid. 6.

non-application of World Bank safeguards in most readiness activities funded by the FCPF, the FAO Guidelines are a primary influence on the development of REDD+ policy and standards on tenure matters. For this reason, it is worth examining their content in some detail, with a view to assessing their conceptual framing and the likely impact of this framing on the interests of indigenous peoples.

### 1. *The FAO Guidelines*

The FAO's policy on indigenous peoples expressly states that '[t]he rationale for engagement with indigenous peoples is grounded in the FAO's mandate to achieve food security for all'.[62] Likewise, the FAO Guidelines are described in REDD+ literature as 'the first broad international consensus by governments, [civil society organizations], the private sector, international organizations and academics on how to govern tenure in a way that increases national food security and promotes sustainable development'.[63] This emphasis on food security and development is significant. As will be discussed below, indigenous interests and land claims may or may not coincide with these goals, in the context of REDD+ or elsewhere.

The FAO Guidelines are pitched at the level of principle, and much of the advice they contain is of a procedural nature. The guidelines urge inclusivity and participation in the design of tenure arrangements. They also emphasise the need for rule-of-law principles to operate in the implementation of property law, including especially via measures to ensure laws are transparent and non-discriminatory, that documentary records of tenure arrangements are accessible, and that dispute-resolution mechanisms are in place. Importantly, the guidelines are not on their face confined to the management of individualised private property, but aim instead to accommodate 'tenure rights in various forms, from limited use to full ownership',[64] and 'all forms of tenure, including public, private, communal, collective, indigenous and

[62] FAO, *FAO Policy on Indigenous and Tribal Peoples* (Food and Agriculture Organization, 2010) 7, www.fao.org/docrep/013/i1857e/i1857e00.htm.

[63] UN-REDD Programme, 'Background Paper for the First Expert Meeting: Options for Addressing Tenure under REDD+', above n 3, 5.

[64] FAO, *Voluntary Guidelines on the Responsible Governance of Tenure of Land, Fisheries and Forests in the Context of National Food Security* (Food and Agriculture Organization of the United Nations, 2012) ('*FAO Voluntary Guidelines (2012)*)', www.fao.org/nr/tenure/voluntary-guidelines/en, 13, para 8.8.

customary'.[65] The key concept of 'tenure' used in the guidelines is correspondingly broad, and is defined as:

> the relationship, whether legally or customarily defined, among people, as individuals or groups, with respect to land ... [as well as] rules invented by societies to regulate behaviour, ... define how property rights to land are to be allocated within societies ... [and] define how access is granted to rights to use, control, and transfer land, as well as associated responsibilities and restraints.[66]

Accordingly, alongside exclusive possessory tenure rights, the guidelines make reference to 'subsidiary tenure rights' (giving the example of 'gathering rights'), as well as to systems of regulation that extend beyond individual property, including 'systems of collective use and management' of publicly owned land and resources, and 'processes of allocation by the State'.[67] It is incumbent on states to 'determine which of the land, fisheries and forests they own or control will be retained and used by the public sector, and which of these will be allocated for use by others and under what conditions'[68] as well as to 'determine whether they retain any form of control over land, fisheries and forests that have been allocated'.[69]

Both the UN-REDD Programme and the FAO acknowledge that tenure is a composite concept comprised of a 'web of intersecting interests'.[70] Several other features of the guidelines are notable. First, the FAO has deployed the concept of a 'bundle of rights' to give substance to the idea of tenure, an explanatory device that typically enumerates rights to use and control land, recover 'rents and profits' from land users, and transfer rights in land, and includes duties incumbent on property holders.[71] The 'bundle of rights' idea is a pillar of Western legal theories of property, used to complicate the classical common law conception of private property as a monolithic and absolute right vested in an individual to the exclusion of all others.[72] Accordingly, the guidelines stress that

[65] Ibid. 2, para 2.4.

[66] FAO, 'Land Tenure and Rural Development' (FAO Land Tenure Series 3, Food and Agriculture Organization of the United Nations, 2002) 7, www.fao.org/docrep/005/y4307e/y4307e00.htm.

[67] FAO, *FAO Voluntary Guidelines (2012)*, above n 64, 12, para 8.3.

[68] Ibid. 13, para 8.5. [69] Ibid. 13, para 8.8.

[70] FAO, 'Land Tenure and Rural Development', above n 66, 7–8. See also, Rothe and Munro-Faure, above n 54, 1.

[71] FAO, 'Land Tenure and Rural Development', above n 66, 10.

[72] See, e.g., William Blackstone, *Commentaries on the Law of England* (Dublin, 6th ed, 1775) (Eighteenth Century Collections Online) vol. 2, 8ff; and, on 'exclusion and

'no tenure right, including private ownership, is absolute'. Instead, all tenure rights are limited by the rights of third parties and by state intervention or regulation in the public interest. Further, all tenure rights are balanced against the duties of the rights holder, and include the duty to 'respect the long-term protection and sustainable use of land, fisheries and forests'.[73] Second, the FAO Guidelines draw on a distinction between 'formal' and 'informal' property rights. Formal rights are those that are 'explicitly acknowledged by the state and which may be protected using legal means'. In the guidelines, formal rights are usually 'statutory rights', while informal property rights 'lack official recognition and protection' and may be either illegal or 'extra-legal'. Customary property rights, in the view of the FAO, are ordinarily informal and extra-legal in character, although the organization concedes that the statutory recognition of customary rights in some countries (and in particular within Africa) has 'blurred' this distinction in practice.[74]

Helpfully, then, the concept of tenure and tenure security used in UN-REDD Programme and FAO advice is not dogmatically exclusive, but rather admits the possibility of multiple property rights coexisting in the same piece of land, an idea that is central to the ways that property rights are conceived as a matter of law, at least in common law jurisdictions. The 'bundle of rights' approach, while attentive to the possibilities of addressing cross-claims by 'layering' different property rights over a single parcel of land, does not on its own provide a way forward when that 'bundle' is contested, wholesale, by claimants who derive their rights from customary legal systems that predate state law. Where customary legal systems are implicated in property disputes, the legal questions become jurisdictional in form, involving conflicts of laws rather than conflicts of rights. In such circumstances, the task is not simply to rank and allocate use and possessory rights over land, but to consider the ways in which customary legal systems and state legal systems can recognise rights sourced in normative systems external to the one in which they are claimed. In settler states, the most difficult property-based conceptual impasses occur where rights holders whose claims are legitimate in one legal system (usually customary law) find that their rights are not recognised in the other, and even more acutely, where third parties have

exclusiveness', Felix S Cohen, 'Dialogue on Private Property' (1954) 9 *Rutgers Law Review* 357, 370–371.

[73] FAO, *FAO Voluntary Guidelines (2012)*, above n 64, 6, para 4.2.

[74] FAO, 'Land Tenure and Rural Development', above n 66, 11.

acquired property rights to the claimed land that *are* recognised in the dominant legal system. The guidelines advise that 'where informal tenure to land, fisheries and forests *exists*, States should acknowledge it in a manner that respects *existing* formal rights under national law'.[75] The real conundrum however, is how to decide that informal tenure 'exists' at all, as a matter of fact and law. This requires an 'act of recognition' in which political or judicial institutions of the state decide to give effect to property rights based in customary law. Having recognised the normative force of customary rights, state institutions must then determine the bases on which those rights can be enforced in state law, and select the rights that can be so recognised and enforced.

One of the most conceptually and politically fraught tasks for settler states and indigenous peoples, then, is to agree on the principles of justice that will govern the recognition of customary property rights and inform the redistribution of formal property rights if this is necessary to resolve conflict between the two. As will be discussed, courts are generally reluctant to order the redistribution of property rights because of the very complex questions of public policy involved. In the common law doctrine of aboriginal title, for example, state-granted or acquired property rights will not be disturbed by indigenous claims, so that where inconsistent grants have been made by the state, preexisting customary title will not find purchase in state law. In some settler states (but not all), where restitution is impossible because the grant of formal rights has 'extinguished' the possibility of asserting customary title in state fora, compensatory processes are provided instead, but these are ordinarily managed as negotiated political processes rather than litigated judicial ones. This draws attention to the fundamental assumptions at play in advice directed towards ensuring the 'security and clarity' of tenure. Security of tenure can be achieved in a variety of ways, and one of those ways is by denying the legal status of customary property rights. The legitimacy of security and clarity depends on the underpinning concepts of distributive and reparative justice by which those acts are to be judged. As Westholm et al. point out, in the context of REDD+, 'clear and secure tenure does not per se imply equity'.[76]

The FAO guidelines acknowledge that property rights are sometimes contested by claimants with equally 'legitimate' but differently justified claims, including competition between claims based on customary law

[75] FAO, *FAO Voluntary Guidelines (2012)*, above n 64, 16, para 10.1 (emphasis added).
[76] Westholm et al, above n 54, 232.

and those based in state law. They also acknowledge that tenure is an aggregate 'bundled' concept that can accommodate multiple usage and possessory rights.[77] However, the concepts of security and clarity emphasised in the guidelines nonetheless drive policy advice towards the 'formalisation' of tenure arrangements. Setting aside the risks posed by state insistence on prioritising state grants over customary rights in formalisation processes, the formalisation of customary rights qua customary rights poses two types of risk to the holders of those rights. The first risk is endogenous, and points to the possibility that the content of customary law itself may not be equitable, in which case recognition may enforce unjust allocations of property within a (formalised) customary system. This possibility receives almost no attention from the UN-REDD Programme and the FAO. Indeed, in the growing body of literature addressing tenure issues in REDD+ policy and programming, very few commentators have questioned the legitimacy of particular forms of property rights supported by customary laws. Yet custom is no guarantee of equity. Westholm et al. have put this problem in blunt terms: 'customary tenure arrangements are often not fair. Social institutions, just like statutory ones, may embody gender inequalities and unequal power relations'.[78]

The second risk is exogenous, and entails the possibility that in formalising customary property rights, state law could compromise, undermine or extinguish some or all of those rights, or deny recognition to them (and so decline to protect or enforce those rights). The UN-REDD Programme's exploratory policies on tenure have noted the danger that customary rights could be compromised or extinguished by processes of statutory formalisation. The UN-REDD Programme has specified that '[g]iving customary and traditional rights statutory legal recognition should neither undermine local tenure systems nor simplify them', and that '[c]apturing their complexity and specificity is necessary for avoiding conflicts among local communities and ensuring that REDD+ is connected to the reality on the ground'.[79] Elsewhere, the

[77] FAO, 'Land Tenure and Rural Development', above n 66, 9. Rothe and Munro-Faure, above n 54, 3: 'Clarifying tenure in the REDD+ context is a complex endeavor, partly because several rights holders may have different rights over the same resources or share the same rights. These rights may also change over time depending on seasons and natural conditions. Primary and secondary rights may exist, and primary rights to trees may be allocated independently of primary rights to lands on which the trees grow'.

[78] Westholm et al, above n 54, 232.

[79] UN-REDD Programme, 'Background Paper for the Expert Meeting: Options for Addressing Tenure under REDD+', above n 3, 13.

FAO has warned that '[i]ronically, attempts to increase the legal security of some may result in others losing their rights', especially use rights, but suggests that this would occur only in 'poorly designed' titling and registration projects, or where the state enforces 'existing rules that prevent access to resources', including forests.[80] Nowhere is it suggested that the loss of property rights is inevitable where formalisation is undertaken with respect to contested land. It remains the case however, that where there are competing claims to land, some claimants will be successful at the expense of others. Attrition of this kind is especially likely where claims are equally legitimate and comparable in extent but premised on different legal systems, so that they are inconsistent and irreconcilable as matters of law and fact. This is significant, because it is the nature of property rights to be permanent and resistant to adjustment or reformulation, especially where they are constitutionally protected against acquisition or diminishment, so that property disputes are more likely than other legal conflicts to be resolved on a 'winner-takes-all' basis. This makes property rights uniquely impervious to methods familiar to human rights law, such as provisions allowing states to impose 'justifiable limitations' on most rights,[81] or judicial methods of 'balancing' or 'proportionality' that are used to rank or reconcile otherwise conflicting rights.

These are serious questions for evolving REDD+ tenure policy. Some commentators have warned, for example, that in the REDD+ context, an emphasis on certainty could result in states asserting greater centralised control over forested land.[82] Others have suggested that the increased value of forested land, along with the prospect of REDD+ payments to governments 'may create a *dis*incentive for forest and conservation and other government authorities to resolve long-standing land disputes in forest areas'.[83] In fact, a 2014 independent evaluation of the UN-REDD Programme included the observation that 'REDD+ and the recent push to secure related carbon rights now threaten to recentralize forest rights and benefits altogether'.[84] This

[80] FAO, 'Land Tenure and Rural Development', above n 66, 20–21.

[81] See Section 1 of the Canadian Charter of Human Rights and Freedoms, which 'guarantees the rights and freedoms set out in [the charter] subject only to such reasonable limits prescribed by law as can be demonstrably justified in a free and democratic society': Canada Act 1982 (UK) c 11, Sch B Pt I. Identical wording is used in the New Zealand Bill of Rights Act 1990.

[82] Sommerville, above n 54, 7; Thompson, Baruah and Carr, above n 14, 36.

[83] Thompson, Baruah and Carr, above n 14, 23.

[84] Frechette, de Bresser and Hofstede, above n 5, 25.

would reduce community control over forested lands, and so increase the likelihood that forest-dependent communities will be denied access to forest resources, and so not receive co-benefits that are allocated in order to reward local contributions to emissions reductions.

These problems require not just the identification of legitimate property rights, but the identification of the right that is to take priority in the event of a conflict between two or more legitimate property rights. The concept of 'legitimacy' is thus the normative fulcrum on which the FAO Guidelines turn, but on its content they are notably silent. States are urged to 'take reasonable measures to identify, record and respect legitimate tenure right holders and their rights, whether formally recorded or not',[85] but the guidelines do not offer a view on the features of land use, or of prior possessory title, that would generate 'legitimate tenure rights' deserving of recognition, nor what to do when these conflict with the legitimate tenure rights of third parties. In short, the guidelines offer advice on the *protection* of legitimate tenure rights, but not on how communities *demonstrate* that they have such rights. The position of non-occupants in cross-claims, in particular, is barely addressed. One tantalising reference specifies that policies governing the allocation of tenure rights 'should specify the means of allocation of rights, such as allocation based on historical use or other means'.[86] In its statements specifically addressing indigenous peoples, however, the guidelines revert to a current-use-based approach, including where claims are made to ancestral lands. The guidelines state that 'recognition should take into account the land, fisheries and forests that are *used* exclusively by a community and those that are shared',[87] and further, that 'where indigenous peoples and other communities with customary tenure systems have legitimate tenure rights to the ancestral lands *on which they live*, States should recognize and protect these rights'[88] – references again containing the implication that 'legitimate tenure rights' are those based on current occupancy or use.

What will be required by states, then, to assist them to take proper account of indigenous property claims, is not a set of guidelines directed to optimal 'outcomes' (security and clarity) but advice on how to assess the legitimacy of the rights that are to be secured and

[85] FAO, *FAO Voluntary Guidelines (2012)*, above n 64, 3. [86] Ibid., 13, para 8.8.
[87] Ibid. 15, para 9.4 (emphasis added). [88] Ibid. 15, para 9.5 (emphasis added).

clarified. Methods and doctrines of this kind have been developed by courts in the settler states, but not, as will be described below, with much assistance from human rights principles. The vastly complicating aspect of REDD+ tenure policy is that, in practice, the recognition of customary law as a source of legally enforceable property rights logically precedes the protection and enforcement of those customary rights in state law. The FAO's references to legitimate 'informal' or 'de facto'[89] rights that 'exist ... in reality' – in contradistinction to formal rights that 'exist because of the formal law'[90] – do little more than beg the question of recognition. By bracketing recognition, the approach taken by the FAO and the UN-REDD Programme in advice on customary tenure raises issues of acute importance for indigenous peoples, because it leans towards the protection of current use, in recognition of the needs-based claims of communities dependent on forest resources for food security. It does not offer a corresponding justificatory basis that could support the claims of prior occupants. It is possible, then, that the methods recommended in the FAO Guidelines could encourage the formalisation of tenure rights that preclude or extinguish indigenous property rights, in furtherance of 'clear and secure' tenure rights for current users.

While the UN-REDD Programme has usefully acknowledged that 'tenure rights are currently not recognised as human rights',[91] its guidance shows that the 'human rights–based approach'[92] it adopts in its work is the primary lens brought to bear on questions of indigenous tenure. This approach adds little to the concept of 'legitimacy' that animates the FAO Guidelines, because it does not contemplate conflict between rights, nor does it address the different logics underpinning property and human rights, respectively. This is demonstrated by the

[89] See Edella Schlager and Elinor Ostrom, 'Property-Rights Regimes and Natural Resources: A Conceptual Analysis' (1992) 68 *Land Economics* 249, 254.

[90] FAO, 'Land Tenure and Rural Development', above n 66, 12.

[91] Rothe and Munro-Faure, above n 54, 5.

[92] A 'human rights-based approach' being one of the five principles underpinning the United Nations Development Assistance Framework (UNDAF), as outlined in Frechette, de Bresser and Hofstede, above n 5 para 53; and the FAO, UNDP and UNEP, *UN Collaborative Programme on Reducing Emissions from Deforestation and Forest Degradation in Developing Countries (UN-REDD): Framework Document* (20 June 2008) 7. See also Forest Carbon Partnership Facility (FCPF) and United Nations REDD Programme (UN-REDD), 'Readiness Preparation Proposal (R-PP)' (Template Version 5, Working Draft, 30 October 2010) 7.

positioning within the guidelines of human and property rights as mutually supportive entitlements:

> Given that all human rights are universal, indivisible, interdependent and interrelated, the governance of tenure of land, fisheries and forests should not only take into account rights that are directly linked to access and use of land, fisheries and forests, but also all civil, political, economic, social and cultural rights.[93]

This ambivalence is also apparent in the guidelines' treatment of indigenous-specific human rights instruments. States are advised that when addressing the tenure rights or claims of indigenous peoples, they must meet their commitments 'as appropriate' under the ILO Convention 169, the CBD and the UNDRIP,[94] and 'should remove and prohibit all forms of discrimination related to tenure rights'.[95] The question we pose in the remainder of this chapter is as follows: to what extent can human rights principles, including principles of equality and non-discrimination, assist indigenous peoples to secure title to their ancestral and traditional territories? In other words, are the types of human rights–based tenure reform anticipated by the UN-REDD Programme and the FAO capable of 'doing justice' to tenure claims based on customary law in circumstances where a state has not yet recognised customary law as a source of legally enforceable property rights? Section C suggests that, in the domestic law of states, the anti-discrimination principles that underpin human rights law can support judicial recognition of customary title, but do not *require* it, and in the absence of such recognition, the legislative allocation of property rights to indigenous groups can be categorised as a form of racially discriminatory preferential treatment. The IACtHR, however, has developed a jurisprudence in which it effectively performs the act of recognition itself in lieu of recognition by the state, and by doing so brings customary property rights within the protection of the equality and property provisions of the American Convention on Human Rights (ACHR). In sum, in order for human rights principles to protect, rather than imperil, indigenous customary property rights, those rights must first be recognised as legally enforceable. The advice offered by the UN-REDD Programme and the FAO does not (so far) guide states in determining when such recognition is required as a matter of justice.

93 FAO, *FAO Voluntary Guidelines (2012)*, above n 64, 7, para 4.8.
94 Ibid. 14–15, para 9.3; see also 22, para 12.7. 95 Ibid. 4–5.

### C. International Human Rights and Property Law: Conceptual Problems

It is worth noting at the outset of this section that none of the major global international human rights instruments recognises a 'right to property' per se. Of the seven major international human rights instruments, only the UNDHR purports to protect property. Article 17 of the UNDHR provides that '[e]veryone has the right to own property alone as well as in association with others' and specifies that '[n]o one shall be arbitrarily deprived of his property'. As will be discussed below, a similar provision appears as Article 21 of the regional ACHR, and as Article 1 of Protocol 1 of the European Convention on Human Rights. These provisions, notably, do not on their face guarantee that a person will own any property at all, or that a person is entitled to be granted property by the state, but rather speak to the equal protection of any property rights a person has already acquired (in other words, property rights recognised in law). As discussed previously, the question of whether property rights can properly be said to 'exist' in the absence of legal recognition is a fraught one. On this conundrum, international human rights law has little to add.

Throughout the FAO Guidelines, states and other actors are urged to respect and protect 'human rights and legitimate tenure rights'.[96] Similarly, the UNREDD Programme policy brief on tenure notes that '[g]eneral principles in international law provide protection [sic] that relate to access to land; e.g. equality and non-discrimination in ownership and inheritance'.[97] Non-discrimination, however, can be a problematic concept for indigenous people and their claims. Put simply, because indigeneity is identified in international and domestic law as a cognate of race, recognition of distinctive property entitlements for indigenous peoples can appear in state law as a form of prohibited racial discrimination. States recognising indigenous property rights are from time to time challenged to show that these acts of recognition do not constitute racial discrimination against non-indigenous peoples or indigenous non-beneficiaries. Section 1 examines the ways that indigenous property claims, including prior possessory claims, have been considered in international law, before turning to a discussion of some of the strategies used in domestic law to overcome or avoid the implication of racial discrimination. Finally, we consider the approach taken by the IACtHR, which has

[96] Ibid. 4. [97] Rothe and Munro-Faure, above n 54, 5.

developed a body of jurisprudence aimed at reconciling principles of equality and non-discrimination with distinctive indigenous rights to ancestral lands, including rights that are not based on current occupancy or use. We take as our starting point the UNDRIP, to which states are referred in the Cancun Safeguards, the FAO Guidelines, the UN-REDD Programme's policy brief on tenure and in the instructions to states applying to the UN-REDD Programme and the FCPF for REDD Readiness funding.

### 1. *Indigenous Property in the UNDRIP*

The UNDRIP is a declaration, and therefore a General Assembly resolution not formally binding on states as a matter of international law, although many of its provisions replicate and augment principles well established in international treaties and in international customary law. The extent to which the UNDRIP extends beyond the previously accepted categories and scope of human rights in international law, and the binding qualities of aspects of the UNDRIP that may be considered 'novel', is a matter of intense scholarly and political debate. The negotiation of the UNDRIP was a long and difficult process, spanning twenty-two years. Some of the most contentious aspects of negotiations centred on the parts of the draft text that addressed indigenous rights to land and natural resources. These debates, and in particular the concerns expressed by states that have recognised historic indigenous property rights, illustrate some of the complexities of reconciling customary and state-based sources of property. The central provision in question is Article 26 of the UNDRIP. This provides that '[i]ndigenous peoples have the right to the lands, territories and resources which they have traditionally owned, occupied or otherwise used or acquired', along with the 'right to own, use, [and] develop' those lands and resources. To comply with Article 26, states are obliged to 'give legal recognition and protection to these lands, territories and resources ... with due respect to the customs, traditions and land tenure systems of the indigenous peoples concerned'.[98] Some states, including the CANZUS states, were concerned that the provision appears to entitle indigenous peoples 'as of right' to lands now owned or occupied by third parties. Canada, for example,

[98] *United Nations Declaration on the Rights of Indigenous Peoples*, GA Res 61/295, UN GAOR, 61st Sess, 107th Plen Mtg, Agenda Item 68, Supp No 49, UN Doc A/RES/61/295 (2 October 2007, adopted 13 September 2007).

argued that the phrase failed to make a necessary distinction between land 'over which indigenous people have or can establish legal rights' and land that they claim but do not (yet) have legal rights to.[99] The United States was similarly of the view that:

> [t]he use of the past tense suggests that lands 'traditionally occupied' at an earlier time – but not now – might be subject to the article. We cannot agree with a blanket statement that all existing legal rights of others to lands 'traditionally occupied' by indigenous persons at some much earlier time should in all cases be set aside regardless of the circumstances.[100]

Similar distributive problems were addressed by CANZUS states in their interventions on UNDRIP draft Article 27 (now Article 28), securing the rights of indigenous peoples to 'the restitution of the lands, territories and resources which they have traditionally owned or otherwise occupied or used' and specifying that, where this is not possible, indigenous peoples shall be provided 'compensation' in the form of 'lands, territories and resources equal in quality, size and legal status'.[101] The United States, for example, expressed its support for 'adequate legal procedures to ensure that claims of non-consensual confiscation or use of the lands or property are fairly resolved', but doubted that restitution was a viable means for resolving all historical land claims, suggesting instead that 'the declaration should call on states to consider the possibility of negotiated settlements, as appropriate, when not otherwise required by law'.[102] Canada likewise suggested that the article should include reference to the need for land-claims settlement processes and noted that alternatives to restitution should be included in the article, alongside market-based compensation.[103] When the four CANZUS states voted against the UNDRIP in 2007, all stated that Article 26 was one of the reasons that they could not endorse the declaration.[104]

[99] Canada, 'Preliminary Comments, Part VI' (1995) doCip Doc 023a (on file with author). See also, Canada, 'Statement on Lands and Resources' (21 October 1999) doCip Doc 014a (on file with author).

[100] United States, 'Preliminary Comments, Part VI (Land and Resources)' (28 November 1995) doCip Doc 030a (on file with author).

[101] UN ECOSOC Commission on Human Rights, *Report of the Sub-Commission on Prevention of Discrimination and Protection of Minorities on Its Forty-Sixth Session*, UN Doc E/CN.4/1995/2 (28 October 1994) ('*Report of the Sub-Commission on Prevention of Discrimination and Protection of Minorities*') 112, art 27.

[102] United States, above n 100.

[103] Canada, 'Preliminary Comments, Part VI', above n 99.

[104] 'DoCip Comparison Table' (on file with the author).

## 2. *Indigenous Property and the UN Human Rights Treaty Bodies*

As noted above, the FAO Guidelines do not address property claims premised on historic or prior customary possession. They also make only a brief and oblique reference to the kinds of restitution that might be sought by dispossessed indigenous communities: '[w]here appropriate, the concerns of indigenous peoples regarding restitution should be addressed in the national context and in accordance with national law and legislation'.[105] This statement is notably less strongly framed than the provisions of the UNDRIP discussed above. It is also less strongly worded than General Recommendation 23 of the CERD Committee, which in 1997 called on state parties:

> to recognize and protect the rights of indigenous peoples to own, develop, control and use their communal lands, territories and resources and, where they have been deprived of their lands and territories traditionally owned or otherwise inhabited or used without their free and informed consent, to take steps to return those lands and territories. Only when this is for factual reasons not possible, the right to restitution should be substituted by the right to just, fair and prompt compensation. Such compensation should as far as possible take the form of lands and territories.[106]

It seems clear that equality-based justifications for human rights are, in some ways, in tension with justifications based on the historic entitlements of indigenous groups. The CERD Committee, for example, has produced a series of general recommendations that deal with the claims of indigenous peoples. These provide some avenues for the assertion of indigenous property rights, but are limited in ways that in most cases make them flawed vehicles for claims based on customary title and prior possession. First, the convention includes provisions that require states to take special measures in order to 'secur[e] adequate advancement of certain racial or ethnic groups or individuals requiring such protection as may be necessary in order to ensure such groups or individuals equal enjoyment or exercise of human rights and fundamental freedoms'.[107] These measures, which the CERD Committee accepts can be equated in

[105] FAO, *FAO Voluntary Guidelines (2012)*, above n 64, 25, para 14.3.

[106] UN Committee on the Elimination of Racial Discrimination, *General Recommendation No 23: Indigenous Peoples*, 52nd sess, UN Doc A/52/18 (18 August 1997) annex V at 122 ('*CERD Committee General Recommendation 23*') para 5.

[107] *International Convention on the Elimination of All Forms of Racial Discrimination* (opened for signature 7 March 1966, 660 UNTS 195, entered into force 4 January 1969) ('CERD') art 1(4); see also, art 2(2).

most cases with 'affirmative action',[108] are subject to important qualifications. They may not 'lead to the maintenance of separate rights for different racial groups and ... they shall not be continued after the objectives for which they were taken have been achieved'.[109] Special measures accordingly provide an equality-based justification for indigenous property rights where these are necessary to alleviate indigenous disadvantage relative to non-indigenous groups and to secure for indigenous peoples the equal enjoyment of fundamental human rights and freedoms. Disadvantage, however, is not always an aspect of indigenous claims-making, nor is the scope of indigenous land claims necessarily tied to the material needs of the claimant community. The requirement of temporality, in addition, is a challenging one for indigenous peoples claiming the restoration of property rights lost as a result of historic injustice. The justificatory basis of such rights does not evaporate when indigenous peoples enjoy equality (however measured) with non-indigenous members of a settler society. Instead, the rights find their basis in measures of continuity that connect contemporary communities with the historic indigenous antecedents who exercised the rights in question, rights that are now claimed by indigenous groups, in perpetuity, as their successors. Since these may be difficult to defend as temporary special measures, they raise again the possibility that protections for indigenous property may be perceived to conflict with norms of non-discrimination in international and domestic law.

This problem was addressed in part by the CERD Committee in its 2009 General Recommendation 32 on special measures, passed two years after the adoption of the UNDRIP. In this recommendation, the CERD draws a distinction between 'special measures', and what it calls 'specific rights' vested in particular groups, including indigenous peoples. Specific rights are not mentioned in the text of the CERD. According to the committee, however, they include:

[108] UN Committee on the Elimination of Racial Discrimination, *General Recommendation No 32: The Meaning and Scope of Special Measures in the International Convention on the Elimination of All Forms [of] Racial Discrimination*, UN Doc CERD/C/GC/32 (24 September 2009) ('*CERD Committee General Recommendation 32*') 4, para 12: 'The term "special measures" includes also measures that in some countries may be described as "affirmative measures", "affirmative action" or "positive action" in cases where they correspond to the provisions of articles 1(4) and 2(2) of the Convention'.

[109] CERD, above n 107 art 1(4).

> the rights of persons belonging to minorities to enjoy their own culture, profess and practice their own religion and use their own language, the rights of indigenous peoples, including rights to lands traditionally occupied by them, and rights of women to non-identical treatment with men, such as the provision of maternity leave, on account of biological differences from men.[110]

These rights, the committee observes, 'are permanent rights, recognized as such in human rights instruments, including those adopted in the context of the United Nations and its specialized agencies'.[111] The purpose of such rights is not to ensure substantive equality by addressing disadvantage, and the CERD Committee's recommendation offers no justification for specific rights other than its reference to rights extant in international legal instruments. Despite its promise, then, these features of the CERD Committee's advice on 'specific rights' make it a rather precarious basis on which to justify the recognition of indigenous property rights in domestic law. Special measures, on the other hand, *have* provided a justification for the recognition of indigenous land rights in Australia, while also placing limits on the scope of that recognition. The Australian example shows, then, the trade-offs that must be contended with when addressing indigenous property claims through the lens of anti-discrimination law. It is discussed here to highlight the awkward interface of human rights and indigenous property claims in the liberal democracies, and to show that indigenous and public actors have been able to navigate a path between these two theoretical frames to secure the recognition of some historic indigenous property rights, albeit in an imperfect and limited way. The methodology developed in Australia can be contrasted with that developed by the IACtHR. Together, the approaches demonstrate two ways in which the tension between equality and property norms can be managed to allow and encourage the recognition of historic indigenous property rights.

### 3. *Indigenous Property and Anti-Discrimination Law in Australia: The Equal Protection of Property Rights (Once Recognised)*

In contrast to the other CANZUS states, in Australia no historic treaties were signed with indigenous peoples and to date no constitutional recognition of indigenous property rights exists. As a consequence,

[110] *CERD Committee General Recommendation 32*, above n 106, para 15.
[111] Ibid., para 15.

federal anti-discrimination law has provided a major vehicle for the assertion and protection of indigenous common law property rights, bringing the competing logics of human rights and property into very sharp relief. The federal Racial Discrimination Act 1975 (RDA), which incorporates the *ICERD* into Australian law, has been of pivotal importance in this enterprise. A central example is the landmark 1992 decision of the Australian High Court, *Mabo v Queensland (No 2)*,[112] which provided the first recognition in Australian common law of property rights sourced in indigenous 'traditional laws and customs'. The High Court recognised that the customary property rights of indigenous Australians had legal effect in common law as 'native title' rights, which survived the acquisition of sovereignty by the British Crown. Native title rights and interests remained intact unless and until they were validly extinguished by a legislative or executive act, abandoned by traditional owners or voluntarily relinquished to the Crown. To be valid, an extinguishment would have to comply with the RDA's prohibition on racial discrimination, and where extinguishment had not occurred, the RDA ensured that rights premised in traditional laws and customs were entitled to be protected to the same degree as other property rights. In the lead judgment, Justice Brennan held:

> Whatever the justification advanced in earlier days for refusing to recognize the rights and interests in land of the indigenous inhabitants of settled colonies, an unjust and discriminatory doctrine of that kind can no longer be accepted. The expectations of the international community accord in this respect with the contemporary values of the Australian people . . . A common law doctrine founded on unjust discrimination in the enjoyment of civil and political rights demands reconsideration. It is contrary both to international standards and to the fundamental values of our common law to entrench a discriminatory rule which, because of the supposed position on the scale of social organization of the indigenous inhabitants of a settled colony, denies them a right to occupy their traditional lands.[113]

The important feature of this application of non-discrimination law to indigenous property claims is that, while equal protection is owed to native title rights *once recognised*, anti-discrimination law, including the RDA, did not compel the *recognition* of those rights. In the logic of native title law, the fact that before 1992 the common law did not recognise 'native title' rights did not amount to discrimination. Similarly, the fact

[112] *Mabo v Queensland (No 2)* (1992) 175 CLR 1. [113] Ibid. 42 (Brennan J).

that non-proprietary aspects of traditional laws and customs continue to be denied legal force in the common law today is not regarded as racially discriminatory. In effect, the lack of recognition means there is no 'right' for anti-discrimination law to protect. The act of recognition effected by the High Court in *Mabo (No. 2)* was an augmentation of Australian common law that was influenced by the development of human rights and anti-discrimination law domestically and internationally, but it was not *required* by that body of law. In fact, in the absence of common law recognition of property rights emanating from customary law, grants of land to indigenous communities can appear as forms of racial discrimination.

In Australia, the interplay of anti-discrimination law and indigenous property rights in the pre-*Mabo (No. 2)* era had a very different character than it does today. A legislative grant of freehold title made by the South Australian government to the Anangu Pitjantjatjara Yankunytjatjara people in 1981, comprised of land in their traditional territory, was considered by the High Court in the landmark case of *Gerhardy v Brown* (1985). The Court found that the South Australian Act was a discriminatory measure in the terms of the RDA, and was saved from invalidation only because it was a temporary 'special measure' designed to overcome indigenous disadvantage, thus falling within the RDA's exemption for such measures.[114] At the time the grant was made, the doctrine of native title did not form part of Australian common law, so that in effect there were no existing property rights that could attract the protection of the RDA. In other words, because of the prohibition against racial discrimination in the RDA, in the eyes of the High Court, the legislation effected a (prima facie) discriminatory *grant* of freehold property rights to the group, rather than operating to extend equal *protection* to customary property rights already in existence. Seven years later, in *Mabo (No. 2)*, the High Court accomplished two tasks: it recognised that traditional laws and customs could generate common law property rights in the form of native title, and it confirmed (in reliance on its majority findings in *Mabo v Queensland (No 1)* (1988)) that by virtue of the RDA, all legal protections provided to persons in respect of their right to 'own and inherit property' now extended to holders of native title.[115]

[114] *Gerhardy v Brown* (1985) 159 CLR 70.

[115] *Mabo v Queensland* (1988) 166 CLR 186, 219 (Brennan, Gaudron and Toohey JJ). Cited, in part, in *Mabo (No 2)* (1992) 175 CLR 1 53 (Brennan J).

Recalling the FAO's emphasis then, on the need to recognise 'legitimate' tenure rights, the use made in the FAO Guidelines of human rights and non-discrimination principles addresses the equal protection of property rights (as in the second branch of the findings in *Mabo (No 2)*), but not *recognition* as the prerequisite of protection. In effect, the FAO Guidelines emphasise the equal treatment of property rights once recognised as 'legitimate', but leave open the question of the criteria on which 'legitimacy' is to be assessed. The application of human rights principles does not assist in this endeavour and, in the absence of recognition, can undermine it.

For this reason, we argue, human rights and non-discrimination principles have not often been used to advance the recognition of indigenous property rights in those countries where other justificatory bases are available. In the other Western settler states – Canada, the United States and New Zealand – historic, indigenous property claims, including those based on common law doctrines of aboriginal title, find purchase in commitments made in historic treaties and in indigenous-specific 'special trust relationships', doctrines of the 'honour of the Crown', and fiduciary responsibilities that attach primarily to the executive branches of settler governments and that guide courts in statutory interpretation.[116] Matters arising from the negotiation of land claims settlements by settler executives are typically regarded as largely non-justiciable political questions.[117] Accordingly, indigenous–state relationships in those countries are increasingly structured by political bargains and characterised by intergovernmental forms of relations, rather than by human rights. Canada, for example, has provided constitutional protection to indigenous property and treaty rights in section 35 of the Constitution Act,[118] notably positioned outside of the human rights protections of the Charter of Human Rights and Freedoms, and 'shielded' from the application of the charter by section 25.[119] The Canadian Supreme Court has

[116] See, e.g., *Nor Anak Nyawai v Borneo Pulp Plantations* [2001] 2 CLJ 769 ('*Nor Anak Nyawai*') (Malaysia); *Kerajaan Negeri Selangor v Sagong Tasi* [2005] 4 CLJ 169 (Malaysia); *Guerin v The Queen* [1984] 2 SCR 335 (Canada); *Manitoba Metis Federation v Canada (Attorney General)* [2013] 1 SCR 623 (Canada); *New Zealand Maori Council v Attorney-General* [1987] 1 NZLR 641 (HC & CA) 664 (New Zealand); *Te Runanga o Wharekauri Rekohu Inc v Attorney-General* [1993] 2 NZLR 301, 304 (New Zealand); *Tsilhqot'in Nation v British Columbia* 2014 SCC 44 (Canada).

[117] But see, for a recent Australian example of judicial attentiveness to intra-group disputes in negotiated settlements, *McGlade v Native Title Registrar* [2017] FCAFC 10.

[118] Canada Act 1982 (UK) c 11, Sch B. [119] Ibid. s 25.

urged the settlement of section 35 claims by negotiation.[120] The non-discrimination provisions of New Zealand's Bill of Rights Act and Human Rights Act have never been used to advance Maori claims in New Zealand's courts,[121] even while Maori have successfully sought urgent action decisions by the CERD Committee on the discriminatory impacts of the New Zealand Foreshore and Seabed Act 2004 (now repealed), a statute that denied to Maori the right to litigate customary title claims to the foreshore.[122] The emphasis on intergovernmental avenues for claims-making in these countries is a reflection of the politically strategic choices of indigenous peoples, as well as traditions of judicial deference to the executive in matters involving indigenous interests, and the acknowledged limitation of non-discrimination norms in the resolution of disputes requiring the redistribution of property and expenditure of public resources.

### 4. *The Inter-American Court Methodology: Equal Recognition of Property Rights*

To this body of domestic law and practice the IACtHR and the Inter-American Commission on Human Rights (IAComHR) have contributed an extensive and growing jurisprudence on indigenous and tribal property rights, predominantly with respect to groups in Latin America. The IACtHR has developed a judicial methodology by which to reconcile and combine the equality and property protections of the ACHR in order to direct states to formalise indigenous and tribal rights to traditional lands. In a series of cases decided in the last fifteen years, the IACtHR and IAComHR have found that the failure of state parties to recognise indigenous customary property rights has breached the non-discrimination and equality provisions of the ACHR (Articles 1(1) and 2), read

[120] See, for example, *Delgamuukw v British Columbia* [1997] 3 SCR 1010 [186]: 'Ultimately, it is through negotiated settlements, with good faith and give and take on all sides, reinforced by the judgments of this Court, that we will achieve what … [is] a basic purpose of s 35(1) – "the reconciliation of the pre-existence of aboriginal societies with the sovereignty of the Crown".' For an example of a successfully litigated aboriginal title claim (the first in Canada), see *Tsilhqot'in Nation v British Columbia* 2014 SCC 44.

[121] New Zealand Bill of Rights Act 1990 (NZ) s 20; Human Rights Act 1993 (NZ) s 21(1)(f). See Fleur Adcock, 'Maori and the Bill of Rights Act: A Case of Missed Opportunities?' (2013) 11 *New Zealand Journal of Public and International Law* 183, 199.

[122] UN Committee on the Elimination of Racial Discrimination, *Decision 1(66): New Zealand Foreshore and Seabed Act 2004*, UN Doc CERD/C/66/NZL/Dec. 1 (27 April 2004) ('*CERD Committee, Decision 1(66)*') para 6.

in conjunction with its Article 21 property guarantee. Article 21 provides in relevant part that 'everyone has the right to the use and enjoyment of his property'.[123] The link between the two parts of the ACHR is summarised by the IACtHR as follows:

> [P]rotection of property under Article 21 of the Convention, read in conjunction with Articles 1(1) and 2 of said instrument, places upon States a positive obligation to adopt special measures that guarantee members of indigenous and tribal peoples the full and equal exercise of their right to the territories they have traditionally used and occupied.[124]

According to the IACtHR, then, the combination of the equality and property protections in the ACHR means that states may not deny the legal force of indigenous property rights to traditionally owned and occupied land where those rights are held in accordance with customary law.[125] In effect, once the IACtHR has deemed the claimed rights to be 'property' in the terms of Article 21, it can further hold that the state cannot deny their legal status, even if those rights are not otherwise given legal effect in domestic law, and even if they have been regarded in state law as 'de facto' or 'informal' property rights only.[126] The novelty of this approach is that it entails the assertion by an international human rights body of the jurisdiction to determine the existence and legal effect of property rights not recognised in state law, as part of assessing a state's performance of its international human rights obligations. In this way, by deciding that customary law generates property rights in state law, the IACtHR has performed an act of recognition equivalent in substance to that undertaken by the Australian High Court in *Mabo (No 2)*.

What might this add to the human rights–based approach currently deployed by the UN-REDD Programme? Importantly for the purposes of this chapter, until recently the jurisprudence of the IACtHR was focused

[123] *Saramaka People v Suriname*, IACHR Series C No 172 (28 November 2007) [110]. See also, *Kuna Indigenous People of Madungandi and Embera Indigenous People of Bayano and Their Members v Panama*, IACHR Report No 125/12, Merits (13 November 2012) 196; *Maya Indigenous Communities of the Toledo District v Belize*, IACHR Report No 40/04, OEA/Ser.L/V/II.122 Doc 5 Rev 1 at 727 (12 October 2004) [193].

[124] *Saramaka v Suriname* IACHR Series C No 172 (28 November 2007) [91].

[125] Ibid. [110]: 'This limitation on the recognition of the legal right of the members of the Saramaka people to fully enjoy the territory they have traditionally owned and occupied is incompatible with the State's obligations under Article 2 of the Convention to give legal effect to the rights recognized under Article 21 of such instrument'.

[126] Ibid. [110].

on the current use and possessory rights of indigenous and tribal peoples, albeit use and possession in accordance with customary law. It has evolved more recently, however, to accommodate the rights of indigenous peoples who are no longer in possession of the lands they claim. This body of law could assist in giving content to the concept of 'legitimacy' that is to trigger recognition in the terms used by the FAO and the UN-REDD Programme where indigenous property rights are not recognised by the state, including in circumstances where claimants have been dispossessed.

The landmark 2001 case of *Mayagna v Nicaragua*[127] addressed the rights of an indigenous community that was in possession of the lands it claimed. Later cases have addressed the claims of indigenous or tribal groups to lands they no longer occupy, in two situations: first, where title to those lands has not been granted to third parties, so the land is effectively state owned (*Moiwana Community v Suriname*),[128] and second, where some of the claimed lands are now held by third parties as private property (*Yakye Axa v Paraguay* and *Sawhoyamaxa v Paraguay*).[129] In situations where formal rights are held over indigenous lands by third parties, the IACtHR has confined itself to acknowledging the need for restitution *of some kind*, including lands in kind or monetary compensation, and has urged procedural fairness in the dispute-resolution processes associated with those claims.[130] In other words, the IACtHR has left decisions on the allocation of competing rights to the state, but has articulated procedural considerations that the state must take into account in the course of such allocation, if it is to comply with its human rights obligations in the terms of the ACHR.

In *Mayagna v Nicaragua*,[131] the IACtHR found that the failure of the Nicaraguan state to 'delimit and demarcate' the lands *possessed* by the Mayagna (Sumo) Awas Tingni people 'violated the right to property protected by Article 21 of the American Convention on Human Rights,

[127] *Mayagna (Sumo) Awas Tingni Community v Nicaragua*, IACHR Series C No 79 (31 August 2001).

[128] *Moiwana Community v Suriname*, IACHR Series C No 124 (15 June 2005).

[129] *Yakye Axa v Paraguay*, IACHR Series C No 125 (17 June 2005) ('*Yakye Axa v Paraguay*'); *Sawhoyamaxa Indigenous Community v Paraguay*, IACHR Series C No 146 (29 March 2006).

[130] 'However, merely abstract or juridical recognition of indigenous lands, territories, or resources, is practically meaningless if the property is not physically delimited and established': *Yakye Axa v Paraguay* IACHR Series C No 125 (17 June 2005) [143].

[131] *Mayagna (Sumo) Awas Tingni Community v Nicaragua*, IACHR Series C No 79 (31 August 2001).

to the detriment of the members of the Mayagna (Sumo) Awas Tingni Community, in connection with articles 1(1) and 2 of the Convention'.[132] According to the IAComHR, the case stands for the proposition that:

> States violate the right to equality before the law, equal protection of the law, and non-discrimination when ... they do not grant indigenous peoples 'the protections necessary to exercise their right to property fully and equally with other members of the ... population'.[133]

The crucial element in this reasoning (to borrow again from the FAO's direction to states to recognise 'legitimate tenure rights') is that the asserted rights were 'legitimate' because the indigenous community in question possessed the land, and possessed it in accordance with customary law. In these circumstances, and 'as a result of customary practices, possession of the land should suffice for indigenous communities lacking real title to property of the land to obtain official recognition of that property, and for consequent registration'.[134] Accordingly then,

> traditional possession of their lands by indigenous people has equivalent effects to those of a state-granted full property title; [and] traditional possession entitles indigenous people to demand official recognition and registration of property title.[135]

In a series of cases decided after 2004, however, the IACtHR has been asked to investigate the question of 'whether possession of the lands by the indigenous people is a requisite for official recognition of property title thereto'.[136] That is, it was asked to identify the attributes of 'legitimate tenure' when an indigenous or tribal community is no longer in possession of the land in

[132] Ibid. [173(2)]. See also *Saramaka People v Suriname* IACHR Series C No 172 (28 November 2007) [91].

[133] *Kuna Indigenous People of Madungandi and Embera Indigenous People of Bayano and Their Members v Panama*, IACHR Report No 58/09, Petition 12.354, Admissibility (21 April 2009) [303]; see also [305]; *Maya Indigenous Communities of the Toledo District v Belize*, IACHR Report No 40/04, OEA/Ser.L/V/II.122 Doc 5 Rev 1 at 727 (12 October 2004) [61], [171].

[134] *Mayagna (Sumo) Awas Tingni Community v Nicaragua*, IACHR Series C No 79 (31 August 2001) [151].

[135] *Sawhoyamaxa Indigenous Community, v Paraguay* IACHR Series C No 146 (29 March 2006) [128]. See also, *Yayke Axa v Paraguay*, IACHR Series C No 125 (17 June 2005) ('*Yayke Axa v Paraguay*') [215], and see [143]: 'a strictly juridical or abstract recognition of indigenous lands, territories or resources lacks true meaning where the property has not been physically established and delimited'.

[136] *Sawhoyamaxa Indigenous Community v Paraguay* IACHR Series C No 146 (29 March 2006) [126]. *Moiwana Community v Suriname* IACHR Series C No 124 (15 June 2005) [134]; *Yayke Axa v Paraguay* IACHR Series C No 125 (17 June 2005) [124]–[131].

question. In the 2005 case of *Moiwana v Suriname*, the IACtHR found that the members of a dispossessed tribal community were 'legitimate owners of their traditional lands' and that their 'traditional occupancy' of the claimed land 'should suffice to obtain State recognition of their ownership'.[137] In that case, the community had been forcibly dispossessed and the claimed land had not been allocated by the state to third parties. In the 2006 case of *Sawhoyamaxa v Paraguay*, the IACtHR further extended its methodology on recognition, holding that the Sawhoyamaxa people had the right to claim the restitution of traditional lands which they no longer possessed, even though some of the land had been granted to third parties,[138] noting that the fact that these lands were in private ownership 'is not in itself an "objective and sufficient ground" barring restitution thereof'.[139] Significantly, the IACtHR articulated a 'connection' test that could allow title to survive even when possession had been lost. Since the unique relationship between indigenous peoples and the land is the 'spiritual and material basis for indigenous identity', according to the IACtHR, the right to claim traditional lands is enforceable 'as long as said relationship exists'.[140] The unique relationship may be expressed by, inter alia: 'spiritual or ceremonial ties; settlements or sporadic cultivation; seasonal or nomadic gathering, hunting and fishing; the use of natural resources associated with their customs and any other element characterizing their culture'.[141] When considering the circumstances in which third-party title could be redistributed to the indigenous claimants, however, the IACtHR drew a distinction between its mandate to make determinations on state compliance with human rights, and the exclusive jurisdiction of the state to make decisions concerning the allocation of property rights, especially in cases involving competing claims:

> [T]he Court cannot decide that Sawhoyamaxa Community's property rights to traditional lands prevail over the right to property of private owners or vice versa, since the Court is not a domestic judicial authority with jurisdiction to decide disputes among private parties. This power is vested exclusively in the Paraguayan State. Nevertheless, the Court has competence to analyse whether the State ensured the human rights of the members of the Sawhoyamaxa Community.[142]

Accordingly, in this case, in order to comply with the ACHR's equality and property protections, the state was required to grant rights to the

[137] *Moiwana Community v Suriname* IACHR Series C No 124 (15 June 2005) [133], [134].
[138] *Sawhoyamaxa Indigenous Community v Paraguay* IACHR Series C No 172 (28 November 2007) [130].
[139] Ibid. [214]. [140] Ibid. [130]. [141] Ibid. [131]. [142] Ibid. [136].

claimed lands or *suitable alternative lands* to the Sawhoyamaxa within three years of the judgment 'irrespective of whether [those rights are] acquired by purchase or by condemnation'.[143] To comply with the order, the state was obliged to 'consider the possibility of purchasing these lands or the lawfulness, need and proportionality of condemning these lands in order to achieve a lawful purpose in a democratic society', and, the IACtHR noted, that if

> restitution of ancestral lands to the members of the Sawhoyamaxa Community is not possible on objective and sufficient grounds, the State shall make over alternative lands, selected upon agreement with the aforementioned Indigenous Community, in accordance with the community's own decision-making and consultation procedures, values, practices and customs.[144]

Thus the allocation of customary lands to third parties did not 'extinguish' the customary title of indigenous peoples, providing that those people have maintained their unique 'spiritual and material' relationship to the land in question.

Finally, in the 2005 case of *Yake Axa v Paraguay*, also involving the claims of a group no longer in possession of their traditional land, the IACtHR outlined the criteria to be addressed 'when indigenous communal property and individual private property are in real or apparent contradiction'[145] so that the property rights of one or the other must be adjusted, finding that

> the American Convention itself and the jurisprudence of the Court provide guidelines to establish admissible restrictions to the enjoyment and exercise of rights, that is: a) they must be established by law; b) they must be necessary; c) they must be proportional; and d) their purpose must be to attain a legitimate goal in a democratic society.[146]

In addition, whatever balance is struck in the case of indigenous claims to traditional land, it must recognise the special significance of land as a basis of indigenous cultural identity, and the control of traditional territory 'as a necessary condition for reproduction of their culture, for their own development and to carry out their life aspirations'.[147] Accordingly, while indigenous property rights need not always take priority over private third-party rights, where there are 'concrete and justified reasons'

[143] Ibid. [215]. [144] Ibid. [212]; see also [214].
[145] *Yayke Axa v Paraguay* IACHR Series C No 125 (17 June 2005) [144].
[146] Ibid. [144]. [147] Ibid. [146].

preventing the state from returning traditional lands to the indigenous community, third-party rights may legitimately be preserved, in which case compensation must be provided to indigenous peoples, in a manner and of a type 'primarily guided by the meaning of the land for them'.[148]

The IACtHR's approach to indigenous prior possessory claims can be summarised as follows. In the absence of evidence of current occupation and use, prior occupancy of the claimed land can suffice to engage the state's obligation to give legal effect to the claimed property rights. Where the land in question has not been allocated to third parties, traditional occupancy in accordance with customary law renders those rights 'legitimate' and deserving of legal protection. Where formal property rights to traditional lands have been vested in third parties, and indigenous claimants have maintained their unique spiritual and material relationship to those lands, their rights are prima facie legitimate and must be given legal effect. In those circumstances, where both third-party and indigenous rights are 'legitimate', the IACtHR will not itself decide on how disputes between those rights are to be resolved. It will, however, find a state to be in breach of the equality and property protections of the ACHR if that state limits the 'legitimate' rights of either party to the dispute, unless such limits are established by law, necessary, proportional, designed to attain a legitimate goal in a democratic society and recognise the special significance of land as a basis of indigenous cultural identity. A refusal to give priority to indigenous rights to traditional lands where there are competing third-party rights over the same land will be in compliance with the ACHR only if there are 'concrete and justified' reasons for that refusal, in which case compensation must be provided in lieu of land in kind or property rights. In summary, the IACtHR has provided an expanded, much more subtle and balanced example of the way that the UN-REDD Programme's human rights–based approach could be brought to bear on the recognition of indigenous rights to forested land, one that deals not just with protection, but also recognition, and which encompasses claims based on prior possession as well as current use.

## D. Conclusion

This chapter has addressed the consideration given to date to indigenous tenure issues in the evolving international discourse on REDD+. It has

[148] Ibid. [149].

pointed out that tenure issues have to date only tentatively been addressed by the two major REDD+ funding bodies – the UN-REDD Programme and the FCPF – and then primarily in criteria used to assess applications for REDD+ Readiness funding, rather than in substantive operational guidance. The World Bank's operational policies do not ordinarily apply directly during the FCPF's funding of REDD Readiness preparations, so that the FAO Guidelines, as endorsed by the UN-REDD Programme, provide the most substantive advice to states to date on how to progress tenure reform in preparation for REDD+ implementation. We have suggested that these guidelines may have the effect of disadvantaging indigenous peoples relative to other property holders because of their emphasis on the current-use and food-security needs of persons and communities occupying or using forested land. Since indigenous claims are more likely than others to be premised on prior possession in accordance with customary law, and because the historic non-recognition of customary law means that those communities are more likely than others to have been dispossessed, an emphasis on current occupation does not capture the circumstances in which many indigenous communities find themselves. Accordingly, the emphasis in REDD+ policy and guidance on the need for 'secure and clear' tenure, and a 'human rights–based' approach to programming, combine to raise the possibility that REDD+ could incentivise states to design tenure arrangements that deny the legal effect of some legitimate indigenous customary rights.

We have suggested that the human rights–based approach to tenure reform so far endorsed by the UN-REDD Programme and FAO is unlikely to provide a way forward on its own in contexts where title and possession are contested, because it imports concepts of racial discrimination that can frustrate the allocation of property rights to indigenous peoples or diminish their extent and duration. The Australian example shows the shortcomings of methods that address indigenous property claims through the lens of human rights law, because this body of law does not contain a justification for the recognition of customary title (although it can encourage such an act of recognition), and is directed instead to the equal protection of property rights *once recognised*. The jurisprudence of the IACtHR, however, shows a promising way forward, since it has directed states to recognise and protect indigenous title based on customary law in fulfilment of the ACHR's property and equality protections, and to resolve disputes with due account of the significance of indigenous ties to traditional land. While this jurisprudence depends for its implementation on state acceptance of the IACtHR's findings on the 'existence' of indigenous property rights as de jure rather than de facto rights, it provides a methodology which mitigates the possibility that recognition could constitute racial discrimination. This set of judicial experiences and findings could usefully inform the development of advice on tenure reform in the REDD+ context, and perhaps help address the bias in existing commentary towards protecting the current users of land by providing a justification for the claims of indigenous peoples who assert title based on prior or historic possession in accordance with customary law.

# 6

# Benefit-Sharing in the REDD+ Regime

## Linking Rights and Equitable Outcomes

LEE GODDEN[1]

The Paris Conference of the Parties to the UNFCCC and its resultant Paris Agreement[2] highlighted that the intersections between the REDD+ regime and the achievement of human rights for indigenous peoples and local forest-dependent communities are of critical importance.[3] As Chapters 4 and 5 demonstrated, there has been progressive recognition of the pressing need for safeguards for these communities, together with calls for the realisation of robust tenure protections and tangible benefits for forest-dependent communities most affected by the implementation of the UN-REDD Programme.[4] The broader concept of benefit-sharing has arisen as a means of providing more defined content to claims by local forest communities and indigenous peoples for equitable participation in the REDD+ regime.[5] The text of the Paris Agreement refers to 'non-carbon benefits', which include the protection of biodiversity and human rights.[6] The indeterminacy as to the composition and character of such

[1] The research and scholarship contributions of Dr Kathleen Birrell to this chapter are gratefully acknowledged.

[2] Conference of the Parties, United Nations Framework Convention on Climate Change, *Adoption of the Paris Agreement: Decision -/CP.21*, UN Doc FCCC/CP/2015/L.9/Rev.1 (29 January 2016) annex, http://unfccc.int/resource/docs/2015/cop21/eng/10a01.pdf.

[3] Stephen Humphreys, 'Editorial: In the Shadow of Paris: Theories of Justice and Principles of Harm' (2016) 7 *Journal of Human Rights and the Environment* 1, 1.

[4] See, for instance, statements from the Indigenous Peoples' Forum on Climate Change (COP 22, 2016): International Indigenous Peoples' Forum on Climate Change, *Resources – 2016: COP 22*, www.iipfcc.org/resources.

[5] REDD+ may be now considered a regime at international law: see Chapters 1–3 and Chapter 9. See also discussion in Margaret A Young, 'REDD+ and Interacting Legal Regimes' in Christina Voigt (ed), *Research Handbook on REDD+ and International Law* (Edward Elgar, 2016) 89, 91, citing Stephen D Krasner, 'Structural Causes and Regime Consequences: Regimes as Intervening Variables' in Stephen D Krasner (ed), *International Regimes* (Cornell University Press, 1983) 1.

[6] *Paris Agreement*, UN Doc FCCC/CP/2015/L.9/Rev.1.

benefits in the REDD+ regime, however, renders their implementation at a local level vulnerable to global political and sectional interests, multinational and national corporate requirements, as well as sovereign nation-state interests, all of which may not coincide with the interests of local forest-dependent communities.[7] In short, it is argued, there exists a disjunction between 'the imperfect democracy of global governance and its implementation nationally'.[8] Benefit-sharing therefore is a critical nexus between the emergent REDD+ regime and the national and local spheres within countries, as benefits are realised most directly at the latter two scales. This chapter examines the interlinking of scales for the implementation of benefit-sharing, while probing the general nature of benefits and benefit-sharing arrangements, as a platform for the more detailed country-based case studies in Chapters 7, 8 and 9. By engaging in a critical examination of the concept of benefit-sharing in REDD+, this chapter also examines how constructs, aligned to deliberative forms of governance such as justice, equity and fairness, are made manifest.

The progressive development of REDD+ and associated schemes over the last decade has been accompanied by increasing concern for the rights of indigenous and other forest-dwelling and forest–dependent communities.[9] In recent years these communities, alongside civil society groups, have advocated for protection of communities and benefit-sharing in REDD+ to redress the impacts of the schemes upon communities, and increasingly, to situate REDD+ projects as a viable alternative to other forms of livelihood for forest-dependent communities. The initial language internationally was that of 'co-benefits', which captured early concerns regarding biodiversity impacts. The UN-REDD Programme, while now acknowledging the need for 'benefits' to include indigenous peoples and forest-dependent communities, still adopts

[7] See, for instance, *People's Agreement of Cochabamba* (24 April 2010), World People's Conference on Climate Change and the Rights of Mother Earth, http://pwccc.wordpress.com/2010/04/24/peoples-agreement. This agreement explicitly condemns REDD and its iterations. See also Kristen Evans, Laura Murphy and Wil de Jong, 'Global Versus Local Narratives of REDD: A Case Study from Peru's Amazon' (2014) 35 *Environmental Science & Policy* 98, 99.

[8] Ibid.

[9] See Chapters 2 and 9. For further discussion of 'justice', see International Bar Association, Climate Change Justice and Human Rights Task Force, *Achieving Justice and Human Rights in an Era of Climate Disruption* (2014) 4. At 3–4, the report acknowledges the imperative to address climate change from the perspective of justice, with a particular focus on indigenous and forest-dwelling communities.

a more constrained scope to the term than the wider concept of benefit-sharing. That concept, as advocated by civil society, seeks to address the potential marginalisation of indigenous and local forest-dwelling communities in the face of an increased commodification and globalisation of environmental governance.[10] Despite greater attention to co-benefits and/or non-carbon benefits, the articulations of 'benefits' in the international agreements for REDD+ lack definition, detail and specificity. In part, it is suggested, this is due to the sensitivity of these matters in terms of national sovereign interests.

Earlier chapters have traced how REDD+ was initially devised to allow forested, developing countries (principally in the Global South) to exchange carbon sequestration credits with developed countries (principally in the Global North) to meet emissions-reduction targets.[11] Conceived as a means of 'draw[ing] developing countries towards emission reductions',[12] the scheme sought to create 'a new economic value for tropical forests'.[13] Some commentators expressed concern that REDD+ was part of a 'wider neoliberal climate regime',[14] whereby global imperatives impact national economies

[10] For discussion of market values and environment, see Melissa Leach and Ian Scoones, 'Political Ecologies of Carbon in Africa' in Melissa Leach and Ian Scoones (eds), *Carbon Conflicts and Forest Landscapes in Africa* (Routledge, 2015) 1, 2; see also Lee Godden, '"Globalized Localisms": Three Phases of International Environmental Governance for Biodiversity Protection' in Shawkat Alam, Natalie Klein and Juliette Overland (eds), *Globalisation and the Quest for Social and Environmental Justice: The Relevance of International Law in an Evolving World Order* (Routledge, 2011) 11.

[11] Conference of the Parties, United Nations Framework Convention on Climate Change, *Report of the Conference of the Parties on Its Thirteenth Session, Held in Bali from 3 to 15 December 2007 – Addendum – Part Two: Action Taken by the Conference of the Parties at Its Thirteenth Session*, UN Doc FCCC/CP/2007/6/Add.1 (14 March 2008) Decision 1/CP.13 ('Bali Action Plan') para 1(b)(iii).

[12] Reducing Emissions from Deforestation in Developing Countries. Approaches to Stimulate Action: Submission from Parties (FCCC/CP/2005/Misc. 1, 11 November 2005) cited in Annalisa Savaresi, 'The Human Rights Dimension of REDD' (2012) 21 *Review of European Community & International Environmental Law* 102, 102.

[13] Leo Peskett, 'REDD+ and Development' in Rosemary Lyster, Catherine MacKenzie and Constance McDermott (eds), *Law, Tropical Forests and Carbon* (Cambridge University Press, 2013) 230, 230. See, for instance, *People's Agreement of Cochabamba* (24 April 2010), World People's Conference on Climate Change and the Rights of Mother Earth, http://pwccc.wordpress.com/2010/04/24/peoples-agreement. This agreement explicitly condemns REDD and its iterations.

[14] For a useful overview, see Julia Dehm, 'Indigenous Peoples and REDD+ Safeguards: Rights as Resistance or as Disciplinary Inclusion in the Green Economy?' (2016) 7 *Journal of Human Rights and the Environment* 170–217; Joanna Cabello and Tamra Gilbertson, 'A Colonial Mechanism to Enclose Lands: A Critical Review of Two REDD+-Focused Special Issues' (2012) 12 *Ephemera* 162, 164.

and local communities as a form of 'globalization of a given localism'.[15] Benefit-sharing in this context may be seen as a 'trade-off' between global climate mitigation objectives and local concerns.[16] Such benefits, while they may 'offset' local impacts by enhancement of community well-being, also may occlude from consideration the thorny issue of whether communities have clear choices about whether or not they participate in schemes.[17]

To date, international guidance offered under the UN-REDD Programme for national implementation of benefits lacks critical specificity as to precisely how such benefits are defined and will flow to indigenous and local forest-dwelling communities in the context of national implementation of REDD+ schemes. Typically, benefits may be monetary payments (results-based payments) or other forms of financial 'benefit'; with both scenarios assuming functioning markets and/or ongoing sources of finance from which these benefits are derived. Alternatively, 'non-carbon' benefits, not directly financial in orientation, may accrue to local communities as a result of their adoption of REDD+ and associated schemes. At the level of national implementation, there are unresolved issues around the willingness and capacity of participating countries to channel benefit flows to subnational groups, such as indigenous communities. Whether countries require 'direction' as to how benefits in REDD+ projects are shared, and with whom, raises sensitive issues around national sovereignty. It highlights the perennial tensions that animate the interaction of global, national and local scales of governance, as well as the multiplicity of normative regimes.

To better understand the tensions around the emergence of benefit-sharing within the REDD+ regime therefore requires examination of the analogous development of ecosystem services,[18] and the financial-transfer and capacity-building models in international environmental law.[19] In turn, this requires reflection on how adequately benefit-sharing

[15] Boaventura de Sousa Santos, *Toward a New Common Sense: Law, Science and Politics in the Paradigmatic Transition* (Routledge, 1995) 263, cited in Peter Fitzpatrick, *Modernism and the Grounds of Law* (Cambridge University Press, 2001) 189.

[16] Karin Bäckstrand and Eva Lövbrand, 'Planting Trees to Mitigate Climate Change: Contested Discourses of Market Efficiency, Planetary Carbon Control and Civic Participation' (2006) 6(1) *Global Environmental Politics* 50, 61.

[17] See Chapters 3 and 4.

[18] For discussion of the rationale for the use of ecosystem services in public policy, see J B Ruhl, 'The Twentieth Annual Lloyd K Garrison Lecture: In Defense of Ecosystem Services' (2015) 32 *Pace Environmental Law Review* 306.

[19] See Chapter 2 for more detail, and see further Alexander Zahar, Jacqueline Peel and Lee Godden, *Australian Climate Law in Global Context* (Cambridge University Press, 2013) ch 8.

can capture and allocate social and cultural benefits as part of an enlarged concept of non-carbon benefits. In this regard, the chapter focuses upon the identification and implementation of co-benefits and the sometimes fraught relationship between benefit-sharing and the 'ownership' of land, resources and carbon rights.

## A. REDD+: Defining Benefits

The Paris Agreement makes reference to existing REDD+ frameworks,[20] which establish generic standards and safeguards for national implementation. These frameworks guide 'the implement[ation] and support, including through results-based payments' of 'policy approaches', 'positive incentives', 'conservation, sustainable management of forests and enhancement of forest carbon stocks in developing countries', as well as potential 'joint mitigation and adaptation approaches for the integral and sustainable management of forests'.[21] These elements are underpinned by the imperative to 'incentiviz[e], as appropriate, non-carbon benefits associated with such approaches'.[22] Despite this text in REDD+ documentation, the international framework presently lacks a solid definition of benefits and rights, and a viable means of situating them within national regulatory frameworks in order to facilitate effective implementation. Such contextualisation, moreover, requires a sustained engagement with local concerns,[23] so that REDD+ policies and guidance developed under the UNFCCC more meaningfully connect to national and local implementation.

In this chapter, there is an examination of how national and regional governance of forest land and resources, tenure security and 'rights' in forests – including prospective carbon rights – under the REDD+ regime are linked to the identification of benefits. Indeed, benefit-sharing is

[20] See the Warsaw Framework: *Decision 9/CP.19*, 'Work Programme on Results-Based Finance to Progress the Full Implementation of the Activities Referred to in Decision 1/CP.16, Paragraph 70', UN Doc FCCC/CP/2013/10/Add. 1 (31 January 2014); *Decision 12/CP.19*, 'The Timing and the Frequency of Presentations of the Summary of Information on How All the Safeguards Referred to in Decision 1/CP.16, Appendix 1, Are Being Addressed and Respected', UN Doc FCCC/CP/2013/10/Add.1 (31 January 2014); *Decision 13/CP.19* and *Decision 15/CP.19*, 'Addressing the Drivers of Deforestation and Forest Degradation', UN Doc FCCC/CP/2013/10/Add. 1 (31 January 2014).

[21] *Paris Agreement*, UN Doc FCCC/CP/2015/L.9/Rev.1, art 5(2).

[22] Ibid.

[23] See Elisa Morgera, 'The Need for an International Legal Concept of Fair and Equitable Benefit Sharing' (2016) 27 *The European Journal of International Law* 353, 356.

premised upon a system which can identify designated individuals or communities that may hold the rights that ground the flow of benefits. Yet all too often these basic aspects are buried under a complex programme of technical requirements and elaborate procedures. Further, to ensure equitable participation in these schemes by forest communities and indigenous peoples (where those groups are identified as the rights holders), the benefits in the REDD+ model should be linked to strong legal protections, whether it be of the land, forest or the carbon-sequestration function itself. Benefit-sharing that ties distributive justice to a precise identification of legal entitlement, however, requires a facilitative consensus as to the meaning of 'ownership' or 'tenure' or 'rights'. For certain forest-based communities and indigenous peoples, the attainment of equitable benefits cannot be tied to security of tenure alone. Tenure questions enliven the issues of equity vis-à-vis third parties (such as governments) and the relevant forest-dependent community. A broader recognition of the complex web of resource and use rights, and attendant governance within a community is needed to achieve equity in terms of benefit-sharing within the members of the group, and also ultimately for effective rights protections.

### B. Benefit-Sharing: 'Fairness' and 'Equity'

The imperatives of distributive justice in benefit-sharing for local and forest-dwelling communities is brought more sharply into focus as the REDD+ regime moves into the national implementation and pilot project phases. Benefit-sharing has been described as comprising 'shared normative elements',[24] from which a working definition might be drawn. Accordingly, rather than a defined model, the concept presents as 'a *frame*' for agreement between parties for cooperation on the basis of perceived benefits.[25] Its multi-tiered elements therefore may include 'the act of sharing, the nature of the benefits to be shared, the activities from which benefit-sharing arise, the beneficiaries, and the teleological connection with equity'.[26] This connection to equity thereby embeds benefit-sharing as integral to addressing 'historical and contemporary patterns of climate injustice'.[27] Indeed by reference to international law, Morgera contends that:

[24] Ibid. 362. [25] Ibid. 356. [26] Ibid. 362.

[27] Anna Grear, 'Towards "Climate Justice"? A Critical Reflection on Legal Subjectivity and Climate Injustice: Warning Signals, Patterned Hierarchies, Directions for Future Law and Policy' (2014) 5(0) *Journal of Human Rights and the Environment* 103, 103.

> benefit sharing should be counted among the specific principles deriving from equity as a general principle of international law, which serves to balance competing rights and interests with a view to integrating ideas of justice into a relationship regulated by international law.[28]

Even though it is stipulated that benefit-sharing must be 'equitable' or 'fair and equitable' under almost all treaties in which the construct appears, its actual scope and content is not determined for the purposes of national implementation.[29] Rather, the term is left open ended, to be identified by reference to the considerations of fairness and equity. These two expressions are given legal form, respectively, as 'the procedural dimensions of justice (fairness)' and 'the substantive dimensions of justice (equity)'.[30] Held in tension, however, the two notions may conflict in practice: 'fairness supports stability within the legal system (predictable and clear procedures), whereas equity tends towards change (recognition or enhanced realization of rights and the (re)allocation of power over resources)'.[31] This dilemma, whereby benefit-sharing is shaped by the two poles of procedural certainty and substantive change is heightened in 'market-based conceptions of distributive equity'.[32] Indeed, local forest communities and indigenous peoples, even if accorded fairness and equity in the REDD+ process, often may still be required to respond to commercialisation pressures and 'pervasive bio-power structures'.[33]

Thus the equitable realisation of benefits in REDD+ is necessarily tied to rights, both procedural and substantive, and requires the full and effective participation of relevant stakeholders, particularly indigenous and local communities,[34] in the design and implementation of REDD+ strategies and action plans.[35] Alongside procedural and normative standards for community participation,[36] these measures extend to the

[28] Morgera, above n 23, 380. [29] Ibid. [30] Ibid. 381. [31] Ibid.

[32] Chukwumerije Okereke, 'Climate Justice and the International Regime' (2010) 1 *WIREs Climate Change* 462, 471.

[33] Lee Godden, 'Death, Desire, Modernity and Redemption: Climate Change and Public International Environmental Law' (2009) 10 *Melbourne Journal of International Law* 543–578, 545.

[34] Conference of the Parties, United Nations Framework Convention on Climate Change, *Report of the Conference of the Parties on Its Sixteenth Session, Held in Cancun from 29 November to 10 December 2010*, UN Doc FCCC/CP/2010/7/Add.1 (15 March 2011), Decision 1/CP.16, Appendix I, para 2(d).

[35] See discussion at UN-REDD Programme, *Stakeholder Engagement: Engaging Indigenous Peoples and Other Forest Dependent Communities*, www.unredd.net/knowledge/redd-plus-technical-issues/stakeholder-engagement.html.

[36] See, Chapter 2 and specifically, UN-REDD Programme, 'UN-REDD Programme Social and Environmental Principles and Criteria (SEPC)' (UNREDD/PB8/2012/V/1, Asunción,

recently formulated Community-based REDD+, an initiative designed to empower and engage local communities through the direct provision of grants for 'community-level projects that complement UN-REDD National Programmes, national REDD+ readiness processes and/or strategies'.[37] This initiative aligns with increasing transnational practice around 'biocultural community protocols',[38] which operate 'at the intersection of communities' customary laws, international environmental and human rights law, national and transnational law'.[39]

Accordingly, forms of benefit-sharing ideally should be locally determined, emphasising community rights rather than the traditional property and/or financial models that privilege bounded individual rights.[40] These flexible types of measures are being adopted more regularly, but concerns about information asymmetries remain acute, especially where FPIC does not guarantee to communities any rights of non-participation in REDD+ projects.[41]

Furthermore, empirical evidence drawn from REDD+ project studies suggests that benefit-sharing often does not match stated project objectives.[42] Indeed, there is a potential for some communities to experience restricted access to forests and resources as a result of REDD+ activities.[43] Further, there is a potential disjunction between the complex indigenous and local forest community systems of tenure and resource rights, and the need for clarity in identifying potential beneficiaries of benefits from REDD+ projects. The current emphasis upon formalisation of tenure may result in perverse outcomes for those indigenous and local groups whose claims to land and resource rights on the basis of historical and cultural associations may not be legally recognised by the nation

Paraguay, 25–26 March 2012) ('UN-REDD Social and Environmental Principles and Criteria'); Forest Carbon Partnership Facility, 'Carbon Fund Methodological Framework' (22 June 2016), www.forestcarbonpartnership.org/carbon-fund-methodological-framework.

37 See UN-REDD Programme, 'Community-Based REDD+', www.unredd.net/community-based-redd.html.

38 See for instance, Daniel Robinson, *Biodiversity, Access and Benefit-Sharing: Global Case Studies* (Taylor and Francis, 2014).

39 See discussion at BeneLex, 'About the Project', www.benelex.ed.ac.uk/about_us.

40 See Kabir Sanjay Bavikatte and Tom Bennett, 'Community Stewardship: The Foundation of Biocultural Rights' (2015) 6 *Journal of Human Rights and the Environment* 7, 7–29.

41 See Chapter 5.

42 Morgera, above n 23, 356. See also Anna Bolin and Dominic Taku Tassa, 'Exploring Climate Justice for Forest Communities Engaging in REDD+: Experiences from Tanzania' (2012) 39 *Forum for Development Studies* 5, 6–7.

43 Savaresi, above n 12, 105.

state.[44] The achievement of distributive justice for indigenous and local communities in benefit-sharing therefore depends upon clear articulation of legal categories of rights, yet these categories also must be capable of accommodating the practices and understandings of indigenous peoples and local forest communities to give effect to legal principles of 'fair' and 'equitable' benefit-sharing in REDD+ policies and guidance.[45]

### *1. Ecosystem Services and Co-Benefits*

As Chapter 2 examined, the models that influenced the UN-REDD Programme and the FCPF[46] had precursors in the afforestation and reforestation projects under the CDM.[47] Similarly, many early ecosystem services projects operative in forests sought to facilitate forest conservation while achieving the benefit of poverty alleviation for forest-based communities.[48] Ecosystem-services models, the REDD+ regime and voluntary carbon markets,[49] all seek to designate and capture the economic value of environmental services, such as clean air and water that are provided to societies.[50]

Growing awareness of the potentially adverse impacts associated with the use of forests simply as 'carbon sinks' led to greater scrutiny.[51] Potential adverse impacts that were identified by communities and civil society included the destruction of biological diversity, the displacement of local communities, the appropriation of indigenous traditional knowledge, and the disruption and alteration of subsistence forest livelihoods.

[44] See Chapter 5 and Kirsty Gover, 'REDD+, Tenure and Indigenous Property: The Promise and Peril of a "Human Rights-Based Approach"' in Christina Voigt (ed), *Research Handbook on REDD+ and International Law* (Edward Elgar, 2016) 249, 269.

[45] See further in Chapters 2, 3 and 4.

[46] FAO, UNDP and UNEP, *UN Collaborative Programme on Reducing Emissions from Deforestation and Forest Degradation in Developing Countries (UN-REDD): Framework Document* (20 June 2008) ('UN-REDD Framework Document').

[47] See Chapter 2 C and Zahar, Peel and Godden, above n 19, 230.

[48] Evans et al, above n 7, 105.

[49] See discussion in Becca Madsen, Nathaniel Carroll and Kelly Moore Brands, *State of Biodiversity Markets Report: Offset and Compensation Programs Worldwide* (Forest Trends, 2010), www.ecosystemmarketplace.com/publications/state-of-biodiversity-markets.

[50] The seminal research is Robert Costanza and Ralph D'Arge, 'The Value of the World's Ecosystem Services and Natural Capital' (1997) 387 *Nature* 253.

[51] Cameron Allen and Stuart Clouth, 'A Guidebook to the Green Economy' (UN Division for Sustainable Development, August 2012); Minh Ha Hoang et al, 'Benefit Distribution Across Scales to Reduce Emissions from Deforestation and Forest Degradation (REDD+) in Vietnam' (2013) 31 *Land Use Policy* 48, 48–60.

The introduction of co-benefits, largely achieved via a 'bottom up' process, was seen to ameliorate or 'offset' impacts, and to provide positive advantages for local communities where REDD+ schemes were implemented.[52] The most tangible benefit is provision of monetary payments to forest-dwelling communities. In the voluntary carbon-trading markets, credits generated by projects that are associated with environmental and community 'non-carbon benefits' often attract a premium price. Donor governments and civil society favour programmes that promote such benefits and, in turn, these schemes typically attract higher multilateral and bilateral funding.[53]

As demonstrated in Chapter 2, the REDD+ regime has evolved rapidly, and the initial, single-purpose, ecosystem-services measure is now a technically and politically complex model.[54] The complexity that now characterises REDD+ may be attributed in part to its engagement with regimes that reflect environmental and development objectives, as outlined in Chapter 2.[55] As REDD+ evolved, there was a parallel convergence of sustainability and human rights regimes in international law that culminated in the Sustainable Development Goals.[56] Moreover, as REDD+ became more aligned to these convergent policy agendas that seek to promote better socio-economic outcomes for communities, it has necessitated consideration of benefit-sharing as integral to 'potential trade-offs between competing values'.[57] These trade-offs, though, are still to be negotiated within a market paradigm, which persists despite the collapse or removal of many national and regional emissions-trading schemes, which were designed as the underpinning economic structure to support REDD+.[58] The regime now relies more heavily on a developed

[52] On bottom-up measures, see Jacqueline Peel, Lee Godden and Rodney J Keenan, 'Climate Change Law in an Era of Multi-Level Governance' (2012) 1 *Transnational Environmental Law* 245. See also UN-REDD Framework Document, above n 46, 5.

[53] Peskett, above n 13, 246. [54] See further in Chapter 2.

[55] See further in Chapter 2, Section B.

[56] *Transforming Our World: The 2030 Agenda for Sustainable Development*, GA Res 70/1, UN GAOR, 70th sess, UN Doc A/RES/70/1 (21 October 2015).

[57] Ingrid J Visseren-Hamakers et al, 'Trade-Offs, Co-Benefits and Safeguards: Current Debates on the Breadth of REDD+' (2012) 4 *Current Opinion in Environmental Sustainability* 646, 649. See also Till Pistorius, 'From RED to REDD+: The Evolution of a Forest-Based Mitigation Approach for Developing Countries' (2012) 4 *Current Opinion in Environmental Sustainability* 638, 640.

[58] Angela Oels, 'Rendering Climate Change Governable: From Biopower to Advanced Liberal Government?' (2005) 7 *Journal of Environmental Policy & Planning* 185, 195.

country donor-funding model. Even so, indigenous peoples and forest-dwelling communities as the recipients of REDD+ benefits are effectively being mobilised within a technocratic and bureaucratic regime that reorganizes the relationship of forest value and community benefits around these market objectives.[59]

The UN-REDD Programme, with its emphasis on reordering forest spaces according to international law and administrative specifications, can be understood from a governmentality perspective whereby communities become framed as project participants. Community identities are set by reference to their capacity to receive payments or benefits.[60] The REDD+ regime prioritises a specialised economic value for forests, and it creates new proprietary categories linked to these values.[61] In turn, the regime constitutes indigenous peoples and forest-dependent communities as extracting value from that forest resource, while community well-being is to be secured by benefits and safeguards. Conflicts concerning the ways in which the economic value or benefits might be 'shared' also raise contentious policy and legal issues.[62] These value contestations may result in the UN-REDD Programme in recipient countries entering a period of stasis or may precipitate a reorganization of the regime objectives and implementation modes – all of which have direct ramifications for the efficacy of democratic experimentalism and the articulation of theories of governance.[63]

[59] This point references the construct of 'governmentality', drawing upon Foucault's historical analysis of the emergence of new sites of power and ordering of life beyond the hierarchical state: see Michel Foucault, *Power/Knowledge: Selected Interviews and Other Writings 1972–1977*, Colin Gordon (ed) (Pantheon Books, 1980), and *Security, Territory, Population: Lectures at the Collège de France, 1977–78* (Palgrave Macmillan, 2007). In Chapter 9, we discuss the potential application of experimentalist governance and the utilisation of normative, soft-law standard setting in the REDD + regime.

[60] Bäckstrand and Lövbrand, above n 16, 62–63, citing Cathleen Fogel, 'The Local, the Global and the Kyoto Protocol' in Sheila Jasanoff and Marybeth Long Martello (eds), *Earthly Politics: Local and Global in Environmental Governance* (MIT Press, 2004) 111.

[61] Benefit-sharing builds on the identification of 'property' in forest resources. For analysis of 'new' property in a new resource, see Kevin Gray, 'Property in Thin Air' (1991) 50 *Cambridge Law Journal* 252.

[62] Andrew McGregor et al, 'Practical Critique: Bridging the Gap between Critical and Practice-Oriented REDD+ Research Communities' (2014) 55 *Asia Pacific Viewpoint* 277, 277.

[63] See further Chapter 9.

## 2. *The Evolution of Benefit-Sharing*

Three phases distinguish the evolution of a non-carbon benefit model within REDD+ beyond the basic idea of payments for forest-based emission-reduction services. The first was the inclusion of non-monetary benefits, while the second phase placed increasing significance on financial and capacity-building aspects 'to generate, commoditise and allocate REDD+ benefits (both carbon and non-carbon)'.[64] As noted in Chapters 3–5, a third, overlapping phase is related to the infusion of policy objectives into the REDD+ programme from other international and transnational legal regimes.[65] This convergence and mixing of policy objectives, however, produces uncertainty as to the foundation for benefit-sharing with indigenous peoples and forest-dwelling communities. Further, the rationales for benefit-sharing often posit direct complementarity between climate-mitigation outcomes and benefits for indigenous and local forest communities.[66] While certain synergies may exist, indigenous peoples and forest-dwelling communities continue to express concern as to whether their unique cultural connection with traditional land and forest places can be adequately recognised and compensated for in benefit-sharing formulas.[67] Other concerns include a general scepticism regarding the extent to which non-carbon benefits will actually flow to forest communities from REDD+ activities.[68] Accordingly, REDD+ negotiations around benefits in international fora

[64] Sophie Chapman, Martijn Wilder and Ilona Millar, 'Defining the Legal Elements of Benefit Sharing in the Context of REDD+' (2014) 8 *Carbon & Climate Law Review* 270, 271.

[65] See discussion in Young, 'REDD+ and Interacting Legal Regimes' in Christina Voigt (ed), *Research Handbook on REDD+ and International Law* (Edward Elgar, 2016) 89.

[66] Inès Ayari, 'Commentary: The Dynamics between Indigenous Rights and Environmental Governance: A Preliminary Analysis and Focus on the Impact of Climate Change Governance through the Reducing Emissions from Deforestation and Forest Degradation (REDD) Programme' (2014) 10 *AlterNative: An International Journal of Indigenous Peoples* 81, 82.

[67] See 'The Anchorage Declaration' (Indigenous Peoples' Global Summit on Climate Change, 24 April 2009) para 2, http://unfccc.int/resource/docs/2009/smsn/ngo/168.pdf. More recently, see Chris Lang, *The Margarita Declaration on Climate Change: 'We Reject the Implementation of False Solutions to Climate Change Such As: Carbon Markets and Other Forms of Privatization and Commodification of Life'* (8 August 2014) REDD-Monitor.org, www.redd-monitor.org/2014/08/08/the-margarita-declaration.

[68] Pistorius, above n 57, 641, 642. See also, Constance L McDermott et al, 'Operationalizing Social Safeguards in REDD+: Actors, Interests and Ideas' (2012) 21 *Environmental Science and Policy* 63.

have been attended by 'long-standing, perhaps intractable, development policy challenges'.[69]

To address the concerns of indigenous peoples and forest-dependent communities, benefits were specifically introduced into the REDD+ regime.[70] As Chapter 2 discusses, the agreements concluded at COP16[71] refer to the expectation that all REDD+ activities should 'enhance other social and environmental benefits'.[72] The agreements situate the attainment of social and environmental benefits alongside moves to incentivise the protection and conservation of natural forests,[73] the encouragement of effective forest governance[74] and the adoption of safeguards.[75] This platform was updated in 2013 by COP17 (Warsaw Framework on REDD+), which seeks to ensure 'results-based actions' prior to receipt of 'results-based finance',[76] and further provides for the transfer of 'results-based payments'.[77] A 2015 Technical Paper prepared by the UN-REDD Programme provides guidance as to

[69] Savaresi, above n 12, 112.

[70] D Rey et al, *A Guide to Understanding and Implementing the UNFCCC REDD+ Safeguards* (ClientEarth, 2013).

[71] Conference of the Parties, United Nations Framework Convention on Climate Change, *Report of the Conference of the Parties on Its Sixteenth Session, Held in Cancun from 29 November to 10 December 2010 – Addendum – Part Two: Action Taken by the Conference of the Parties at Its Sixteenth Session*, UN Doc FCCC/CP/2010/7/Add.1, Decision 1/CP.16 ('The Cancun Agreements: Outcome of the Work of the Ad Hoc Working Group on Long-Term Cooperative Action under the Convention'), paras 71, 73. See further in Chapters 2, 3 and 4.

[72] *Cancun Agreements*, UN Doc FCCC/CP/2010/7/Add.1, Appendix I, para 2(e). See further in Chapters 2, 4, 5.

[73] Ibid.

[74] Sarah Marlay, 'Clarifying the Role of Non-Carbon Benefits in REDD+' (Policy Paper, Environmental Defense Fund, July 2013), http://blogs.edf.org/climatetalks/files/2013/08/Non-Carbon-Benefits-Policy-Paper.pdf; *Cancun Agreements*, UN Doc FCCC/CP/2010/7/Add.1, Decision 1/CP.16, s III(C).

[75] Ibid., Decision 1/CP.16, Appendix I, paras 2(c) and 2(d). See further Chapters 3 and 4, and discussion in Annalisa Savaresi, 'The Legal Status and Role of Safeguards' in Christina Voigt (ed), *Research Handbook on REDD+ and International Law* (Edward Elgar Publishing Limited, 2016) 126, 129. Appendix I to the *Cancun Agreements* outlines the safeguards to be promoted and supported when undertaking REDD+ activities. *Cancun Agreements*, UN Doc FCCC/CP/2010/7/Add.1, Decision 1/CP.16, Appendix I, para 2.

[76] 'Work Programme on Results-Based Finance to Progress the Full Implementation of Activities Referred to in Decision 1/CP.16, paragraph 70', above n 20, paras 3, 4.

[77] *Report of the Conferences of the Parties to Its Nineteenth Session, Held in Warsaw from 11 to 23 November 2013 – Addendum – Part Two: Action Taken by the Conference of the Parties at Its Nineteenth Session*, UN Doc FCCC/CP/2013/10/Add.1, Decision 10/CP.19, para 2.

how nation states should address mechanisms to secure land tenure in developing national strategies or action plans designed to deliver benefits to communities.[78] As noted in earlier chapters, the effective realisation of non-carbon benefits is seen as requiring a supporting complex of normative safeguards, such as transparency in governance[79] and FPIC.[80] To the extent that both safeguards and benefit-sharing increasingly are directed toward securing fairness and equity for indigenous and forest-dependent communities, a convergence in the REDD+ regime has been occurring, with explicit consolidation of the two strands in the Paris Agreement.

## 3. *Benefit-Sharing in REDD+ and Parallel Models*

Benefit-sharing is not limited to the REDD+ regime. Forms of benefit-sharing with indigenous peoples and local forest communities is well established in many transnational environmental, energy and resources sectors, as well as in certain intellectual property regimes.[81] Benefit-sharing operates at many levels, from transnational financial allocations by multinational companies to local communities, to redistributions of income within nations by means of various taxation, subsidies and infrastructure programmes for indigenous peoples and local communities.[82]

### (a) Private Law Ordering

Benefit-sharing models in environmental and resource-extraction sectors most recently have been influenced by collaborative and deliberative governance models,[83] as well as corporate social responsibility principles.

[78] See Emelyne Cheney et al, 'REDD+ Safeguards Information Systems: Practical Design Considerations' (UN-REDD Programme Technical Brief 01, October 2015) 1.

[79] *Paris Agreement*, UN Doc FCCC/CP/2015/L.9/Rev.1 art 6(2).

[80] Robert Fisher and Rosemary Lyster, 'Land and Resource Tenure: The Rights of Indigenous People and Forest Dwellers' in Rosemary Lyster, Catherine MacKenzie and Constance McDermott (eds), *Law, Tropical Forests and Carbon: The Case of REDD+* (Cambridge University Press, 2013) 187.

[81] Elisa Morgera, 'The EU and Environmental Multilateralism: The Case of Access and Benefit-Sharing and the Need for a Good-Faith Test' (2014) 16 *Cambridge Yearbook of European Legal Studies* 109.

[82] See generally Lila Barrera-Hernández et al (eds), *Sharing the Costs and Benefits of Energy and Resources Activity: Legal Change and Impact on Communities* (Oxford University Press, 2016).

[83] Elinor Ostrom, 'Polycentric Systems for Coping with Collective Action and Global Environmental Change' (2010) 20 *Global Environmental Change* 550.

These principles emphasise that corporations and organizations involved in resource extraction (including in forests) require a 'social licence to operate'.[84] These systems of private law ordering hinge around contracts and agreements, together with associated normative framework. They create complex interrelationships between private international law, multinational corporations, indigenous and local communities, and national and provincial governments. Benefit-sharing concepts centred on the UN-REDD Programme and the World Bank's FCPF have many similarities to parallel models in environmental conservation agreements that have been in place for several decades.[85]

In transnational mining and resource extraction, benefit-sharing has a very long and, at times, very troubled history.[86] Benefit-sharing arrangements typically occur as part of broader agreement making between government, industry and indigenous peoples and/or local communities.[87] Despite recent increases in the range of payments and enlargement in the scope of benefits flowing to communities in many countries across the globe, many indigenous peoples and local communities may still fail to realise positive outcomes from these agreements. International civil society and NGOs have developed quite extensive guiding principles and/or protocols for these benefit-sharing arrangements.[88] Agreement making with indigenous and local communities from which these benefits derive has largely proceeded either as a form of private international law, an assemblage of soft law practices or as mandated within national laws. International public law has proved less amenable to formalising benefit-sharing, with some notable exceptions.

### (b) The Nagoya Protocol: Access and Benefit-Sharing

Principles and protocols in public international law that encompass benefit-sharing with indigenous peoples and local communities *within*

[84] Maureen Tehan et al, 'Sharing Land and Resources: Modern Agreements and Treaties with Indigenous People in Settler States' in Marcia Langton et al (eds), *Settling with Indigenous People* (Federation Press 2006) 1.

[85] Philippe Sands and Jacqueline Peel, *Principles of International Environmental Law* (Cambridge University Press, 3rd edn, 2012) 665.

[86] Marcia Langton and Judy Longbottom (eds), *Community Futures, Legal Architecture: Foundations for Indigenous Peoples in the Global Mining Boom* (Routledge, 2012).

[87] Lee Godden and Lily O'Neill, 'Benefit Agreements with Indigenous Communities: The Native Title Act in Australia' in Lila Barrera-Hernández et al (eds), *Sharing the Costs and Benefits of Energy and Resource Activity: Legal Change and Impact on Communities* (Oxford University Press, 2016) 135.

[88] Chapman et al, above n 64, 272.

nations are not typical. As noted in Chapter 3, a prominent exception is the access and benefit-sharing arrangements under the Nagoya Protocol to the *Convention on Biological Diversity* (CBD).[89] The CBD requires that, subject to national legislation, a state party shall respect, preserve, and maintain the traditional knowledge and ways of life of indigenous peoples and local communities when accessing genetic resources.[90] The protocol established rules governing access to genetic resources in situ (i.e. within sovereign national territory) and for the distribution of benefits to indigenous communities,[91] that is, as a form of 'biocultural right' for communities. The access and benefit-sharing model is particularly relevant to the REDD+ regime as many genetic resources are located within forests.

The protocol was devised amid concerns that multinational corporations could obtain genetic resources for commercial use and develop patents without the provision of benefits to 'provider' countries. Indigenous representatives also raised concerns about the uncompensated use of traditional knowledge in the identification of and access to genetic resources.[92] The protocol requires nation-state parties to take legislative, administrative and policy measures to ensure that indigenous and local communities receive fair and equitable benefits from the utilisation of genetic resources.[93] It encourages parties to adopt codes of conduct, guidelines and best practice in relation to access to genetic resources and benefit-sharing.[94] Robinson has found, through detailed case studies ranging from Africa to Oceania, that the implementation of access and benefit-sharing agreements is often patchy. Often there is limited national compliance in adopting protocols and difficulties in ensuring that corporations comply when pursuing patents.[95] Thus,

[89] See *Nagoya Protocol on Access to Genetic Resources and the Fair and Equitable Sharing of Benefits Arising from Their Utilization to the Convention on Biological Diversity*, opened for signature 2 February 2011, UN Doc UNEP/CBD/COP/DEC/X/1, annex I (entered into force 12 October 2014) art 1 ('Objective').

[90] *Convention on Biological Diversity*, opened for signature 5 June 1992, 1760 UNTS 79 (entered into force 29 December 1993) art 8(j). For discussion of the benefit-sharing arrangements, see Sands and Peel, above n 85, 456–459.

[91] Elisa Morgera, Matthias Buck and Elsa Tsioumani, *The 2010 Nagoya Protocol on Access and Benefit-Sharing in Perspective: Implications for International Law and Implementation Challenges* (Brill Nijhoff, 2012).

[92] Sands and Peel, above n 85, 464.

[93] *Nagoya Protocol*, UN Doc UNEP/CBD/COP/DEC/X/1, art 5(2).

[94] *Nagoya Protocol*, UN Doc UNEP/CBD/COP/DEC/X/1, art 20.

[95] Robinson, above n 38.

despite the existence of the protocol as an innovative sharing arrangement, significant inequalities remain between those organizations seeking access and those seeking benefit-sharing,[96] signalling a note of caution for the prospects of equitable benefit-sharing in the REDD+ regime.

### 4. *Non-Carbon Benefits in REDD+*

Benefit-sharing within the UN-REDD Programme and the FCPF, while exhibiting features common to many international and transnational ecosystem-service models, has characteristics that are distinctive to the climate change mitigation regime. Groups advocating effective cultural and economic protections for indigenous peoples and local forest-dwelling communities in the REDD+ context consistently urge the development of a clearly articulated but specific framework for benefit-sharing within the regime.[97] The REDD+ regime, though, remains 'open to interpretation by different actors and valuable to each for different reasons',[98] including in relation to defining non-carbon benefits. The gradual transition to regard in situ REDD+ activities as a source of shared benefits therefore has necessitated a complex reframing of regime objectives.[99] This progressive reframing is clearly evident in the international guidance developed for countries that are implementing REDD+. Progressively, the modes of guidance have come to focus more upon equity and socio-ecological outcomes than technical forest evaluation and mapping, monitoring and reporting requirements – although these remain prominent.

At a practical level, the achievement of equitable and effective outcomes in the REDD+ regime for local communities will depend upon the

[96] Morgera, above n 81, 129.

[97] Anne M Larson, 'Forest Tenure Reform in the Age of Climate Change: Lessons for REDD +' (2011) 21 *Global Environmental Change* 540, 541. See also James Anaya, *Report of the Special Rapporteur on the Rights of Indigenous Peoples*, UN Doc A/HRC/21/47 (6 July 2012) paras 52, 62–71.

[98] Harro van Asselt and Constance L McDermott, 'The Institutional Complex for REDD+: A "Benevolent Jigsaw"?' in Christina Voigt (ed), *Research Handbook on REDD+ and International Law* (Edward Elgar, 2016), citing Constance L McDermott et al, 'Operationalizing Social Safeguards in REDD+: Actors, Interests and Ideas' (2012) 21 *Environmental Science and Policy* 63, 64.

[99] Melissa Leach and Ian Scoones, 'Political Ecologies of Carbon in Africa' in Melissa Leach and Ian Scoones, *Carbon Conflicts and Forest Landscapes in Africa* (Routledge, 2015) 1, 3–4.

design and implementation of its benefit-sharing mechanisms, and these measures will need to operate across multiple levels of governance.[100] The initial strategic-planning phase is the point at which many of these strategic and design choices will be made.

## C. REDD+ 'Readiness' and Benefit-Sharing

The readiness phase in the REDD+ regime is also the point at which tensions may surface in respect of the need to achieve equitable benefit-sharing for indigenous peoples and local forest communities while simultaneously maintaining integrity for nation-state sovereignty. These potential tensions underscore the complex legal challenges involved.[101] Moreover, significant policy choices around the identification of benefits, and their allocation and distribution at a national level, can be obscured by the discourse of readiness. The policy choices involved are illuminated when considering what is comprised by 'readiness'.[102] The preparation for REDD+ governance by participant countries includes stakeholder participation and consultation processes, the need to determine clear and secure land and forest tenure rights, conflict resolution measures, transparent and accountable systems to manage revenues, transparent and comprehensive systems for non-carbon monitoring, institutional coordination and policy coherence across sectors affecting forests, institutional coordination across levels of government for forest management and mechanisms for equitable benefit distribution.[103]

Similarly, guidance on strategic planning and benefit-sharing has been developed by the FCPF. As Chapter 2 noted, the fund requires countries to have a benefit-sharing plan when applying to its carbon fund,[104] but its composition may be flexible at the readiness phase. In the voluntary carbon market, there are relevant standards for project development – for example, the Verified Carbon Standard, and the Climate, Community

[100] Pham Thu Thuy et al, 'Approaches to Benefit Sharing: A Preliminary Comparative Analysis of 13 REDD+ Countries' (Working Paper 108, Center for International Forestry Research, 2013).

[101] For example, *Cancun Agreements*, UN Doc FCCC/CP/2010/7/Add.1, Decision 1/CP.16, para 72.

[102] Chapter 2 has more fully described the readiness elements.

[103] Lauren Goers Williams, 'Putting the Pieces Together for Good Governance of REDD+: An Analysis of 32 REDD+ Country Readiness Proposals' (Working Paper, World Resources Institute, March 2013).

[104] See FCPF, Carbon Fund Methodological Framework, above n 36.

and Biodiversity Standards that, in effect, accredit particular benefit-sharing arrangements. The FCPF and the UN-REDD Programme also support countries in achieving 'readiness' by formulating initial strategy proposals, developing governance institutions to manage programmes, and building capacity to implement REDD+ activities.[105]

Accordingly, many participant countries have introduced policy platforms and legal reforms, including reforms to land law, in the 'readiness' phase. Participant nations must be judged by reference to centralised criteria as sufficiently 'ready' to receive 'benefits', notwithstanding the global climate mitigation outcomes being pursued. Pilot projects also have proliferated, some in advance of full readiness preparation, and at times despite unresolved land-tenure matters and, in some instances, despite a lack of meaningful consultative measures.[106] Further, the 'vague and hortatory' language of many benefit-sharing instruments[107] may constrain the realisation of tangible benefits.

## D. Defining Benefits Independently of Context

As noted, there is no commonly agreed definition of 'benefit' under the REDD+ frameworks. The uncertainty is compounded by the proliferation of terms over time: 'co-benefits', 'multiple benefits', 'non-carbon benefits' or 'non-monetary benefits',[108] with more precise definitions now considered to be specific to national and subnational contexts.[109] Non-carbon benefits arising from REDD+ schemes may now include not only direct payments to affected communities, but also wealth creation, population security, and individual or community empowerment.[110] Related environmental benefits may comprise improved environmental sustainability, such as water

[105] Matthew Sommerville, 'Land Tenure and REDD+: Risks to Property Rights and Opportunities for Economic Growth' (Issue Brief, USAID, February 2013).

[106] David Takacs, 'Environmental Democracy and Forest Carbon (REDD+)' (2014) 44 *Environmental Law* 71.

[107] Savaresi, above n 12, 112.

[108] Chapman, Wilder and Millar, above n 64, 271.

[109] See *Report of the Conference of the Parties on Its Twenty-First Session, Held in Paris from 30 November to 13 December 2015*, UN Doc FCCC/CP/2015/10/Add.3 (29 January 2016) Decision 18/CP 21. This decision, for example, explicitly states that non-carbon benefits cannot be universally defined: '*Recognizes* that non-carbon benefits associated with the activities referred to in decision 1/CP.16, paragraph 70, are unique to countries' national circumstances, in accordance with national sovereignty, legislation, policies and priorities'.

[110] Kathleen Lawlor et al, 'Community Participation and Benefits in REDD+: A Review of Initial Outcomes and Lessons' (2013) 4 *Forests* 296, 298.

regulation and erosion control. Relevant governance benefits can extend to security of land tenure, increased transparency in government and local participation in the management of forest resources.[111] In short, REDD+ 'benefits' have come to function as a new nexus of green governance. While the governance scope of REDD+ has expanded, the proposition that REDD + schemes alone can generate benefits *independently* of in situ local economies and social systems, however, is challenged by much research on REDD+ implementation.

## E. Benefit-Sharing Arrangements

Specific arrangements for benefit-sharing are not set out in the UN-REDD Programme frameworks.[112] Thus, different benefit-sharing arrangements exist in REDD+ participant countries, reflecting the diverse views about the intended outcomes of benefit-sharing and divergent bases for sharing.[113] If we examine the possible rationales for benefit-sharing, several significant questions arise, for instance: Is benefit-sharing to be premised only upon the identification of 'legal' or 'formalised rights and interests'? Is its purpose to compensate vulnerable communities in forests? Does it reward sustainable forest management irrespective of legal rights? Are benefits to be measured by reference to economic-efficiency outcomes?[114] Each rationale may require a different benefit-sharing arrangement in response. Nonetheless, the specific allocation of benefits to indigenous peoples and local communities is most frequently linked to a clear identification of the legal interests involved. This linkage reinforces the importance of securing forest tenure and forest rights for communities as an important first step.[115] Furthermore, experience with the implementation of REDD+ projects suggests that,

[111] Marlay, above n 74, 1.

[112] See 'Methodological Issues Related to Non-Carbon Benefits Resulting from the Implementation of the Activities Referred to in Decision 1/CP.16, paragraph 70' above n 109.

[113] Thuy et al, above n 100, 7–15. See also Lea Fortmann et al, 'Incentive Contracts for Environmental Services and Their Potential in REDD' (Policy Research Working Paper 6829, World Bank, April 2014).

[114] Ibid. 10.

[115] Alain Karsenty, Aurélie Vogel and Frédéric Castell, '"Carbon Rights", REDD+ and Payments for Environmental Services' (2014) 35 *Environmental Science & Policy* 20, 27: the authors note that reforming land-tenure codes to allow individuals, families and communities to claim collective tenure rights on the land and the trees is vital to fairness in REDD+. See also generally Chapter 5 and, in respect of specific case studies, see Chapters 7–9.

even in one project, a multiplicity of rationales and types of benefit-sharing arrangements is typical.[116]

A typology of benefit-sharing mechanisms that have been implemented in the forest sector identifies six options: fund-based approaches, forest-concession agreements, market-based (performance-based) instruments, access and benefit-sharing, community-based natural resource management and joint forest management.[117] Many countries with substantial forests have adopted a range of these measures prior to being involved with REDD+.[118] Further, REDD+ participant countries are able to choose the manner in which legal interests in the schemes (including 'carbon rights') are designated under national laws, as well as the management of benefit-sharing within national programmes.[119] Thuy et al. suggest that there is significant uncertainty in the various models studied in their research regarding who is eligible for benefits from REDD+.[120] They also contend that, while REDD+ may deliver benefits to communities, project implementation is often suboptimal due to 'conflicting legal provisions, overlapping mandates and inconsistent implementation among government agencies, weak law enforcement, limited funding and staffing and lack of transparency, unchecked corruption and elite capture'.[121]

These findings align with research on case studies of benefit-sharing across forest-resource sectors that reveal mixed outcomes for the indigenous peoples and local communities involved. In interviews conducted with local communities to examine REDD+ project benefits in Eastern Brasilia, Cromberg et al. found that there was a lack of confidence in direct cash payments, and that communities expressed a preference for investment in improved production systems, which value local practices.[122] The interviews revealed that community members raised questions about the value and equity of proposed payments.[123] An important factor in securing equitable benefits was increased local community capacity and participation in forest

[116] For an analysis of REDD+ as efficient 'economic' payments, see Karsenty, Vogel and Castell, above n 115; Thuy et al, above n 100, 10.

[117] Thuy et al, above n 100, 13.

[118] This is the situation for Malaysia (Chapter 7), Papua New Guinea and Vanuatu (Chapter 8), and even in northern Australia (Chapter 10).

[119] Sophie Chapman et al, 'A Legal Perspective of Carbon Rights and Benefit Sharing under REDD+: A Conceptual Framework and Examples from Cambodia and Kenya' (2015) 9 *Carbon & Climate Law Review* 143, 144.

[120] Thuy et al, above n 100, 10.

[121] Ibid. 12.

[122] Marina Cromberg, Amy E Duchelle and Isa de Oliveira Rocha, 'Local Participation in REDD+: Lessons from the Eastern Brazilian Amazon' (2014) 5 *Forests* 579, 592.

[123] Ibid.

governance that incorporated 'local knowledge, land use practices and aspirations'.[124] Support for traditional knowledge practices and robust internal governance mechanisms for communities can be important components in the success of benefit-sharing arrangements under REDD+ models as the PNG, Vanuatu, Malaysia and northern Australian case studies demonstrate.

A study examining REDD+ pilot projects in forests in Nepal surveyed a community forest-management approach.[125] The project focus was 'to develop and demonstrate an innovative benefit-sharing mechanism for REDD+ incentives, as well as institutionally and socially inclusive approaches to local forest governance'.[126] The study demonstrated enhanced levels of carbon sequestration commensurate with improved community forest management.[127] Non-carbon benefits accruing to such communities were described as 'pro-poor and livelihood improvement activities', and 'income-generating activities', such as high-value agriculture, business, vocational skills and training.[128] Furthermore, financial benefits were distributed amongst participants not only on the basis of emissions-reduction performance, but also on the basis of social need by marginalised groups.[129] Effective and inclusive stakeholder engagement was promoted by REDD+ in the face of 'social traditions that militate against inclusion of the poor and underprivileged'.[130] The case studies in later chapters of this book also highlight the importance of ensuring REDD+ benefit-sharing mechanisms can deal with intra-group equity concerns in order to guard against REDD+ exacerbating inequalities existing within local communities.[131] In sum, the still-limited empirical evidence to date around REDD+ implementation, and the analysis in the case study chapters, reveals patchy success for benefit-sharing under REDD+-style projects unless robust enabling conditions for indigenous people and local communities exist in the implementing countries.[132]

[124] Cromberg et al, above n 122, 579. See also Ashwini Chhatre and Arun Agrawal, 'Trade-Offs and Synergies between Carbon Storage and Livelihood Benefits from Forest Commons' (2009) *PNAS* 106 17667–17670.

[125] Shanti Shrestha, Bhaskar Singh Karky and Seema Karki, 'Case Study Report: REDD+ Pilot Project in Community Forests in Three Watersheds of Nepal' (2014) 5 *Forests* 2425, 2425.

[126] Ibid. [127] Ibid. 2435. [128] Ibid. 2432. [129] Ibid. 2433. [130] Ibid. 2436.

[131] Susan Chomba et al, 'Roots of Inequity: How the Implementation of REDD+ Reinforces Past Injustices' (2016) 50 *Land Use Policy* 202, 211.

[132] Joyeeta Gupta et al, 'The Future of Forests' in Joyeeta Gupta, Nicolien van der Grijp and Onno Kuik (eds), *Climate Change, Forests and REDD: Lessons for Institutional Design* (Routledge, 2013) 229, 248.

### 1. *National and Regional Governance of Forest Land and Resources*

Central to robust enabling conditions for equitable benefit-sharing in the REDD+ regime is governance that provides an effective and fair means to resolve contested questions concerning 'ownership' and 'rights'. Legally, the issue of 'ownership' is complex and multi-layered. It transcends national and regional governance of forest resources, public and private controls of forested areas subject to REDD+ projects, and 'designation' of the legal interests or carbon rights that are the reference point from which REDD+ monetary benefits flow. The lack of express guidance on ownership and rights at the national and subnational levels under the UN-REDD Programme may signify not only a desire to retain flexibility to accommodate individual country circumstances, but also a reluctance on the part of international organizations to override national sovereignty on contentious matters concerning national controls over land and forest 'ownership'.

A fundamental tenet of national sovereignty is the power of national governments – and regional governments pursuant to specific constitutional arrangements, such as federalism – to exercise controls in the 'public interest' over public lands and natural resources, including forests.[133] Within most national legal systems, the public lands upon which many forests are located are vested in governments – essentially, a constitutional form of nation-state 'ownership'. Governments typically exercise controls over the forests through the grant of forestry leases or concessions to private individuals and corporations. For many indigenous and local communities occupying and using the same forest areas, therefore, these lands and forests are held by governments – national or regional – and also potentially are subject to private or commercial forestry operations and agricultural uses, as well as unsanctioned uses such as illegal logging.

Within this multi-scalar model, indigenous peoples and local forest-dwelling communities may hold rights to the land and forest resources recognised and protected by the national legal system, such as statutory land and resource rights.[134] Furthermore, or in the alternative, communities and individuals may hold rights enforceable only within the relevant indigenous or customary legal system (which may be termed 'informal' by reference to national legal systems).[135] Simultaneously these

[133] This power is referred to as the 'vesting' of land and resources in the state.
[134] These issues are developed further in Chapter 5 and Chapters 7–9.
[135] See further in Chapter 5.

informal 'customary rights' may constitute a land claim in respect of those public lands and forests. As Chapter 5 detailed, the Paris Agreement provides guidance for REDD+ activities and asserts the need for the parties' adherence to the UNDRIP, which articulates indigenous rights to traditional ownership of land and resources and associated rights to self-determination.[136] In other instances, indigenous groups may already hold statutory land rights or have considerable legal and constitutional protections for customary land holding. It is within this multilayered governance structure that issues of tenure security and safeguards arise, which are pertinent to benefit-sharing. Indeed, in some ways 'safeguards' may be conceived as procedural and legal prerequisites to the crystallisation and realisation of benefits for indigenous and local forest-dwelling communities under REDD+. Therefore, even though there are references dating back to the Cancun Agreements for the need for countries engaged in REDD+ activities to address land tenure and security in forests, these requirements should be seen as existing against a backdrop of myriad complexities surrounding tenure and 'ownership'.[137]

## 2. *Tenure Security and Benefit-Sharing*

The difficulties of aligning tenure and rights with the flow of benefits to indigenous and local communities are further compounded by treating tenure as one dimensional. Tenure, too, is a multivariate term.[138] One estimate says, 'Indigenous peoples and local communities claim on a customary basis at least 50 per cent of the world's lands, but legally own just 10 per cent, and exercise some formal rights to use or manage an

[136] *United Nations Declaration on the Rights of Indigenous Peoples*, GA Res 61/295, UN GAOR, 61st sess, 107th plen mtg, Agenda Item 68, Supp No 49, UN Doc A/RES/61/295 (2 October 2007, adopted 13 September 2007) art 26, paras 2–3. At para 2, the Declaration provides the right to 'own, use, develop and control the lands, territories and resources that [indigenous peoples] possess by reason of traditional ownership or other traditional occupation or use'. See also discussion in Handa Abidin, *The Protection of Indigenous Peoples and Reduction of Forest Carbon Emissions: The REDD-Plus Regime and International Law* (Brill Nijhoff, 2015) 66–67. The International Labour Organization, *Convention Concerning Indigenous and Tribal Peoples in Independent Countries*, opened for signature 27 June 1989, 1650 UNTS 383 (entered into force 5 September 1991) (ILO Convention 169) also provides the right to 'ownership and possession of the peoples concerned over the lands which they traditionally occupy', and other associated rights (art 14); see, more broadly, arts 13–19.

[137] Larson, above n 97, 542, notes that secure tenure does not necessarily 'guarantee' improved forest protection by local communities.

[138] See further Chapter 5.

additional 8 per cent.'[139] Given this disparity, the focus of concern amongst communities, the NGO sector and many international organizations has been upon the establishment of formalised tenure arrangements to underpin benefit-sharing in the REDD+ regime.[140] Further, it has been argued, biodiversity conservation as a form of non-carbon benefits is integrally linked to the maintenance of cultural diversity and the protection of the land and resource rights of indigenous peoples and local communities.[141]

Tenure is often used interchangeably (and often somewhat inaccurately) with property rights in much REDD+ literature. The meaning of property, like tenure, varies according to legal context.[142] Moreover, indigenous and local community rights, as Chapter 5 discusses, are often communal in nature.[143] The degree of security afforded such 'rights' frequently turns upon the extent to which the land and resource rights of groups have been recognised and/or formalised by the governing state.[144] Acceptance of 'customary' rights can be highly variable.[145] For some indigenous peoples and local community groups, constitutional protections are in place; in other situations, the assertion of land and resource claims generates intense, even violent, conflict.[146] Accordingly,

[139] Rights and Resources, 'Toward a Global Baseline of Carbon Storage in Collective Lands: An Updated Analysis of Indigenous Peoples' and Local Communities' Contributions to Climate Change Mitigation' (Rights and Resources Initiative, Woods Hole Research Center, and World Resources Institute, 1 November 2016) 5, http://rightsandresources.org/wp-content/uploads/2016/10/Toward-a-Global-Baseline-of-Carbon-Storage-in-Collective-Lands-November-2016-RRI-WHRC-WRI-report.pdf.

[140] See, for example, 'UN-REDD Highlights Tenure as a Success Factor for REDD+' (2013) (36) *UN-REDD Programme Newsletter*.

[141] 'Toward a Global Baseline of Carbon Storage in Collective Lands', above n 139, 6.

[142] See Chapter 5 and Joyeeta Gupta, Nicolien van der Grijp and Onno Kuik (eds), *Climate Change, Forests and REDD: Lessons for Institutional Design* (Routledge, 2013) esp 91, 194–198 (9.4), 250; Kathleen Birrell, Lee Godden and Maureen Tehan, 'Climate Change and REDD+: Property as a Prism for Conceiving Indigenous Peoples' Engagement' (2012) 3 *Journal of Human Rights and the Environment* 196.

[143] See comparative discussion of 'communal lands and individual ownership' in Lee Godden and Maureen Tehan, *Comparative Perspectives on Communal Lands and Individual Ownership: Sustainable Futures* (Routledge, 2010) Chapter 1.

[144] Rosemary Lyster, 'REDD+, Transparency, Participation and Resource Rights: The Role of Law' (2011) 14 *Environmental Science & Policy* 118.

[145] See Martijn Wilder, Sophie Chapman and Arjuna Dibley, 'Special Issue: The Legal Aspects of REDD+ Implementation: Translating the International Rules into Effective National Frameworks' (2015) 9 *Carbon & Climate Law Review* 99.

[146] See, for example, Mas Achmad Santosa, Josi Khatarina and Aldilla Stephanie Suwana, 'Indonesia REDD+: Beyond Carbon, More Than Just Forest' in Paul Martin et al (eds), *The Search for Environmental Justice* (Edward Elgar, 2015) 158, 174.

determining 'tenure' in REDD+ should be treated with caution, since a constellation of rights, 'formal' and 'informal', may exist in relation to the land, forest, carbon sequestered or emissions avoided. This complexity renders a benefit-sharing model founded only upon 'formalised' tenure rights an important first step, but it may obscure a wider web of community interests and uses of forests that also may ground a benefit flow. In particular, many women may have informal use or gathering rights in forests as opposed to land rights.[147]

Fisher and Lyster also urge clarification of 'tenure', but direct their query to whether the rights of indigenous peoples and communities over forests (and thus benefits) should depend on 'land tenure', or whether a forest 'resource' tenure may be more suitable.[148] They advocate exploration of resource-tenure models in REDD+ as a means to secure rights over forests affected for forest-dependent communities.[149] This form of 'tenure' does not accord with land 'ownership',[150] but is similar to statutory 'tenures', such as mining leases, forestry concessions and fishing licences, which also do not confer ownership of the underpinning land or sea. Any 'resource tenure' that grounds benefit-sharing mechanisms, however, still requires an effective means of enforcement within existing land and resources laws that govern forest activities in countries where REDD+ is implemented. For indigenous peoples and local communities, this model also runs the risk of reducing their rights to only usufructuary interests. Within national legal systems, typically such interests are perceived to be of lesser priority than proprietary rights. At the very least, if this model is adopted, benefits that are linked to these 'tenures' should be enforceable beyond an immediate contractual payment for services model.[151]

The final issue for consideration in a multilayered legal structure for benefit-sharing is whether legal rights to stored carbon (or the capacity to reduce emissions) should be recognised and accorded legal status, independently of land holding.[152] Legally, it is possible to recognise rights to

147 See further in Chapters 5 and 8.

148 Fisher and Lyster, above n 80, 194–195, citing Rosemary Lyster, 'The New Frontier of Climate Law: Reducing Emissions from Deforestation (and Degradation)' (2009) 26 *Environmental and Planning Law Journal* 417.

149 Ibid. 150 Fisher and Lyster, above n 80, 197.

151 For discussion, see Thuy et al, above n 100, 13–21.

152 Australia and New Zealand, for example, have adopted legislation that recognises 'carbon rights' independently of landholding and forests. See Zahar, Peel and Godden, above n 19, ch 10.

carbon sequestration and/or emissions-reduction capacity independently of the land, vegetation or soil in which the carbon is stored. Indeed, an ecosystem-services model facilitates this very type of division of forests into separate pockets of economic value and associated rights. Despite being technically and legally feasible, however, significant debate exists as to how 'carbon rights' should be implemented, enforced and made fungible, while still advancing benefits for local community interests. Very few countries participating in REDD+ have legislated to give effect to independent carbon rights.[153] Defining the legal nature of carbon sequestration or emissions-reduction capacity may prove a facilitative step for benefit-sharing,[154] but the process must be carefully examined to ensure that it provides effective and equitable outcomes for forest-dependent communities.

The success of the REDD+ regime in achieving fair and equitable benefit-sharing therefore depends in specific ways on robust but inclusive tenure forms and their recognition by national and subnational governments. Despite the growing volume of research addressing the compound issues of tenure, 'ownership' of forests and benefit-sharing, definitive models remain elusive.[155] Existing models in national jurisdictions such as those outlined in Chapters 7–9 may be instructive for the development of legal frameworks that can account for multilayered rights, and which may then better inform benefit-sharing arrangements. Further, there may be some advantages to the creation of specialist, statutory forest-resource tenure that integrates with existing land laws.[156] Nonetheless, as Chapter 9 suggests, the adoption of a model for indigenous and local community rights to forests is possible but should operate within a system that has other protections in place for communities, as well as opportunities for indigenous and local communities to actively govern those areas of traditional land and resources.

To date, few mechanisms under REDD+ decisively identify or assign legally enforceable rights for indigenous and local forest-dwelling communities at a subnational level. Part of the difficulty lies in the shifting nature of what is being 'valued' as the REDD+ regime has expanded in

[153] Thuy et al, above n 100, 31.

[154] Chapman, Wilder and Millar, above n 64, especially 274.

[155] Thuy et al, above n 100.

[156] See Chapter 9 and 10, and Lee Godden et al, 'Reducing Emissions from Deforestation and Forest Degradation in Developing Countries (REDD): Implementation Issues' (2010) 36 *Monash University Law Review* 139, 156.

scope and purpose. Accordingly, the following questions arise: Do the rights relate solely to technical and environmental outcomes gauged by measuring avoided emissions? Do they comprehend an equity-based claim to an entitlement based on forest occupancy? Should they compensate those whose 'labour' is invested in securing the reduced emissions? Most benefit-sharing mechanisms 'value' these 'rights' only in a high level or generic way, so a proliferation of benefit-sharing mechanisms within participating countries has resulted. Moreover, even when advocated as strategies for advancing distributive-justice outcomes, tenure reform and the adoption of forest tenure 'rights' may, in fact, consolidate and justify the hierarchical reorganization of value and governance over forests and their inhabitants that is put into play by the REDD+ regime. Further, while the promotion of rights-based approaches may facilitate short-term outcomes for communities within REDD+ projects, it may not counter longer-term inequalities as the regime reorganizes governance at a local level within countries as they implement REDD+.

## F. Conclusion: Implementing Fairness and Equity in Benefit-Sharing

Within the REDD+ regime, the practical implementation of fair and equitable benefit-sharing turns upon a global regime with its attendant legal and policy frameworks being exercised in a manner which is responsive to the concerns of local communities. REDD+ has the potential to both ameliorate and exacerbate existing causes of inequality for indigenous and forest-dwelling communities. Indeed, Atela et al. argue:

> Equitable and pro-poor benefit-sharing are necessary but not sufficient for effective REDD+ implementation unless national institutions are reformed and global carbon pricing harmonized with local livelihood needs.[157]

Indeed, the REDD+ regime may support livelihoods but may also import structural inequalities that further disadvantage indigenous and local communities.[158] The risks are greater where national governments fail to distribute benefits fairly, or where effective governance of benefit-

[157] Joanes O Atela et al, 'Implementing REDD+ at the Local Level: Assessing the Key Enablers for Credible Mitigation and Sustainable Livelihood Outcomes' (2015) 157 *Journal of Environmental Management* 238.

[158] Peskett, above n 13, 231.

sharing is absent.[159] These factors are crucially important to ensuring equity in the national implementation of REDD+, and to the future evolution of benefit-sharing arrangements.

This chapter has considered the derivation of concepts of benefit-sharing and non-carbon benefits in REDD+, and the implications of these frameworks for indigenous and local forest-dwelling communities. Defining 'benefit-sharing' as a 'frame' for the distribution of perceived benefits, simultaneously grounded in procedural 'fairness' and substantive 'equity',[160] the chapter has identified a variety of practical conflicts arising in the national and local implementation of the REDD+ regime. We have examined the changing nature and scope of the ecosystem services model underpinning REDD+, of which benefit-sharing is an integral component. This chapter also has examined the links between the identification of safeguards, tenure and 'rights' in the REDD+ regime and the distribution of benefits at national and subnational levels in terms of the ramifications for indigenous peoples and local forest-dependent communities. Fair and equitable benefit-sharing remains an issue of pressing concern for the international community, for each REDD+ participant country and for each participant community. Moreover, given the global scope of the regime, transnational responsibility and/or multilateral responsibility for ensuring the flow of benefits to indigenous and local communities must be considered.[161] Indeed, the extent to which the objectives of fairness and equity in benefit-sharing are achievable for indigenous and local communities depends in specific ways upon the mutual constitution of international legal principles defining benefit-sharing with national legal regimes, as informed by the robust engagement and participation of indigenous peoples and local communities in REDD+ governance.

[159] Léa Bigot et al, 'Case Study: Vietnam' in Joyeeta Gupta, Nicolien van der Grijp and Onno Kuik (eds), *Climate Change, Forests and REDD: Lessons for Institutional Design* (Routledge, 2013) 99, 113.

[160] Moregera above n 23, 380.

[161] Morgera, above n 81, 141.

# PART III

# Country Examples

# 7

# Malaysia and the UN-REDD Programme

## Exploring Possibilities for Tenure Pluralism in Forest Governance

LEE GODDEN[1]

Malaysia is home to tropical forests rich in biodiversity, as well as diverse indigenous peoples who are heavily dependent on forest resources, particularly in East Malaysia. Indigenous peoples and forest-dependent communities within Malaysia have struggled for protection of the forests and their livelihoods over many years. Malaysia joined the UN-REDD Programme in May 2012.[2] A significant reason for this engagement with the international regime was the widening of the REDD programme under the UNFCCC process to include sustainable forest management and enhancement of carbon forest stocks.[3] To date, Malaysia is not a member of the FCPF.[4] The UN-REDD Programme framework in Malaysia has developed alongside existing systems of forest-resource governance and national timber-certification schemes in various parts of the country, exemplifying a domestic form of regime interaction.[5] Moreover,

[1] The research and scholarship of Professor, Dr Fadzilah Majid Cooke, Universiti Malaysia Sabah, Malaysia, that contributed to this chapter is gratefully acknowledged, as well as her valuable assistance in fieldwork, and in commenting upon earlier drafts of the chapter.

[2] 'New UN-REDD Programme partner countries can benefit from receiving targeted support from the UN-REDD Global Programme – Support to National Actions, as well as unique knowledge sharing opportunities, facilitated through UN-REDD Programme events and the Programme's online collaborative workspace': UN-REDD Programme, 'Malaysia and Uganda Join the UN-REDD Programme' (June 2012) *UN-REDD Programme Newsletter*, www.un-redd.org/Newsletter29/Malaysia_Uganda_Join_UN-REDD/tabid/104318/Default.aspx).

[3] The degree of involvement prior to 2012 had regional variation; see, for example, Frederick Kugan and Heidi William, 'The Long and Bumpy Road to REDDINESS – Why Bother?' (Sabah Forestry Headquarters, Sandakan Sabah, Malaysia, 2011) (copy on file with author).

[4] See Forest Carbon Partnership Facility (2015), www.forestcarbonpartnership.org.

[5] For analysis of the interaction, see Jessica Rae, Mahala Gunther and Lee Godden, 'Governing Tropical Forests: REDD+, Certification and Local Forest Outcomes in Malaysia' (2011) 7(2) *Macquarie Journal of International and Comparative Environmental Law* 40.

despite the range of sustainable forestry initiatives, Malaysian forests remain at high risk of deforestation.[6]

Malaysia, although not a high-profile UN-REDD Programme participant, is a significant case study for examining the effectiveness of REDD+ planning, governance and implementation processes when engaging with indigenous peoples and forest-dwelling communities. The importance of the case study is underlined by reference to its significance for climate mitigation outcomes, given the levels of loss of intact forest landscapes[7] and the conversion of land to palm oil plantations.[8] Malaysia has a mixed history in dealing with its forest-dependent communities, and although customary law governing land is recognised in varying degrees, there is strong regional variation in the extent to which local community rights in forest lands have been formally instituted.[9] In a country where indigenous peoples have often been marginalised and their claims to customary land and forest resources strongly contested,[10] REDD+ safeguards around FPIC, together with measures to accord tenure security and benefit-sharing to communities, potentially offer an important means for redress.

Despite such potential, there are challenges in instituting comprehensive adoption of safeguards in the Malaysian context, and in facilitating the flow of benefits to forest-dwelling communities from projects to be implemented under the UN-REDD Programme. Accordingly, this chapter delineates pertinent legal, economic and demographic factors relating to forests, and outlines the relevant indigenous groups, forms

[6] Malaysia's deforestation rate has accelerated over time: see UN-REDD Programme, 'Partner Countries: Malaysia' (2012), www.un-redd.org/RegionalActivities_AsiaPacific_New/tabid/104258/AsiaPacific_Malaysia/tabid/106577; UN-REDD Programme collaborative site for Malaysia: UN-REDD Programme, 'Partner Countries: Malaysia' (2017), www.unredd.net/index.php?option.com

[7] Global Forest Watch, 'Country Profile: Malaysia', www.globalforestwatch.org/country/MYS. For a forest-layer analysis of Malaysia, see the interactive map at Global Forest Watch, www.globalforestwatch.org/map/7/4.96/114.57/MYS-1 ('Malaysia Data').

[8] Food and Agriculture Organization of the United Nations (FAO), 'Global Forest Resources Assessments' (2015), www.fao.org/forest-resources-assessment/en; Food and Agriculture Organization of the United Nations (FAO), 'Global Forest Resources Assessment 2015: Country Report – Malaysia' (2015) 16, 49, 68.

[9] Carol Yong, 'Logging in Sarawak and the Rights of Sarawak's Indigenous Communities' (JOANGOHutan, 2010) 6, www.pro-regenwald.de/docs/logginginsarawak_joangohutanreport.pdf.

[10] See, e.g., Yogeswaran Subramaniam, 'Orang Asli, Land Rights and the Court Process: A "Native Title" Lawyer's Perspective' in Kirk Endicott (ed), *Malaysia's Original People: Past, Present and Future of the Orang Asli* (NUS Press, 2016) 423, 425.

of land law and tenure, and history of land-rights claims before considering the adoption of a UN-REDD Programme 'readiness' framework in the country. The chapter concludes by considering whether the UN-REDD Programme can bring normative coherence to the diverse legal frameworks and plural systems governing forests to provide an effective impetus for stronger protections for forest-dependent communities in Malaysia. It reflects upon the 'vertical relationship' between the international regime and the Malaysian sovereign state in terms of the normative leverage that the REDD+ model might offer in advancing human rights and tenure-security outcomes at a domestic level.

## A. REDD+ in Malaysia

Multidimensional challenges face the UN-REDD Programme in Malaysia. Many are integral to the power structures around forest and land governance systems that have become entrenched over long periods, as well as the importance of forests to diverse stakeholders within the country. Malaysia, located in the tropical forest belt, is a federal nation comprising a multi-ethnic society. Its British colonial legal past is mixed with Islamic law principles governing Muslim groups in West Malaysia and the coastal groups of East Malaysia, and with customary law governing the indigenous peoples of interior Sabah and Sarawak. This legacy has led to a mixed land-law system which combines elements of statutory codes, customary *(adat)* law and common law native title. Matters relating to indigenous tenure security and rights to forest areas must be negotiated across a complex and interwoven legal system for land and resource management and biodiversity conservation.[11] Malaysia has a chequered history in terms of human rights, and uneven recognition of customary land tenures.[12] Although there has been a history of

[11] Malaysia, for example, has ratified the World Heritage Convention. Two World Heritage sites (natural heritage) are located on the island of Borneo: UNESCO: World Heritage Centre, 'The States Parties: Malaysia' (2015), http://whc.unesco.org/en/statesparties/my. See also Global Forest Watch, above n 7, for a list of the international environmental conventions to which Malaysia is a party.

[12] Carol Yong, 'Forest Governance in Malaysia: An NGO Perspective' (JOANGOHutan, 2006) 13, www.fern.org/sites/fern.org/files/media/documents/document_3754_3755.pdf; Amity Doolittle, 'Native Customary Land Rights in Sabah, Malaysia 1881 – 2010' in Marcus Colchester and Sophie Chao (eds), *Divers Paths to Justice: Legal Pluralism and the Rights of Indigenous Peoples in Southeast Asia* (Forest Peoples Programme and Asia Indigenous Peoples Pact, 2011) 81, 82. Subramaniam, above n 10.

customary tenure formalisation, many outstanding indigenous land claims remain.[13] Access to forests for diverse uses by local communities, such as swidden cultivation, is under threat in many regions.[14] In tandem, the country is experiencing strong economic growth, but drivers of national productivity, such as timber and oil palm production, typically depend on the exploitation of forest lands often held as customary tenures, highlighting the insecurity of lands held under customary claims.[15] The country's record of SFM and its national forest certification system stand alongside extensive deforestation and illegal logging, especially in Sarawak.[16]

Nonetheless, upon entering the UN-REDD Programme, Malaysia moved quickly into a REDD 'readiness stage', given its level of scientific expertise and experience in forestry management.[17] Malaysian states, such as Sabah, had undertaken considerable preparatory REDD+ work before the country formally entered the scheme. Initially, the emphasis was on technical aspects of forest carbon sequestration, such as forest carbon mapping in the early phases of the UN-REDD-Programme.[18] The degree of involvement of Malaysia's indigenous peoples in the early stage was constrained, although consultative fora with indigenous peoples were organized in peninsular Malaysia and Sabah.[19]

[13] Ramy Bulan, 'Indigenous Land Rights in Malaysia' in J Ronald Engel, Laura Westra and Klaus Bosselmann (eds), *Democracy, Ecological Integrity and International Law* (Cambridge Scholars, 2010) 171; Human Rights Commission of Malaysia (SUHAKAM), *Report of the National Inquiry into the Land Rights of Indigenous Peoples* (2013).

[14] Ole Mertz et al, 'The Last Swiddens of Sarawak, Malaysia' (2013) 41 *Human Ecology* 109, 110.

[15] Fadzilah Majid Cooke, 'In the Name of Poverty Alleviation: Experiments with Oil Palm Smallholders and Customary Land in Sabah, Malaysia' (2012) 53 *Asia Pacific Viewpoint* 240, 244.

[16] Yong, above n 12. Alison Hoare, 'Illegal Logging and Related Trade: The Response in Malaysia – A Chatham House Assessment' (Research Paper, Energy, Environment and Resources, January 2015), www.chathamhouse.org/publication/illegal-logging-and-related-trade-response-malaysia#.

[17] 'REDD in Malaysia' (2017) The REDD Desk: A Collaborative Resource for REDD Readiness, http://theredddesk.org/countries/malaysia.

[18] Interview with staff from the Forest Research Institute (Kuala Lumpur, Malaysia, July 2012) (copy on file with author).

[19] Frederick Kugan, 'Sabah's Roadmap on REDD+ Strategy' (paper presented at National Inception Workshop on REDD+: Building Capacity for Decision Makers and the Way Forward, Sunway Putra Hotel, Kuala Lumpur, 13–14 February 2012). See also interview with Frederick Kugan (Sabah Forestry Department Sandakan, Malaysia, July 2012) (copy on file with author).

Malaysia is home to diverse indigenous peoples. On the Malay Peninsula, where most Malaysians live, the indigenous peoples are a minority, known as the Orang Asli. The diversity across Malaysia amongst its indigenous peoples and variation in the extent of legal recognition and formalisation of their rights to land and forest resources represent a significant challenge for developing locally responsive REDD+ governance and highlight the critical nexus between measures by the nation state that determine indigenous 'identity' and the extent of effective inclusion and representation for indigenous peoples in forest-related decision-making.[20] The UN-REDD Programme in Malaysia was preceded by several international initiatives which sought to ensure that law and policy directed to environmental sustainability took due cognisance of the potential impacts on local communities;[21] even so, the record in forestry is patchy.[22] This earlier history may contribute to the wariness of Malaysia's indigenous peoples regarding international programmes that promise to safeguard their interests while delivering global outcomes.[23]

Within the REDD+ regime, as discussed, safeguards for forest-dependent indigenous peoples and local communities around FPIC have been developed and nations directed to consider principles for securing land tenure and equitable benefit-sharing.[24] As a part of the REDD+ documentation, nations must provide comprehensive planning for how indigenous peoples are to be consulted; integral to these procedures is the requirement for FPIC.[25] Nation states are requested to put in place arrangements with respect to tenure security and co-benefits. Malaysia, in its 'readiness' phase, stated support for developing guidelines for

[20] See further in Chapter 4.

[21] Victoria Tauli-Corpuz, United Nations Special Rapporteur on the Rights of Indigenous Peoples, *Expert Testimony of the UN Special Rapporteur on the Rights of Indigenous Peoples, Victoria Tauli-Corpuz, before the Inter-American Court of Human Rights (IACHR) on the Case of Kaliña and Lokono Peoples vs the Government of Surinam* (3 February 2015), http://unsr.vtaulicorpuz.org/site/index.php/en/statements/56-testimony-case-surinam.

[22] Anthony Hall, 'Forests, Conservation and Climate Change in Latin America' in *Forests and Climate Change: The Social Dimensions of REDD in Latin America* (Edward Elgar, 2012) 10.

[23] See Rae, Gunther and Godden, above n 5. See also Chapter 2, Sections A and B.

[24] See Chapters 2 and 3; UN ECOSOC Permanent Forum on Indigenous Issues, *Indigenous People's Rights and Safeguards in Projects Related to Reducing Emissions from Deforestation and Forest Degradation*, UN Doc E/C.19/2013/7 (5 February 2013) ('*Indigenous People's Rights and Safeguards in Projects Related to Reducing Emissions from Deforestation and Forest Degradation*').

[25] See further in Chapter 4.

safeguards but appears to have taken some time to agree on actual mechanisms.[26] More recently, it appears that Malaysia has formalised its position on REDD+ safeguards and developed a results-based action plan covering REDD+ safeguards, forest governance and forest management.[27]

Malaysia's position on FPIC principles for consultation with indigenous peoples may be influenced by its experience with consultations in developing forest certification standards.[28] This certification process was characterised by strained engagement and some conflict with forest-dependent communities, particularly where claims based on indigenous rights to land and resources were not recognised.[29] As earlier chapters discuss, without wider state acknowledgment of customary tenure, and resolution of historic claims to forests, significant barriers remain to the effective implementation of REDD+.

### 1. *Malaysia as a Governance-Interaction Case Study*

Malaysia is an important choice for a case study as the introduction of REDD+ exemplifies 'regime interaction' at horizontal and vertical scales. The UN-REDD Programme developed under the auspices of the UNFCCC, but it necessarily intersects with other international regimes.[30] At the international or 'horizontal' level, the adoption of REDD+ in Malaysia is characterised by the convergence of interests in forest governance across a range of international agencies. Agencies such

[26] There is a lack of documentation on these measures on the UN-REDD Programme collaborative site for Malaysia: UN-REDD Programme, 'Partner Countries: Malaysia' (2017), www.unredd.net/index.php?option=com_docman&Itemid=134&view=list&slug=malaysia-2398. See also 'REDD in Malaysia', above n 17; Dato' Dr Abdul Rahman, E Philip and K A Pan, 'REDD Plus Framework for Malaysia', www.cbd.int/doc/meetings/for/wscbredd-apac-01/other/wscbredd-apac-01-malaysia-en.pdf.

[27] Extensive research failed to locate a public record copy of the government documents, but a media release refers to submission to the UNFCCC Secretariat at COP22 in Marrakesh, 14 November 2016: Ministry of Natural Resources and Environment (NRE), 'Malaysia Hosted REDD+ Day at COP-22, Marrakech, Morocco' (Media Statement, 15 November 2016).

[28] Adrian Wells, Thang Hooi Chiew and Chen Hin Keong, 'Systems for Verification of Legality in the Forest Sector, Malaysia: Domestic Timber Production and Timber Imports' (Country Case Study 8, VERIFOR, 2008) 82. This early report notes that the most complex challenge lies with aboriginal land rights in land and forest resources.

[29] Yong, above n 12, 28.

[30] Annalisa Savaresi, 'The Role of REDD in the Harmonisation of Overlapping International Obligations' in Erkki J Hollo, Kati Kulovesi and Michael Mehling (eds), *Climate Change and the Law* (Springer, 2013) 391, 397.

as the FAO, UNEP and UNDP all have a presence within the country. Biodiversity protection and SFM are promoted by a web of international organizations working with national government agencies and NGOs.[31] At a vertical scale, the governance of tropical forests is increasingly shaped by bilateral, state-to-state arrangements, as well as nonstate institutions such as forest-certification bodies and the NGO sector.[32] In Sabah, for example, Japan has been an important donor of technical expertise to the Sabah Forestry Department.[33]

Internationally, adaptive management concepts have been progressively introduced into forest resource management with the recognition that forest systems, as complex systems require a more sophisticated model of multidimensional environmental governance than previously acknowledged.[34] The recognition of environmental systems as dynamic has provided the impetus for legal models to embrace regulatory approaches that allow for greater institutional flexibility, as well as those which recognise a diverse range of legitimate actors.[35] Although the inclusion of indigenous peoples and local communities as significant 'actors' in forest sustainability regimes has had a chequered history, there is growing acceptance of the need for legal mechanisms to be cognisant and inclusive of indigenous and customary resource rights.[36] Recent attention to FPIC principles has arisen in the context of developments

[31] Malaysia attracts much online attention from civil society 'forest watch' groups: see, e.g., World Resources Institute, 'Risk Tool: Malaysia', Forest Legality Initiative, www.forest legality.org/risk-tool/country/malaysia.

[32] H O Mohd Shahwahid, 'Forest Certification in Malaysia' in Benjamin Cashore et al (eds), *Confronting Sustainability: Forest Certification in Developing and Transitioning Countries* (Yale School of Forestry and Environmental Studies, 2006) 69, 71.

[33] Interview with Kugan, above n 19.

[34] Emily Boyd et al, 'Exploring Development Futures in a Changing Climate: Frontiers for Development Policy and Practice' (2009) 27 *Development Policy Review* 659.

[35] Joanne Scott, 'The Multi-Level Governance of Climate Change' in Paul Craig and Gráinne de Búrca (eds), *The Evolution of EU Law* (Oxford University Press, 2nd edn, 2011) 805.

[36] See, e.g., International Union of Forest Research Organizations (IUFRO), United Nations Development Programme (UNDP/UN-REDD), International Union for the Conservation of Nature (IUCN), 'Social Safeguards: Protecting the Rights and Interests of Indigenous Peoples and Forest-Dependent Communities in REDD+' (Programme, Forest Day, Durban, 4 December 2011), www.iufro.org/download/file/7930/4887/redd-social-safe guards-forum_doc; John Parrotta, 'Social Safeguards: Protecting the Rights and Interests of Indigenous Peoples and Forest-Dependent Communities in REDD+' (Keynote Remarks, Forest Day 5, Discussion Forum, 4 December 2011). For discussion around land tenure and collective interests, see Lee Godden and Maureen Tehan (eds), *Comparative Perspectives on Communal Lands and Individual Ownership: Sustainable Futures* (Routledge, 2010).

such as dams and logging (both legal and illegal) on customary lands, and areas that potentially are open to historic-based claims.[37]

The other critical nexus of regime interaction is between REDD+ and human rights.[38] Malaysia's relationship with its indigenous peoples has attracted international human rights concerns[39] and been the subject of a national inquiry.[40] Accordingly, the interface between these various organizations with differing objectives and operational frameworks will influence the national implementation of the UN-REDD Programme. Civil society groups such as Transparency International have been significant in providing oversight of the international legal and financial initiatives for forest sustainability, including the adoption of REDD+ models. The civic sector typically operates in an intermediary, regulatory space and provides an important network function, at times, linking international agencies and national governmental organizations.[41]

Moreover, as the most recent in a series of international initiatives, REDD+ may subsume 'long-standing, perhaps intractable, development policy challenges within its remit in order to incentivise climate change mitigation through bio-sequestration in the forestry sector'.[42] Malaysia's government, however, shares features with similar 'post-colonial, developmentalist' regimes in South-East Asia, as it is often suspicious of challenge and streamlines policy-making so that the state leaves little space for oppositional activity.[43] All the same, the influence of civil society in Malaysia is growing.[44] Malaysia's profile as a timber exporter and oil palm producer has focused civil society attention on its forests[45] and the impacts of forest projects on indigenous peoples living within forests. Human rights advocacy for more stringent consultative standards

[37] S Robert Aiken and Colin H Leigh, 'Seeking Redress in the Courts: Indigenous Land Rights and Judicial Decisions in Malaysia' (2011) 45 *Modern Asian Studies* 825, 828.

[38] See Chapter 4. [39] Aiken and Leigh, above n 37, 826.

[40] Human Rights Commission of Malaysia (SUHAKAM), above n 13.

[41] Fadzilah Majid Cooke, 'Recent Development and Conservation Interventions in Borneo' in Fadzilah Majid Cooke (ed), *State, Communities and Forests in Contemporary Borneo* (Australian National University Press, 2006) 4.

[42] Simon West, '"Command Without Control": Are Market Mechanisms Capable of Delivering Ecological Integrity to REDD?' (2010) 6 *Law, Environment & Development Journal* 298, 301.

[43] Meredith L Weiss and Saliha Hassan, 'Introduction' in Meredith L Weiss and Saliha Hassan (eds), *Social Movements in Malaysia: From Moral Communities to NGOs* (Routledge, 2002) 1.

[44] Aiken and Leigh, above n 37, 828.

[45] Majid Cooke, 'Recent Development and Conservation Interventions in Borneo', above n 41, 5–8.

has previously impacted Malaysia's national forest governance models,[46] although with limited practical devolution of government decision-making. As an important node of vertical integration, civil society groups bring an orientation toward democratic and collaborative forms of forest governance.[47] The work of NGOs in facilitating collaborative responses has converged with social movements from within Malaysia that are actively questioning the control over forest resources exercised by the state, particularly in Sabah and Sarawak.[48]

### 2. *REDD+: Competing Opportunities*

Another reason for examining REDD+ implementation in Malaysia is the research that is emerging on land-use transitions[49] and the opportunity costs associated with the adoption of REDD+ projects.[50] Through East Malaysia, in particular, the expansion of oil palm production has provided competing opportunities for local customary lands.[51] Typically, the viability of REDD+ projects will depend on generating sufficient payments to communities to offset revenue from alternative land-use activities such as timber extraction, oil palm production and small-scale agriculture.[52] A recent study concluded that '[b]ased on the relatively modest profits from forest carbon financing compared to the profits from oil palm and timber plantations, REDD+ will remain ill-suited to slowing these intensive industries across the region'.[53] These findings are significant when examining the plural normative context in which

[46] Lyn Teck Wyn, 'Critical Review of the Forest Regulatory Framework and Its Implementation in Malaysia' in Henry Scheyvens (ed), *Critical Review of Selected Forest-Related Regulatory Initiatives: Applying a Rights Perspective* (Institute for Global Environmental Strategies, 2011) 51.

[47] See discussion of democratic experimentalism in Chapter 10.

[48] Fadzilah Majid Cooke, 'Constructing Rights: Indigenous Peoples at the Public Hearings of the National Inquiry into Customary Rights to Land in Sabah, Malaysia' (2013) 28(3) *SOJOURN: Journal of Social Issues in Southeast Asia* 512, 533.

[49] Saiful Arif Abdullah and Adnan A Hezri, 'From Forest Landscape to Agricultural Landscape in the Developing Tropical Country of Malaysia: Pattern, Process, and Their Significance on Policy' (2008) 42 *Environmental Management* 907.

[50] Victoria Graham et al, 'A Comparative Assessment of the Financial Costs and Carbon Benefits of REDD+ Strategies in Southeast Asia' (2016) 11(11) *Environmental Research Letters* 4022.

[51] Fadzilah Majid Cooke, Sumei Toh and Justine Vaz, 'Community-Investor Business Models: Lessons from the Oil Palm Sector in East Malaysia' (International Institute for Environment and Development, 2011) 13–14, http://pubs.iied.org/pdfs/12570IIED.pdf.

[52] Graham et al, above n 50, 2. [53] Ibid. 9.

REDD+ will need to operate with respect to the benefits it may bring for local forest communities.

Historically, many communities combined shifting cultivation (swidden or slash and burn) with hunting and collection of forest resources, together with small-scale cash crops.[54] The decline of swidden, an important source of subsistence for local communities that assists in maintaining food security, could be accelerated due to pressures from competing land uses and from carbon sequestration. Significantly, if the UN-REDD Programme incorporated traditional practices, it could rehabilitate views about the utility of swidden cultivation (such land is often regarded as 'idle').[55] There is potential for swidden burning practices to constitute a form of REDD+-type activity.[56]

Having sketched the challenges facing REDD+ in Malaysia, the following sections examine how the UN-REDD Programme might negotiate its path forward, in conjunction with forest-dependent indigenous and local communities.

## B. The Malaysian Forests and Peoples

### 1. *A Federal Nation*

The initial carbon-sequestration RED scheme was conceived as 'an agreed performance-based mechanism operating at the national level in participating countries'.[57] The expanded REDD+ model now accepts that relevant activities take place at multiple levels. The implementation of REDD+ at a subnational level can even involve 'partnerships between state and provincial administrations in developed and developing countries'.[58] Subnational implementation of the UN-REDD Programme is critical for federal nations such as Malaysia.

54 Aiken and Leigh, above n 37, 831.

55 Majid Cooke, 'In the Name of Poverty Alleviation', above n 15, 244. Amity A Doolittle, 'Imagining New Environmental Futures: Alternative Strategies for Natural Resource Governance' in *Property and Politics in Sabah, Malaysia: Native Struggles over Land Rights* (University of Washington Press, 2005) 152.

56 See Chapter 8 and Stephanie Niall et al, 'Climate Change and REDD+: Integrating Customary Fire-Management Schemes in East Malaysia and Northern Australia' (2013) 28 *SOJOURN: Journal of Social Issues in Southeast Asia* 538, 558.

57 *Indigenous People's Rights and Safeguards in Projects Related to Reducing Emissions from Deforestation and Forest Degradation*, above n 24, 3.

58 Ibid. 5.

Malaysia's federation comprises thirteen states and three federal territories. Sabah, Sarawak and Labuan are located on the island of Borneo, known collectively as East Malaysia, while the remainder forms West or Peninsular Malaysia. Article 74(2) of the Federal Constitution of Malaysia specifies that the state (provincial) governments have jurisdiction over lands, waters and natural resources, including forests and agricultural production. As forest lands are vested in the respective states, this provision empowers each state to independently regulate forests through specific regional laws, and to grant timber concessions and other land and resource development permits to third parties. Indigenous communities, however, may have or lay claim to customary tenures over such forests.[59] The largest concentration of forest-dependent communities occurs within East Malaysia where many people still live in their traditional territories.[60] Displacement of many indigenous peoples from these traditional forest areas also has occurred.[61]

Forest laws comply in varying degrees with national policy directions, although there is significant autonomy for the forestry departments of Sabah and Sarawak.[62] Certain national legislation, such as that mandating environmental impact assessment, can override state legislation.[63] As a practical matter, this national oversight can be contested as the East Malaysian state governments wield considerable political power.[64] Given the importance of logging and land-use development to regional economies, especially in East Malaysia,[65] the state forestry departments play a significant role in REDD+ implementation.[66]

[59] Aiken and Leigh, above n 37, 839.

[60] Yong, above n 12, 11. For Orang Asli, see Kirk Endicott, 'Introduction' in Kirk Endicott (ed), *Malaysia's Original People: Past, Present and Future of the Orang Asli* (NUS Press, 2016) 1, 4–6.

[61] Aiken and Leigh, above n 37, 839; Sheema A Aziz et al, 'Why Conservationists Should Be Concerned about Natural Resource Legislation Affecting Indigenous Peoples' Rights: Lessons from Peninsular Malaysia' (2013) 22 *Biodiversity and Conservation* 639, 647.

[62] Robert Stephen Milne, *Malaysia–New States in a New Nation: Political Development of Sarawak and Sabah in Malaysia* (Frank Cass, 1974). See more recently World Resources Institute, above n 31.

[63] Yong, above n 9, 61–64.

[64] K S Jomo and Wee Chong Hui, 'The Political Economy of Malaysian Federalism: Economic Development, Public Policy and Conflict Containment' (2003) 15(4) *Journal of International Development* 441.

[65] Majid Cooke, Toh and Vaz, above n 51, 14.

[66] In Sabah, for example, the REDD+ roadmap is coordinated through the Sabah Forestry Department.

### (a) Governing Forest in Malaysia

Many of the current structures for forest governance in Malaysia have origins in the colonial period. In Borneo, the introduction of the plantation economy by the North Borneo Company set the foundation for the later vesting of the land and resources in the British Crown, and subsequently in the respective Malaysian state governments. Since independence, Malaysia has had several forest sustainability programmes, and developed its own timber-certification system.[67] The nation also negotiated the application of standards prescribed by the Forest Stewardship Council (FSC). The outcomes of SFM in Malaysia are a litmus test for elsewhere in Asia. In opening its Malaysian office in 2014, the FSC identified the main challenges for the FSC system 'will be around building awareness of and value for high conservation value species, indigenous peoples' and forest workers' rights and good forest governance'.[68] Given such normative objectives, REDD+ initiatives may draw Malaysia further into the web of international multilateral and bilateral forestry-governance arrangements, although the overarching relationship between REDD+ and SFM is progressively being integrated.[69]

## 2. *Malaysia's Forests*

Malaysia contains some of the world's most valuable tropical forests. At the UN Conference on Environment and Development at Rio in 1992, Malaysia pledged to maintain at least 50 per cent of perpetual forest and tree cover in the country. Malaysia's Second National Communication to the UNFCCC reported that 56 per cent of the country was covered in forests in 2000, with 55 per cent remaining in 2007.[70] Forest cover in 2012 was estimated at 53 per cent.[71] While Malaysia consistently

[67] Rae, Gunther and Godden, above n 5, 45–46, 57–58.

[68] Forest Stewardship Council, 'Forest Stewardship Council Opens Doors to Sustainable Forest Management in Malaysia' (19 November 2014), http://ic.fsc.org/fsc-press-releases.638.979.htm.

[69] See, e.g., the role of REDD+ in facilitating forest assessments: Food and Agriculture Organization of the United Nations (FAO), 'Global Forest Resources Assessment 2015: How Are the World's Forests Changing?' (2015), www.fao.org/resources/infographics/infographics-details/en/c/325836.

[70] 'National REDD+ Readiness in Malaysia' (2015) The REDD Desk: A Collaborative Resource for REDD Readiness, http://theredddesk.org/countries/initiatives/national-redd-readiness-malaysia.

[71] Figures are challenged by third-party estimates. Yong notes that there is a lack of good-quality data about the definition of a forest, the extent of forest cover, loss of forest,

reports little change in its forest cover, other commentators suggest that 'natural' or primary forests in Malaysia have experienced major deforestation and degradation.[72] Efforts have been made to address this, for example in Sabah, where ecosystem-service models have been trialled via forest carbon projects since 1992.[73] Forest lands remain vitally important to Malaysia. The Malaysian forestry sector plays a pivotal role in economic growth.[74]

### (a) Defining Forests in Malaysia

Malaysia broadly defines 'forest', and the designation includes some areas of oil palm, rubber and tree crops. This is significant in the Malaysian context where commercial 'forest' plantation areas are expanding. Malaysia's forested land is predominantly Crown land, that is, land vested in the relevant state government – although this legal position may not preclude native customary tenures and native title claims.[75] Thus 'forests' comprise Permanent Reserved Forest, state land forests, national parks, and wildlife and bird sanctuaries.[76] Some forest land is reserved for future development purposes.[77] Permanent Reserved Forest includes protection forests, production forests, amenity forests or research forests.[78] Malaysia reported its total forested area as 21,019 million hectares.[79] There are also small areas of privately owned plantation forests.[80] Malaysia claims that its rate of deforestation has remained

(un)sustainable yields' (Yong, above n 12, 7). See also Food and Agriculture Organization of the United Nations, 'Global Forest Resources Assessment 2010: Country Report – Malaysia' (2010), www.fao.org/docrep/013/al558E/al558e.pdf. The FAO estimates that, in 2014, of the total country land area of 32,855 thousand hectares the forest cover comprised 22,180.8 thousand hectares.

72 See Global Forest Watch, above n 7.

73 Kugan and William, above n 3, 2.

74 Julia McMorrow and Mustapa Abdul Talip, 'Decline of Forest Area in Sabah, Malaysia: Relationship to State Policies, Land Code and Land Capability' (2001) 11(3) *Global Environmental Change* 217, 217.

75 Ramy Bulan, 'Native Title in Malaysia: A Complementary Sui Generis Proprietary Right under the Federal Constitution Commentary' (2007) 11 *Australian Indigenous Law Review* 54, 54.

76 'REDD in Malaysia', above n 17; Food and Agriculture Organization of the United Nations (FAO), 'Country Report – Malaysia' (2015), above n 8, 6–8.

77 M Samsudin et al, 'Reducing Emissions from Deforestation and Degradation (REDD+) – Malaysia' (paper presented at International Technical Seminar on REDD+, Tokyo, 16–17 February 2011).

78 Food and Agriculture Organization of the United Nations (FAO), 'Country Report – Malaysia' (2015), above n 8.

79 Detailed statistical information can be found at ibid.

80 Ibid.

fairly stable but, as noted, this has been contested. Indeed, a study released in 2013 suggested that 'Malaysia had the world's highest rate of forest loss between 2000 and 2012'.[81] Malaysia's forest loss was partly offset by gains in vegetation cover from natural recovery, reforestation and the establishment of industrial plantations.

The security of tenure and the spectrum of forest resource rights for indigenous communities vary widely across these 'forests'.[82] Further uncertainty occurs in relation to land available for 'future development'. Practically, this may be where swidden cultivation operates or where local communities access forest resources to supplement subsistence agriculture.[83] As the Committee on World Food Security noted:

> Tackling the issue of land and resource tenure is one of the prerequisites for mechanisms such as REDD+ to have positive impacts not only on reducing emissions from the forestry sector but also in reducing poverty and achieving food security.[84]

### (b) Forests as an Economic Driver in Malaysia

Malaysia exhibited rapid growth, first transforming its economy toward export-oriented manufacturing,[85] then later through export commodities such as timber and oil palm production.[86] The general increase in living standards has largely bypassed Malaysia's indigenous peoples.[87] Malaysia's push towards modernisation, driven by a 'top down', planned

[81] Rhett A Butler, 'Malaysia Has the World's Highest Deforestation Rate, Reveals Google Forest Map' (15 November 2013), http://news.mongabay.com/2013/1115-worlds-highest-deforestation-rate.html.

[82] See generally Human Rights Commission of Malaysia (SUHAKAM), 'National Inquiry: Suhakam Inquiry Highlights Issues Faced by Sabah's Indigenous People' (11 January 2014), www.suhakam.org.my/national-inquiry.

[83] Bulan, 'Indigenous Land Rights in Malaysia', above n 13, 171; Jefferson Fox, Jean-Christophe Castella and Alan D Ziegler, 'Swidden, Rubber and Carbon: Can REDD+ Work for People and the Environment in Montane Mainland Southeast Asia?' (2014) 29 *Global Environmental Change* 318, 318.

[84] 'In a landmark decision on 11 May, 2012, the Committee on World Food Security (CFS) endorsed a set of global guidelines that will help governments safeguard the rights of people to own or access land, forests and fisheries': UN-REDD Programme, 'Reports & Analysis: Committee on World Food Security Launches Groundbreaking Guidelines on Land Tenure' (June 2012) *UN-REDD Programme Newsletter*, www.un-redd.org/Newsletter28/Impact_of_NewLandTenureGuidelines/tabid/104166/Default.aspx).

[85] Harold Brookfield (ed), *Transformation with Industrialization in Peninsular Malaysia* (Oxford University Press, 1994) preface.

[86] Majid Cooke, Toh and Vaz, above n 51, 14–15.

[87] Aiken and Leigh, above n 37, 829.

development approach, has had social and environmental costs.[88] Timber is an important revenue basis but 'it is ironic that the people who live in and depend on the forests are amongst the poorest in Malaysia'.[89] Palm oil is now a major economic catalyst for 'development' in Malaysia – as well as a significant contributor to forest deforestation and degradation, and to increasing global GHG emissions.[90] Oil palms not only are mainly grown in large plantations, but also by smallholders – including indigenous and local forest-based communities. Government agencies promote smallholder oil palm production as a means of poverty alleviation.[91] The pressure from oil palm production is neatly explained by reference to Sabah:

> Since oil palm already occupies 90% of land planted with industrial crops any future expansion will either use up the remaining 10% of areas under crop or expand into lands not under industrial crops such as those claimed under customary rights that are largely used for subsistence agriculture by indigenous communities. Much of the land claimed under customary rights is not yet titled.[92]

These trends highlight the challenges facing REDD+ in Malaysia, and in redressing social and cultural disadvantages and endemic poverty and achieving sustainable development goals. Yet the simplistic view that native peoples are passive players in forces that government and international agencies set in motion needs qualification as many groups actively seek economic opportunities or resist projects inconsistent with cultural values.[93] Nonetheless, land tenure security issues remain critical to the degree of autonomy that indigenous peoples may exercise in challenging 'top down development'.[94]

### (c) Deforestation and Forest Degradation

Malaysia has a long record of involvement in SFM, commencing with the National Forestry Policy 1978 that set out principles for SFM, forest harvesting, regeneration, rehabilitation and management of non-wood forest products. The National Forestry Act 1984 applies to Malaysian

[88] Majid Cooke, 'Constructing Rights', above n 48, 531. [89] Yong, above n 12, 20.
[90] Varsha Vijay et al, 'The Impacts of Oil Palm on Recent Deforestation and Biodiversity Loss' (2016) 11(7) *PLoS ONE* e0159668.
[91] Majid Cooke, 'In the Name of Poverty Alleviation', above n 15, 241.
[92] Majid Cooke, Toh and Vaz, above n 51, 15.
[93] Doolittle, 'Imagining New Environmental Futures: Alternative Strategies for Natural Resource Governance', above n 55, 163.
[94] Majid Cooke, 'In the Name of Poverty Alleviation', above n 15, 245.

Peninsula states, but Sabah and Sarawak continue to use their respective forestry regulations. A major strategy in SFM was the Malaysian Timber Certification Scheme.[95]

Despite such regulations, there are conflicting views about deforestation in the country, especially in East Malaysia. It had a historically low emissions baseline from deforestation when compared to surrounding tropical forest nations. When attention was first drawn to forest decline many years ago, shifting cultivation was identified as the central reason for forest loss. Shifting or swidden cultivation involves a 'land use system that employs a natural or improved fallow phase, which is longer than the cultivation phase of annual crops, sufficiently long to be dominated by woody vegetation, and cleared by means of fire'.[96] Amongst traditional communities in Malaysia, swidden cultivation is much more than simply a land-use practice; it represents a complex series of rules and norms highly adapted to the environment in which it is practiced.[97] Swidden agriculture, once stigmatised as destructive of forests, may now be 'recognized for its potential for resource conservation'.[98] As noted in Chapter 9 there is potential for forms of traditional 'burning practices', such as those involved in swidden cultivation, to constitute REDD+ activities.

Malaysia also has instigated measures over recent decades to address illegal logging, forest degradation and widespread forest clearance.[99] Malaysia has committed to biodiversity and forest conservation initiatives,[100] such as the Heart of Borneo project.[101] Yet Malaysian forests continue to be heavily exploited by the commercial timber industry,[102]

[95] Robin A Lewis and Stacey R Davis, 'Forest Certification, Institutional Capacity, and Learning: An Analysis of the Impacts of the Malaysian Timber Certification Scheme' (2015) 52 *Forest Policy and Economics* 18.

[96] Ole Mertz et al, 'Swidden Change in Southeast Asia: Understanding Causes and Consequences' (2009) 37 *Human Ecology* 259, 261.

[97] Wil de Jong, 'Developing Swidden Agriculture and the Threat of Biodiversity Loss' (1997) 62 *Agriculture, Ecosystems & Environment* 187, 188; P J A Kleinman, D Pimentel and R B Bryant, 'The Ecological Sustainability of Slash-and-Burn Agriculture' (1995) 52 *Agriculture, Ecosystems & Environment* 235, 236.

[98] de Jong, above n 97, 196. [99] McMorrow and Talip, above n 73.

[100] Annawati van Paddenburg et al, 'Heart of Borneo Investing in Nature for a Green Economy', WWF HoB Global Initiative (2012), http://awsassets.wwf.org.au/downloads/fl013_heart_of_borneo_investing_in_nature_for_a_green_economy_jan12.pdf.

[101] Marc Ancrenaz et al, *Strategic Plan of Action (Sabah): The Heart of Borneo Initiative (2014–2020)* (Sabah Forestry Department and WWF-Malaysia, 2013).

[102] Nigel Dudley, Jean-Paul Jeanrenaud and Francis Sullivan, *Bad Harvest? The Timber Trade and the Degradation of the World's Forests* (Routledge, 2014).

with illegal logging remaining an entrenched problem[103] despite government remediation efforts through certification.[104] Forests continue to be subject to pressures from plantation agriculture.[105]

Malaysian timber certification schemes have also been criticised for failing to effectively resolve conflicts relating to customary land and forest-resource claims.[106] Strong tensions remain between forest governance mechanisms and indigenous peoples as evident in the national inquiry conducted by the Malaysian Human Rights Commission (SUHAKAM), which commenced in 2011.[107]

With these factors at play in Malaysia, the UN-REDD Programme represents an important opportunity to address external and internal pressures precipitating forest decline; respond to environmental, social and cultural impacts; and assist global climate mitigation efforts. While these opportunities are apparent, REDD+ frameworks in Malaysia will require a responsive set of safeguards and robust measures around tenure and co-benefits to effectively address the multiple factors confronting forest-dependent communities in Malaysia.

## C. Indigenous Peoples and Forest-Dependent Communities

The patchwork of formal and informal customary rights to lands and forest resources in Malaysia is a complex pattern, further complicated by a plural legal system.[108] The legal system reflects the legacy of a colonial past,[109] but also the diverse cultural origins of Malaysians. Indigenous peoples in Malaysia are officially collectively designated as Orang

[103] Brendan Fisher, David P Edwards and David S Wilcove, 'Logging and Conservation: Economic Impacts of the Stocking Rates and Prices of Commercial Timber Species' (2014) 38 *Forest Policy and Economics* 65.

[104] 'Despite certification making some headway with illegal logging, it does not address the problem of conversion of forests outside the forest reserves to non-forest uses': Mohd-Shahwahid, above n 32, 72.

[105] Majid Cooke, 'In the Name of Poverty Alleviation', above n 15, 240–253.

[106] Rae, Gunther and Godden, above n 5, 60–61; see also 57.

[107] See, e.g., Human Rights Commission of Malaysia (SUHAKAM), above n 13, 91–30 [6.57]–[6.69] ('Agropolitan and MESEJ Projects for Hardcore Poor'), 93–4 [6.70]–[6.76] ('Communal Titles').

[108] Amanda Whiting, 'In the Shadow of Developmentalism: The Human Rights Commission of Malaysia at the Intersection of State and Civil Society Priorities' in C Raj Kumar and D K Srivastava (eds), *Human Rights and Development: Law, Policy and Governance* (LexisNexis, 2006) 324.

[109] Doolittle, 'Native Customary Land Rights', above n 12, 83.

Asal.[110] Other studies differentiate between the Orang Asli (0.6 per cent of the peninsular population) and 'Natives'. 'Natives', as identified in the Malaysian Constitution, are located in the states of Sabah and Sarawak.[111]

While indigenous groups and local communities who traditionally rely on forest resources comprise approximately 12 per cent of Malaysia's population, indigenous peoples (Natives) constitute 50–60 per cent of the population in Sabah and Sarawak.[112] As of the 2010 census, twenty-eight indigenous groups formed the majority of the population in Sarawak.[113] Inland indigenous peoples are collectively known as Orang Ulu, while more coastal peoples, comprising Iban and Bidayuh peoples, are collectively known as Dayak. The term 'Dayak' has at times been used to include the Orang Ulu group. The use of these two groupings is flexible as they represent officially imposed categories which overlay the classifications adopted by indigenous peoples themselves to acknowledge their own ethnic identities.[114] The two groupings comprise roughly half of the population of Sarawak.[115] Officially, Sabah has more than thirty known indigenous groups, such as the Kadazan Dusun, Murut and Orang Sung, again comprising half of the population.[116] In Sabah, a native community is defined as any group, the majority of whom are natives and who live under the jurisdiction of the local authority or of a native chief or headman.[117]

[110] See, e.g., South East Asia National Human Rights Institutions Forum (SEANF) and Human Rights Commission of Malaysia (SUHAKAM), *Project on Good Practices in Promoting and Protecting the Rights of Indigenous Peoples: Good Practices in Malaysia* (2014) 5–11.

[111] Federal Constitution (Malaysia) art 161A.

[112] Bulan, 'Native Title in Malaysia: A Complementary Sui Generis Proprietary Right under the Federal Constitution Commentary', above n 75, 54.

[113] Human Rights Commission of Malaysia (SUHAKAM), above n 13, 15.

[114] Tan Chee-Beng, 'Indigenous People, the State and Ethnogenesis: A Study of the Communal Associations of the "Dayak" Communities in Sarawak, Malaysia' (1997) 28(2) *Journal of Southeast Asian Studies* 263.

[115] Eleanor Goroh, 'Malaysia' in Kathrin Wessendorf (ed), *Yearbook – The Indigenous World* (International Work Group for Indigenous Affairs, 2011) 280, 280, www.iwgia.org/regions/asia/malaysia.

[116] According to the Sabah Museum, there are seventy-two ethnic and sub-ethnic groups, but according to the federal government list, there are less than thirty. See Human Rights Commission of Malaysia (SUHAKAM), above n 13, 13.

[117] Bulan, 'Native Title in Malaysia: A Complementary Sui Generis Proprietary Right under the Federal Constitution Commentary', above n 75, 58. Local authorities in Sabah are established under the Native Court Ordinance 1993.

In Peninsular Malaysia, the Orang Asli[118] have distinctive laws, culture and identity, and comprise eighteen tribal groups, some having a culture dating back more than 20,000 years that included not only hunting and gathering activities in forests, but forms of trade in forest resources.[119] The Orang Asli historically have been marginalised, with some groups displaced from forest areas and resettled or existing at the fringe of forest development projects.[120]

The Malays, Natives and Aborigines (Orang Asli) are classified as 'bumiputra' – sons of the soil.[121] Embedded in the 'New Economic Policy' of 1971, the classification 'bumiputra' confers positive discrimination status,[122] but the general consensus of research is that the policy has not contributed significantly to advancing prosperity for Malaysia's indigenous groups. Different identity criteria distinguish 'natives' between Sarawak and Sabah,[123] but in each instance the status remains important in terms of providing access to customary land holding. A person can gain native status upon meeting certain criteria, such as assimilation into the culture.[124] The nexus between indigenous identity and customary tenure was examined in Chapter 5, and it remains vital in Malaysia as

> [t]he application of customary laws as a 'system of personal laws' means that the question of Indigenous or native identity is crucial to determining entitlement to rights based on customary laws, native laws and custom.[125]

[118] See, e.g., Aiken and Leigh, above n 37, 830. [119] Aziz et al, above n 61, 641.

[120] Yong, above n 9, 21; see, e.g., Signe Howell, 'Continuity through Change: Three Decades of Engaging with Chewong: Some Issues Raised by Multitemporal Fieldwork' in Kirk Endicott (ed), *Malaysia's Original People: Past, Present and Future of the Orang Asli* (NUS Press, 2016) 57, 70.

[121] Ramy Bulan, 'Indigenous Peoples and the Right to Participate in Decision-Making in Malaysia' (paper prepared for International Expert Seminar on Indigenous Peoples and the Right to Participate in Decision-Making, Chang Mai, Thailand, 20–22 January 2010) 2.

[122] Yong, above n 12, 15.

[123] See Ramy Bulan, 'Native Status and the Law' in Min Aun Wu (ed), *Public Law in Contemporary Malaysia* (Longman, 1999) 248.

[124] Human Rights Commission of Malaysia (SUHAKAM), above n 13, 12 [2.10]–[2.11] (Sabah), see also 14 [2.16]; 14 [2.17] n 14 (Sarawak); Bulan, 'Native Title in Malaysia: A Complementary Sui Generis Proprietary Right under the Federal Constitution Commentary', above n 74, 58.

[125] Bulan, 'Native Title in Malaysia: A Complementary Sui Generis Proprietary Right under the Federal Constitution Commentary', above n 75, 56.

### 1. *Indigenous Peoples' Land Rights in Malaysia*

Within Malaysia, property rights are formally protected against acquisition by the state in the federal Constitution.[126] In addition, there are various protections for customary land rights within federal and state land legislation such as Native Reserves.[127] Since 1997, the courts in Malaysia have recognised a form of common law native title, influenced by the Australian High Court decision in *Mabo (No 2).*[128] Groups must establish a connection to the land claimed over a long period of time and the process of proving title is time consuming and difficult.[129] In turn, the pre-existing customary law or *adat* regime coexists with common law and statutory law.[130] Customary law and *adat*, which is passed down from generation to generation, govern land and resources, often entailing communal forms of use. Generally, customary rights do not include the right to alienate land, and individual rights include forest areas – and evidence of foraging, hunting and gathering on that land.[131] While the common law doctrine of native title is now settled law in Malaysia,[132] there are many claims pending, based on indigenous peoples' historic use and occupancy of forests.[133] Native title rights are recognised by the courts conjunctively with the constitutional law and statutory frameworks.[134] Various statutes regulating natural resources and conservation also impinge upon indigenous customary rights to forest lands and resources in Malaysia.[135]

[126] According to article 13(1), 'No person shall be deprived of property save in accordance with law.' Article 13(2) states that 'No law shall provide for the compulsory acquisition or use of property without adequate compensation': Federal Constitution (Malaysia).

[127] Aboriginal Peoples Act 1954 (Revised 1974) (Malaysia); Land Ordinance 1930 (Revised 1996) (Sabah) (Malaysia); Land Code 1958 (Sarawak) (Malaysia).

[128] Ramy Bulan, 'Ripples of Mabo: Aboriginal and Native Customary Land Rights in Malaysia' in Toni Bauman and Lydia Glick (eds), *The Limits of Change: Mabo and Native Title 20 Years On* (AIATSIS, 2012) 41, 41.

[129] Subramaniam, above n 10.

[130] Bulan, 'Native Title in Malaysia: A Complementary Sui Generis Proprietary Right under the Federal Constitution Commentary', above n 75, 73.

[131] *Adong bin Kuwau v Kerajaan Negeri Johor* [1997] 1 MLJ 418; *Nor Anak Nyawai v Borneo Pulp Plantations* [2001] 2 CLJ 769.

[132] Bulan, above n 75, 67.

[133] Bulan, above n 128, 45.

[134] Bulan, above n 75, 64.

[135] Regarding the Orang Asli, see Aziz et al, above n 61, 641.

Despite various types of legal recognition of indigenous rights, state governments retain much underlying control of Malaysia's forested lands. In practice, indigenous customary land and resource rights are often contested, displaced or delayed due to bureaucratic inefficiencies. Nonetheless, as Doolittle notes in respect of Sabah:

> [r]egardless of whether the state criminalizes, selectively codifies, or invents customary law, local institutions of access to and ownership of natural resources often persist in everyday practice, with or without official recognition from the state.[136]

Thus, indigenous peoples' rights to forest lands in Malaysia (including swidden areas) must be understood within complex, multifaceted, statutory and customary law systems that govern lands and forests. Capturing this coexistence may prove difficult in terms of translating this system into the model of security of tenure that is proposed for REDD+.

### (a) A Legacy Land Law

Adding to the complexity are the strong regional variations in land laws for Sabah, Sarawak and Peninsular Malaysia which reflect both the current political differences and the colonial past of each region.[137] Calls for tenure security and clarity that are now being raised as a prerequisite for implementation of REDD+ confront a legacy in Malaysia where, despite a system of legal pluralism, historically there was limited security for customary law in many parts of the country. Further, significant aspects of customary rights to the use of forests were omitted from the customary systems that were formally 'recognised' in colonial statutes. In Sabah, for example, land law systems put in place under the North Borneo Chartered Company brought 'the logic of Western-based property rights and economic rationality [which] replaced indigenous rights to resources and local patterns of resource use'.[138] Even when land was reserved for the use of 'natives', these protections were often ignored to facilitate the plantation agricultural system.[139] Moreover, as Doolittle notes, the statutory land law entrenched a system of paternalistic control that still continues, serving as 'one of the largest obstacles to natives gaining title to their land in the 21st century where the ruling elite are

136 Amity A Doolittle, *Property and Politics in Sabah, Malaysia: Native Struggles over Land Rights* (University of Washington Press, 2005) 24.

137 Majid Cooke, Toh and Vaz, above n 51, 11.

138 Doolittle, above n 136, 29.

139 Ibid. 5.

able to place their land claims at the forefront of the application process overriding pre-existing native claims'.[140]

### (b) Statutory Formalisation of Indigenous Tenure and Rights

The statutory system, with its colonial origins, remains the primary process for giving formal legal effect to customary tenures in Malaysia. The express statutory recognition of indigenous peoples' rights in forest areas is a matter for each Malaysian state to determine – thus indigenous peoples' rights to forest lands across Malaysia are accorded varying levels of formalisation, tenure security and recognition of communal title. Mechanisms such as native reserves have been limited in their practical implementation.[141] Majid Cooke suggests that the land 'codes' enabled a centralisation of forest governance and effected the transfer of control over forests from decentralised local community groups to the bureaucratic state.[142] The following provides a snapshot of relevant land laws.

(i) **Peninsular Malaysia** A range of land legislation and forest governance measures impinge upon the Orang Asli in Peninsular Malaysia. Although forest lands are formally under state control, the federal government has wider powers in Peninsular Malaysia than in Sabah and Sarawak. The National Land Code 1965, which applies only in Peninsular Malaysia, governs land tenure and institutes a land-title registration system. It does not provide for indigenous communal titles, although it contains an ambiguous savings provision for customary tenure.[143] The Aboriginal Peoples' Act 1954 provides for 'Aboriginal Areas' and the gazettal of 'Aboriginal Reserves'. The gazettal of Orang Asli traditional areas is incomplete and inefficient.[144] Permanent Reserved Forest cannot be established in 'Aboriginal Reserves', but the Act allows permanent reserved forest in 'Aboriginal Areas',[145] albeit with a first right of Orang

[140] Doolittle, 'Native Customary Land Rights', above n 12, 81.

[141] Amity A Doolittle, 'Redefining Customary Law in Govuton' in *Property and Politics in Sabah, Malaysia: Native Struggles over Land Rights* (University of Washington Press, 2005) 67, 70–78.

[142] Majid Cooke, Toh and Vaz, above n 51, 15.

[143] 'Except in so far as it is expressly provided to the contrary, nothing in this Act shall affect the provisions of- (a) any law for the time being in force relating to customary tenure': National Land Code 1965 (Malaysia) s 4(2).

[144] Aziz et al, above n 61, 641.

[145] See *Koperasi Kijang Mas v Kerajaan Negeri Perak* [1991] CLJ 486.

Asli to gather forest products.[146] In 2011, Aboriginal Areas located in forest/wildlife reserves or national parks covered 9,873 hectares.[147]

Even in gazetted areas, Orang Asli have insecure title to ancestral land (in effect the rights constitute tenancies at will) and allocated land can be resumed by the state.[148] In 1996, there was the first successful native title claim to protect Aboriginal customary interests in forests – *Adong bin Kuwau v Kerajaan Negri Johor.*[149] Yet, 'the federal government, along with certain of the state governments, has continued to adopt an adversarial approach to indigenous land issues'.[150] In contrast, the 'native peoples' of Sabah and Sarawak have relatively greater, although still limited, statutory formalisation of customary lands and rights.[151]

(ii) **Sarawak** In Sarawak, the complex land and natural resource laws express the legacy of the Brooke (North Borneo Company) and British colonial regimes; a post-independence system with some codification of indigenous communal rights, together with recent native title claims. Following the cession of Sarawak in 1946, the colonial government enacted the Land (Classification) Ordinance 1948. Under the Ordinance, natives were entitled to occupy interior area land (where many ancestral territories were located), but were deemed licensees of the Crown. The legal effect was the removal of the proprietary character of these rights from people who for generations had occupied that land.[152] As a mere licence, it suggests that 'natives had only a usufructuary right with no kind of ownership'.[153] Post-independence statutory regimes have continued this vulnerability of indigenous communal rights within Sarawak. This situation poses a significant challenge for REDD+ projects in terms of establishing 'tenure security' within Sarawak.

The 1958 Sarawak Land Code recognises native communal reserves and native customary rights. Native Communal Reserves occur on Crown lands. Indigenous groups who claim native customary rights must

[146] Chip Fey, 'Review of Legal Frameworks for Community-Based Natural Resource Management in Selected Asian Countries (Draft)', World Agroforestry Centre (September 2007) 51; Aziz et al, above n 61, 648.

[147] Wyn, above n 46. [148] Aiken and Leigh, above n 37, 834. [149] [1997] 1 MLJ 418.

[150] Aiken and Leigh, above n 37, 825. [151] Ibid. 834.

[152] Ramy Bulan, 'Native Customary Land: The Trust as a Device for Land Development in Sarawak' in Fadzilah Majid Cooke (ed), *State, Communities and Forests in Contemporary Borneo* (ANU ePress, 2006) 48, 48, http://press.anu.edu.au/titles/asia-pacific-environment-monographs/borneo_citation.

[153] Ibid.

have held these rights prior to 1958 – with obvious difficulties in bringing evidence to prove the rights. Further, the group must obtain a permit from the relevant superintendent of land. Section 5(2) of the Land Code sets out six ways to establish native customary rights, including by felling virgin jungle and occupation of the land, planting of land with fruit trees and occupation of cultivated land. The Sarawak Forest Ordinance 1954 governs native customary rights in forest areas, and prohibits indigenous people from creating customary rights in forest reserves and state lands without authorisation from the district officer. Indigenous communities can apply to register their respective rights but registration must occur within sixty days from the gazettal of forest areas, otherwise the rights are extinguished.[154] This policy has been strongly criticised as allowing insufficient time for claims registration.[155] Where native customary rights have been established, companies exploiting forest resources via a concession from the state must compensate for any violation or disturbance of traditional rights or agricultural areas belonging to local communities.[156]

The landmark native title case in Sarawak, *Nor Anak Nyawai*,[157] involved the plaintiff's claim that a pulp and paper company had trespassed and damaged their ancestral land. The High Court ruled that the plaintiffs had preexisting rights, evidenced by continuous occupation, which had not been extinguished by the Sarawak Land Code. This decision was overturned by the Court of Appeal[158] on the basis that there was insufficient evidence of actual, traditional cultivation or 'temuda' in the disputed area.[159] As Bulan notes:

> Limiting occupation to actual settlement and separating it from the surrounding lands on which rests a native community's traditional culture and livelihood strikes at the very basis of claims based on customary practice.[160]

The need to consider both formal cultivated areas and rights to surrounding forest areas as a package is an important consideration when developing measures around tenure security for REDD+ projects. It draws attention to the need to consider a spectrum of customary forest

[154] Sarawak Forest Ordinance 1954 (Malaysia), s 6.
[155] Wells, Chiew and Keong, above n 28, 15. [156] Ibid. 16.
[157] *Nor Anak Nyawai v Borneo Pulp Plantations* [2001] 2 CLJ 769.
[158] *Superintendent of Land & Surveys, Bintulu v Nor Anak Nyawai* [2006] 1 MLJ 256.
[159] Bulan, 'Native Title in Malaysia: A Complementary Sui Generis Proprietary Right under the Federal Constitution Commentary', above n 75, 67.
[160] Ibid.

rights from those akin to 'title' to use-based resource rights when REDD+ projects are contemplated. Further, if the current orientation towards formalisation of land tenure rather than use rights (or resource tenures) remains the dominant way of conceiving security of tenure, it may be difficult to include many customary access and resource rights in Malaysian forest areas within REDD+ models. The difficulties are exacerbated as the Sarawak Land Code and Forest Ordinance have been amended to restrict customary rights.[161] The Land Surveyors Ordinance 2001 was enacted to criminalise community mapping[162] and to restrict their admissibility as evidence.[163] These changes make it difficult to ascertain whether customary forest rights exist, and may limit the effective acknowledgment of such rights in timber certification and REDD+ frameworks.

The expansion of oil palms represents another challenge. To date, much of the expansion in Sarawak has taken place in state forests (which may well be a source of subsistence for forest-dependent communities), but increasingly plantations are being established on areas of native customary rights, where communities traditionally practice semi-subsistence cultivation.[164] To be effective in these circumstances, REDD+ frameworks must effectively include customary rights while also competing against the expansion of palm oil.

**(iii) Sabah** Land law within Sabah, like Sarawak, reflects a colonial legacy that has been largely retained within the post-independence statutory frameworks.[165] In Sabah, the plurality of legal modes for determining entitlements to land and forest resources has been more clearly acknowledged than in other parts of Malaysia (even if customary modes were often displaced in the push to develop Sabah). Moreover, the initial exclusions of various types of native customary rights from the process of

[161] In 1996, a number of amendments were made to the Land Code 1958 to shift the burden of proof to the native claimant to prove the existence of customary rights, and to insert a presumption that the state owned the land without the presence of native customary rights until those rights are proved: Ramy Bulan and Amy Locklear, *Legal Perspectives on Native Customary Land Rights in Sarawak* (Suruhanjaya Hak Asasi Manusia, 2008).

[162] Community mapping uses traditional knowledge and anthropological evidence to establish the patterns of traditional customary use of areas and/or resources. The mapping is often used as evidence in native 'title' claims.

[163] Yong, above n 9, 26.

[164] Astrid Oberborbeck Andersen et al, 'Negotiating Development Narratives within Large-Scale Oil Palm Projects on Village Lands in Sarawak, Malaysia' (2016) 182 *The Geographical Journal* 364, 365.

[165] Doolittle, 'Native Customary Land Rights', above n 12, 88.

formalisation, and the incomplete demarcation of customary lands in the early colonial period, effectively limited the extent of customary tenure that was included in statutory land laws. Statutory formalisation was principally directed to 'native' land kept under permanent cultivation.[166] This orientation meant that the characteristic, communal, swidden customary rights were not easily accommodated. Increasing formalisation was also part of a strategy to derive a revenue basis from native lands as formalisation was the basis of land taxes. Some colonial administrators sought to insulate native lands from being alienated in commercial market processes, as the forest lands were progressively developed. The mechanisms adopted by the administrators included introducing measures such as the inalienability of native lands, except to the Crown.[167]

The Sabah Land Ordinance 1930 currently regulates native tenure in Sabah. Section 65 defines customary tenure as land in lawful possession by natives either in continuous occupation or cultivation for three or more consecutive years or held by title. This interest confers 'a permanent heritable and transferable right of use and occupancy in his land' subject to certain conditions.[168] 'Undocumented' customary tenures are required to be registered, while section 69 sets out requirements for making a claim over state land. Claims are limited to twenty hectares of land, with cultivation conditions and rent payments.[169] The procedure is marked by a lengthy claims process. Any untitled land is regarded as state land. A claim for native customary rights can be made in Sabah in circumstances that include: land possessed by customary tenure; land planted with fruit trees; isolated fruit trees, sago, rotan or other plants of economic value; grazing land; land that has been cultivated or built on within three years; and burial grounds or shrines.[170] When native customary rights are established, either monetary compensation or a grant of the land to the claimant occurs with a title issued.[171]

Many natives still do not have title to land which has been in possession and use by their families for generations.[172] Some people have unsuccessfully tried to formally register their land claims as native title over many years without success.[173] The backlog of native title claims constitutes a major source of conflict.[174] The large number of unsettled

[166] Ibid. [167] Ibid. 93.

[168] Land Ordinance 1930 (Sabah) (Malaysia) s 66. Conditions pertain to continuing to cultivate land and communal obligations.

[169] Ibid. ss 70–71. [170] Ibid. s 15. [171] Ibid. s 16.

[172] Doolittle, 'Native Customary Land Rights', above n 12, 95. [173] Ibid. 94. [174] Ibid. 95.

claims increases the potential for overlapping claims and disputed boundaries between possible REDD+ project participants, with consequences for payment of benefits. In turn, the overlaps make the adoption of FPIC measures quite problematic as it is unclear which groups or people should be consulted. This raises the thorny question of whether REDD+ safeguards are applicable to unsettled claims based on historic use or whether the principles are limited to currently recognised customary tenures. Further, even if the safeguards and calls for tenure security leverage an increase in the settlement of claims, the extent of the claims backlog makes the task very challenging.

The Sabah Forest Enactment 1968 does not allow for native title in Permanent Reserved Forest. Under Section 14 of the Act, indigenous communities can register certain rights and privileges over forest reserves within three months from gazettal of the forest area. Rights are extinguished if not registered and subject to cancellation if unused for three years. Additionally, Sabah's Sustainable Forest Management Licence Agreements oblige forestry-licence holders to respect the rights and privileges included in the customary law gazettal process, including rights to gather forest products for subsistence and the designation of community forestry areas. These offer useful models for consideration in the implementation of the UN-REDD Programme. These types of provisions were included to assist the transition of the SFM licence agreements towards FSC certification in Sabah. These obligations are unique to the state of Sabah and are not mandated in forestry concession licenses in Peninsula Malaysia or Sarawak.[175]

### (c) REDD+ and Communal Title

One way in which the State of Sabah ostensibly addressed the backlog of titling for customary tenures was to develop the concept of communal titles in 2011.[176] Communal titles under the Land Ordinance are a 'development intervention' operating over large areas of customary tenure.[177] Under this scheme, a communal title ties land use within the customary title lands to the expansion of commercial palm oil

[175] Fey, above n 146, 54.

[176] Land Ordinance native communities and village groups can apply for a communal title, rather than individuals seeking title; however, once granted, there is no 'power of sale' and rent may be levied by the minister: Land Ordinance 1930 (Sabah) (Malaysia) s 76.

[177] Doolittle, 'Native Customary Land Rights', above n 12, 99–100.

production.[178] Communal titles enable the government agency, the Sabah Land Development Board, to enter into palm oil joint ventures with indigenous communities.[179] Communal titles are inalienable and title is held by the collector of land revenue, as trustee for the 'natives' concerned.

As oil palm plantations are a strong driver of development, the measure was part of a policy designed to move rural peoples away from subsistence agriculture and its incidental poverty.[180] Results have been very mixed. An empirical study of the introduction of oil palm plantations found that, 'incorporation into the market economy through cash crop production has diversified local economies and improved the socio-economic status of some communities in Sabah, but also has caused the violation of rights of access to the means of livelihood'.[181]

If the UN-REDD Programme is to be successfully implemented in Sabah, clearly it will need to offset the opportunity costs for indigenous peoples generated by oil palm development, address security of customary tenure and provide sufficient economic incentives for local communities to commit forest lands to these 'passive uses' for long periods.[182] Tenure security will need to be matched by economic certainty for communities. In Sabah, the introduction of REDD+ represents another facet of a series of international and national programmes that have, as part of their guiding rationale, the development of land and natural resources to achieve benefits for local communities. Increasingly central to the international REDD+ regime is the view that for local forest communities and indigenous peoples to secure benefits from REDD+ projects there must be clarification of tenure arrangements.[183] Yet, as Doolittle highlights, land law since the colonial period has been as much a source of the dispossession of local communities as it has been a means of securing protections.[184] Accordingly, REDD+ frameworks should be cognisant of the potential for tenure-security measures to have adverse

[178] Majid Cooke, 'In the Name of Poverty Alleviation', above n 15, 241.

[179] Majid Cooke, Toh and Vaz, above n 51, ch 4.

[180] Majid Cooke, 'In the Name of Poverty Alleviation', above n 15, 240. [181] Ibid. 241.

[182] Andersen et al, above n 164, 368–369; see also 370–371.

[183] Ann-Kristin Rothe and Paul Munro-Faure, 'Tenure and REDD+: Developing Enabling Tenure Conditions for REDD+' (Policy Brief, Issue No 6, UN-REDD Programme, 2013), http://theredddesk.org/resources/tenure-and-redd-developing-enabling-tenure-conditions-redd.

[184] Doolittle, 'Native Customary Land Rights', above n 12, 94–95, 99.

impacts and to cut across local practices, as well as for these programmes to be points of mediation and resistance for local communities.[185] This is not to indicate that securing indigenous and customary rights to land and forest resources should not be pursued. Rather, there needs to be renewed emphasis upon understanding the fine-grained nature of customary tenure and rights in forests, as well as the myriad and often partial nature of their recognition within the mainstream legal system. The process may even require a transformation of property relationships between the state and local communities; a situation that brings into issue the wider questions of national sovereignty and the normative impetus of international law and guidance expressed through the REDD+ regime. Malaysia is at a critical point in this process as it consolidates its readiness framework, completes technical reference measures and develops guidelines for safeguards.

## D. REDD+ in Malaysia

Malaysia signed the UNFCCC in 1993 and ratified the convention in July 1994.[186] After formally joining the UN-REDD Programme in 2012, it received supporting funds towards the preparation of a REDD Readiness Plan with contributions made by the Malaysian government.[187] The initial National REDD+ Readiness in Malaysia Programme, which finished in 2013, did not involve a REDD+ project.[188] Instead, it was directed towards addressing deforestation, forest degradation, the enhancement of carbon stocks, conservation and SFM. The Peninsular, Sabah and Sarawak forest departments took the lead with technical assistance by the Forest Research Institute Malaysia and with expertise also provided by the World Wildlife Fund for Nature.

[185] Lisa Ogle, 'Free, Prior and Informed Consent for REDD+ in the Asia-Pacific Region: Lessons Learned, Challenges and Recommendations' (UN-REDD Programme, August 2012) 7.

[186] Sabah Forestry Department and WWF (Malaysia), 'REDD+: Developing Sub-National Strategy in the State of Sabah, Malaysia', Sabah Forestry Department, Sandakan, Malaysia (2011) (copy on file with author).

[187] The relevant financing agreements are RD3569 US$543,000: Funding Institution United Nations Development Programme (Malaysia), and RD3570 US$131,000: Funding Institution Government of Malaysia: 'National REDD+ Readiness in Malaysia', above n 70.

[188] Ibid.

### 1. *Overview*

The UNDP supported Malaysia in the development of a national framework for the policy, legal and institutional components of REDD+. This involved setting up a sustainable funding mechanism for REDD+ and improving stakeholder participation in the UN-REDD Programme. Malaysia is a programme partner and official observer to the UN-REDD Programme Policy Board. The country is taking a phased approach, commencing with readiness, including development of the national REDD+ strategy and capacity building, although some pilot projects have been initiated.[189] Malaysia is focussing its efforts in REDD+ on Permanent Reserved Forest and protected areas. Co-benefits (environmental) are promoted through enhancing natural forests and biodiversity and degraded forest carbon stocks, and '[l]ocal community participation will be included where applicable'.[190] Working groups were established to set emission baselines; measurement, reporting and verification (MRV); institutional arrangements; governance; arrangement for payment of benefits; and capacity building.[191] Malaysia also has signed a Voluntary Partnership Agreement with the EU Forest Law Enforcement, Governance and Trade (FLEGT) initiative, but there are fears that verification work may cost more than benefits from such partnerships.[192]

A major part of the 'readiness stage' relates to providing carbon baselines for forest areas. The Forest Research Institute Malaysia has undertaken the scientific component of the development of baselines in conjunction with regional forestry agencies.[193] A number of international organizations, including the Forestry and Forest Product Research Institute of Japan and the International Tropical Timber Organisation, have supported these efforts. Malaysia appears to have moved steadily along this 'readiness' path with an announcement around reference levels at COP22.[194]

189 Rahman, Philip and Pan, above n 26. 190 Ibid. 191 Ibid.

192 Benjamin Cashore and Michael W Stone, 'Can Legality Verification Rescue Global Forest Governance? Analyzing the Potential of Public and Private Policy Intersection to Ameliorate Forest Challenges in Southeast Asia' (2012) 18 *Forest Policy and Economics* 13, 18.

193 Interview with staff from the Forest Research Institute, above n 18; see also the Forest Research Institute Malaysia, 'FRIM, FFPRI Share Research Findings REDD+' (6 February 2013) Ministry of Natural Resources and Environment, www.frim.gov.my/frim-ffpri-share-redd-research-findings.

194 Malaysian submissions to the UNFCCC on the REDD+ web platform are on forest reference emissions level and/or forest reference level (not updated since 2015); REDD+

Malaysia's REDD policy developments are coordinated on a national level, but use a 'nested' approach, where accounting for carbon and credits is to occur at both national and subnational levels. At the subnational level, Sabah has been taking an active lead in readiness activities. Sabah has instituted a REDD+ road map outlining the steps needed to become REDD+ ready.[195] The Sabah Forestry Department hosted a workshop in August 2011 as part of the Heart of Borneo Initiative,[196] with the assistance of the World Wildlife Fund,[197] to facilitate stakeholder consultation on the proposed roadmap. The roadmap forms the basis for the state's subnational strategy,[198] and includes plans for sourcing financing, defines FPIC and outlines pilot projects.[199] In contrast to Sabah's activity, progress in Sarawak has been hindered not only by land claims but by a downturn in forest productivity.[200]

The Malaysian position on adopting safeguards, tenure security for forest communities and co-benefits remains formative. In 2012, it was stated that:

> As the National REDD+ Strategy is under development, issues relating to rights and tenure for REDD+ have yet to be fully considered. The National REDD+ Strategy covers how safeguards and benefit sharing systems will be formed.[201]

Resolution of these issues is a crucial component of REDD+ if the broader socio-economic and human rights objectives that are now associated with the programme are to be achieved.[202] The extent to which Malaysia is able to develop viable safeguards and measures in relation to customary tenure security, resource rights and benefit-sharing – and to have them agreed upon and ultimately adopted – will be critical to the

Web Platform, 'Submissions: Malaysia', United Nations Framework Convention on Climate Change, http://redd.unfccc.int/submissions.html?country=mys. The latest submission, 'Modified Submission on Proposed Reference Level', is undated but was likely drafted in late 2015 or early 2016.

195 Sabah's REDD+ Roadmap provides that an FPIC guideline and a benefit distribution system should be agreed upon by 2014: 'REDD in Malaysia', above n 17.

196 'Tackling Climate Change through Sustainable Forest Management and Community Development in Malaysia' (2012) The REDD Desk: A Collaborative Resource for REDD Readiness, http://theredddesk.org/countries/initiatives/tackling-climate-change-through-sustainable-forest-management-and-community.

197 Ancrenaz et al, above n 101.

198 Joe Leong, 'Sabah Moves Forward with REDD+ Road Map' *Borneo Post*, 24 August 2011, www.theborneopost.com/2011/08/24/sabah-moves-forward-with-redd-road-map-latest.

199 Interview with Kugan, above n 20.

200 Cashore and Stone, above n 192, 18.

201 'REDD in Malaysia', above n 18.

202 Ogle, above n 185, 31.

successful implementation of the UN-REDD Programme. While workshops and readiness-phase documents in Malaysia recognise the importance of these elements for implementation of the programme, no formal agreement on REDD+ safeguards has been published to date.[203] As a Malaysian submission to the UN-REDD Programme indicates, the safeguards are being built on national systems. A 'Safeguard Information System' will be designed to be more transparent than existing systems, and is to be developed in phases.[204] The submission also lists challenges in developing the system, including engaging relevant stakeholders, linking subnational systems with national ones, resourcing and funding for the safeguards system, and 'information assemblage into a transparent, consistent and comprehensive manner'.[205]

Malaysia is a signatory to the CBD and the Cartagena Protocol,[206] but has not ratified the Nagoya Protocol.[207] These earlier multilateral agreements are to provide guidance to support the implementation of social and environmental safeguards for the REDD+ regime, which in Sabah, are identified as central to the state-level implementation framework.

[203] The authors note a Malaysian government presentation at COP22 in Marrakesh that was reported in the media; the official documentation is not publicly available as of 8 February 2017: Ministry of Natural Resources and Environment (NRE), above n 27.

[204] Subsidiary Body for Scientific and Technological Advice, United Nations Framework Convention on Climate Change, *Report of the Subsidiary Body for Scientific and Technological Advice on Its Thirty-Eighth Session, Held in Bonn from 3 to 14 June 2013*, UN Doc FCCC/SBSTA/2013/3 (29 August 2013) para 29. See also Virginia Wiseman, *UNFCCC Publishes Country Views on REDD+ Safeguard Information Systems* (12 November 2014) IISD/SDG Knowledge Hub, http://sdg.iisd.org/news/unfccc-publishes-country-views-on-redd-safeguard-information-systems/?rdr=climate-l.iisd.org.

[205] Subsidiary Body for Scientific and Technological Advice, United Nations Framework Convention on Climate Change, *Views on Experiences and Lessons Learned from the Development of Systems for Providing Information on How All the Safeguards Are Being Addressed and Respected and the Challenges Faced in Developing Such System: Submissions from Developing Country Parties*, UN Doc FCCC/SBSTA/2014/MISC.6 (6 November 2014) 'Paper No 5: Malaysia', 25.

[206] *Convention on Biological Diversity* (opened for signature 5 June 1992, 1760 UNTS 79, entered into force 29 December 1993) ('Convention on Biological Diversity'); *Nagoya Protocol on Access to Genetic Resources and the Fair and Equitable Sharing of Benefits Arising from Their Utilization to the Convention on Biological Diversity* (opened for signature 2 February 2011, UN Doc UNEP/CBD/COP/DEC/X/1, annex I, entered into force 12 October 2014).

[207] 'Malaysia – Country Profile', Convention on Biological Diversity, www.cbd.int/countries/default.shtml?country=my; 'Parties to the Nagoya Protocol', Convention on Biological Diversity, www.cbd.int/abs/nagoya-protocol/signatories.

## 2. Pilot Programmes

Pilot programmes are regarded as the forerunner to subsequent implementation of REDD+ projects in Malaysia. Sabah has taken initiatives in this regard. The state has experience in forest carbon projects, as the host of the world's first tropical improved forest-management project.[208] The project in the Lahad Datu district has been in operation since 1992. More recently, Sabah received funding from the European Union to tackle climate change through sustainable forest management and community engagement in a project from 2011–2013. The project aimed to enhance institutional governance capacity for REDD+ and focused on three pilots, in partnership with local communities, in the areas of Kg Gana and Lingkabau Forest Reserve, Kinabatangan River Corridor and Kinabalu Ecolinc Zone.[209]

Working in partnership with local communities was an acknowledged priority, as evident in the Kinabalu Ecolinc Project, which was initiated by Sabah Parks to improve ecological connectivity between Kinabalu Park and Crocker Range Park.[210] Approximately 11,000 indigenous peoples live within the Ecolinc Zone.[211] Sabah Tourism, Culture and Environment Minister Datuk Masidi Manjun described the community-focused nature of the project in *The Borneo Post*:

> The government prefers that this initiative not involve any gazetting of a new forest reserve ... we would rather take it as an effort on the part of the village people themselves to look after their own village surrounding environment, with the government assuming the role of a facilitator. We trust the village people because they have been living there for generations.[212]

The EU project-funding document for the pilot projects, however, listed barriers preventing effective engagement with local communities in sustainable forest management. These included current agency policies and

[208] 'First Forest Project on Sabah, Borneo Registered under the VCS' (6 September 2011) Forest Carbon Portal: Tracking Terrestrial Carbon, www.forestcarbonportal.com/news/first-forest-project-sabah-borneo-registered-under-vcs.

[209] 'Tackling Climate Change through Sustainable Forest Management and Community Development in Malaysia', above n 196.

[210] The Board of Trustees of the Sabah Parks, 'Kinabalu Ecolinc Project', www.sabahparks.org.my/index.php/discover-us-3/kinabalu-ecolinc-project.

[211] Murib Morpi, 'Ecological Linkages to Connect National Parks' *Borneo Post*, 22 July 2011, www.theborneopost.com/2011/07/22/ecological-linkages-to-connect-national-parks.

[212] Ibid.

practices, as well as insecure land tenure and forest-resource access rights.[213] Another project in Sabah is an UNDP and GEF-supported initiative on biodiversity conservation in multiple-use forest landscapes.[214] The implementing partners are the Malaysian Ministry of Natural Resources and Environment, Sabah State Economic Planning Unit and Sabah Forestry Department. The project aims to develop an integrated management strategy to mainstream biodiversity, ecosystem functions and resilience, while enabling ongoing sustainable uses.[215] It involves a pilot programme in Southeast Sabah in partnership with Forest Management Unit licensees.[216] Related initiatives include Geoscience Australia support for the provision of satellite data for carbon accounting through provision of data estimating areas of deforestation and degradation.[217] Another EU-supported project involves eight South-East Asian countries (based in the Global Environment Centre in Malaysia) addressing the sustainable management of peatland forests. USAID has supported a project around concessions and forest management in East Malaysia.[218] A growing number of studies are directed at resolving the economic barriers for successful implementation of REDD+ projects.[219]

## E. Conclusion

The increasing number of pilot projects in states, such as Sabah, and the growing sophistication and comprehensiveness of the programmes for meeting the technical requirements for the implementation of the

[213] 'Tackling Climate Change through Sustainable Forest Management and Community Development in Malaysia', above n 196.

[214] United Nations Development Programme (UNDP), 'Biodiversity Conservation in Multiple-Use Forest Landscapes in Sabah, Malaysia' (2012) UNDP in Malaysia, Singapore & Brunei Darussalam, www.my.undp.org/content/malaysia/en/home/operations/projects/environment_and_energy/80468_forestlandscapes.html.

[215] Global Environment Facility (GEF) and United Nations Development Programme (UNDP), 'Biodiversity Conservation in Multiple-Use Forest Landscapes in Sabah, Malaysia' (United Nations Development Programme County: Regional (Sabah, Malaysia) – Project Document, rev 14 March 2012), http://info.undp.org/docs/pdc/Documents/MYS/00060158_4186_FINAL_Sabah_Prodoc.pdf.

[216] 'Tackling Climate Change through Sustainable Forest Management and Community Development in Malaysia', above n 196.

[217] Ibid. [218] Ibid.

[219] See, e.g., Ammar Abdul Aziz et al, 'Using REDD+ to Balance Timber Production with Conservation Objectives in a Mangrove Forest in Malaysia' (2015) 120 *Ecological Economics* 108.

UN-REDD Programme signal a growing momentum for the programme to move to the next stage in Malaysia.[220] Significant issues in critical areas, such as tenure security and the allocation of benefit payments, remain to be resolved. The general perspective Malaysia may take to meet the challenges around implementing REDD+ can be gleaned from a document prepared for Sabah REDD+ Readiness planning:

> REDD+ is based on a simple and appealing idea, but turning the appealing idea into implementation is much more complex. Challenges highlighted by sceptics in designing and implementing and monitoring an economically environmentally and socially sustainable mechanism are surmountable by developing an understanding of the science, methods and possible outcomes of REDD+ relevant to their work and influence.[221]

In this comment Malaysia's orientation towards solving the challenges of REDD+ based primarily on technical 'know how' is apparent.[222] Malaysia does indeed have considerable expertise in forest governance; the nation is familiar with bio-sequestration integrity processes (such as MRV), and the country has long-standing involvement in SFM and forest certification. It has a relatively robust economy (in part due to forest exploitation) and more resources than many other tropical forested countries – but it is taking a cautious approach to REDD+.

On the other hand, the nation faces a legacy of customary land rights and land law issues and there is a record of conflict with indigenous, forest-dependent communities, particularly in East Malaysia. The systems for recognition of indigenous rights in forest areas are complex and variable across the country. Such systems do not reflect the full spectrum of customary uses of the forests – with swidden cultivation at particular risk. These factors will pose significant challenges for 'designing and implementing and monitoring an economically environmentally and socially sustainable mechanism' where REDD+ can be effectively implemented in a culturally appropriate manner. For REDD+ to be effective in Malaysia, it will require a broadening of the process and practices entailed in collaborative forest governance. Robust governance will require meeting the pluralist challenges of recognising and securing lands claimed under

[220] It is noted that official documentation on UN-REDD Programme websites still indicates that Malaysia is in the readiness phase.

[221] Sabah Forestry Department and WWF (Malaysia), above n 186.

[222] See, e.g., Mui-How Phua et al, 'Estimating Logged-Over Lowland Rainforest Aboveground Biomass in Sabah, Malaysia Using Airborne LiDAR Data' (2016) 27(4) *Terrestrial, Atmospheric & Oceanic Sciences* 481.

customary rights and/or forest resource rights to support indigenous and local communities in sustainable forest governance in the face of significant development pressures.

Further, although Malaysia has voted in favour of the UNDRIP,[223] it is unclear whether these principles will be actively translated into the Malaysian national REDD+ policy setting.[224] Indigenous communities have made extensive complaints to the Human Rights Commission of Malaysia (SUHAKAM) about infringement of their customary land rights.[225] Yet the position with respect to many local community land claims continues a pattern of state and federal government reluctance to effectively incorporate indigenous peoples in decision-making and to fully acknowledge the customary rights of those communities.[226] While there is political representation for customary groups in Malaysia and relatively strong traditional authority structures in place, Malaysia's indigenous peoples 'struggle to maintain their identity and control over their land and resources in the face of growing encroachment on their territories'.[227] International human rights principles incorporated into the REDD+ regime, such as FPIC, can play a pivotal role. The effective translation into REDD+ of measures such as tenure security and equitable benefit-sharing for forest-dependent communities in Malaysia will require robust national adoption across many institutional, economic and social spheres, not least within the existing land law and forest resource management structures to support meaningful consultation and justice in the implementation of the REDD+ scheme.

[223] *United Nations Declaration on the Rights of Indigenous Peoples,* GA Res 61/295, UN GAOR, 61st Sess, 107th Mtg, Agenda Item 68, Supp No 49, UN Doc A/RES/61/295 (2 October 2007, Adopted 13 September 2007). See generally United Nations Office of the High Commissioner for Human Rights, Declaration on the Rights of Indigenous Peoples, www.ohchr.org/EN/Issues/IPeoples/Pages/Declaration.aspx.

[224] Colin Nicholas, Jenita Engi and Teh Yen Ping, *The Orang Asli and the UNDRIP: From Rhetoric to Recognition* (Center for Orang Asli Concerns, 2010).

[225] National Human Rights Commission of Malaysia (SUHAKAM), 'Background Paper: SUHAKAM's National Inquiry into the Land Rights of Indigenous Peoples in Malaysia' (May 2011), www.suhakam.org.my/wp; Human Rights Commission of Malaysia (SUHAKAM), above n 13.

[226] Majid Cooke, 'Constructing Rights', above n 48, 517.

[227] Bulan, 'Indigenous Peoples and the Right to Participate in Decision-Making in Malaysia', above n 121, 1.

# 8

# REDD+ in Melanesia

## Papua New Guinea and Vanuatu

LEE GODDEN[1]

### A. Melanesia Overview

Papua New Guinea (PNG) and Vanuatu are two Melanesian countries that exemplify the opportunities that REDD+ can provide to smaller nation states that are heavily dependent on the forest/land use sector for national economic development, but also the complexities and difficulties that such nations may face in successfully implementing the programmes. In both nations, customary landowners constitute the majority of the population; a situation that reconstitutes the 'tenure' question in REDD+ as one directed more closely to effective governance of tenure arrangements rather than tenure security itself. The two case studies therefore offer a useful counterpoint to the growing number of studies of REDD+ implementation that typically look to countries in South America and Africa to examine REDD+ safeguards and tenure compliance at national level. PNG also offers an instructive example of the long-term trajectory for REDD+ implementation as that country had a prominent role in bringing REDD+ to international attention. Further, the subsequent developments in the two Melanesian countries are worthy of attention because they demonstrate an emergent national legal and policy platform to support REDD+, albeit one that must navigate the complexity and uncertainty of aligning customary law and authority structures to that regime at the local community level.

In the ten years following the UNFCCC conferences in which PNG, in particular, advocated for REDD+ (see Chapter 2), there have been wide-ranging policy discussions and new domestic governance arrangements. PNG is a partner country of the UN-REDD Programme and a country

[1] The research contributions of Stephanie Niall, Elizabeth McPherson, Cate Read and Carly Godden to this chapter are gratefully acknowledged.

participant of the World Bank's Forest Carbon Partnership Facility (FCPF), and has national FPIC Guidelines. Vanuatu is a country participant of the FCPF but has not received UN-REDD Programme funds, and is yet to address FPIC guidelines in its REDD+ planning process. Both countries are members of the Pacific Islands Regional Policy Framework for REDD+,[2] and both are piloting REDD+ carbon sequestration projects. Further, both countries mention REDD+ as a mitigation action under their 'nationally determined contributions' for the Paris Agreement, although they stress that REDD+ implementation requires considerable international assistance.[3]

These Melanesian countries provide an interesting context for examining the relationship between REDD+ and indigenous communities in developing countries. Despite their colonial legacy, neither PNG nor Vanuatu is a country with an indigenous or tribal minority, and customary land holding attracts constitutional protection. Non-indigenous residents comprise a tiny minority of the population and are not represented in national democratic governance. As such, this chapter points to the particular challenges raised in REDD+ planning and implementation in countries where the vast majority of local inhabitants are members of rural and tribal communities governed by their own customary laws but dependent on forested lands for their livelihoods and cultural continuity.[4]

With this in mind, Section A of this chapter examines the historical context and potential success of REDD+ in PNG and Vanuatu, demonstrating the importance of REDD+ in meeting climate-mitigation targets while managing the challenges presented by customary tenure systems, the pressures of socio-economic development and limited governance

[2] Secretariat of the Pacific Island Community, 'Pacific Islands Regional Policy Framework for REDD+' (Supported by the SPC/GIZ Regional Project, 'Climate Protection through Forest Conservation in Pacific Island Countries', 2013), http://theredddesk.org/countries/policies/pacific-islands-regional-policy-framework-redd.

[3] Papua New Guinea, 'Intended Nationally Determined Contribution (INDC) under the United Nations Framework Convention on Climate Change', submitted 24 March 2016, available at NDC Registry (Interim), www4.unfccc.int/ndcregistry/Pages/All.aspx; Republic of Vanuatu, 'Intended Nationally Determined Contribution (INDC)', submitted 21 September 2016), available at NDC Registry (Interim), www4.unfccc.int/ndcregistry/Pages/All.aspx.

[4] Cf Taiji Fujisaki et al, 'Does REDD+ Ensure Sectoral Coordination and Stakeholder Participation? A Comparative Analysis of REDD+ National Governance Structures in Countries of the Asia-Pacific Region' (2016) 7 *Forests* 195.

capacity.[5] Section B assesses the use made to date of UN-REDD Programme and FCPF readiness funding, where planning has mostly focused on technical accounting and reporting requirements, with limited progress until recently towards resolving tensions around how FPIC should operate with respect to customary land-tenure arrangements. Section C outlines the 'no regrets' approach to REDD+ Readiness and implementation in the Pacific.[6] Section D discusses the entrenched problems of forest-resource development, the land tenure arrangements in PNG and Vanuatu that developed to make customary interests 'legible' to third-party interests to facilitate development, the subsequent reforms and the implications for REDD+. Experiences in PNG and Vanuatu serve as a caution in respect of the complexities and difficulties inherent in providing for 'secure and clear' tenure as a prerequisite for REDD+ implementation.[7] Section E discusses the development of FPIC principles in each country, while Section F examines NGO funded pilot projects that are similar to REDD+ in both countries as providing instructive examples for moving forward with REDD+ via an experimental process of 'learning by doing'.

## B. Indigenous People and Customary Tenure in PNG and Vanuatu

In PNG, the land rights of the customary owners were not extinguished and were recognised by the colonial state. Post-independence, the PNG nation has constitutional powers to deal with land and resources, although only 3 per cent of the landmass in the nation is 'alienated land' held by the PNG state or privately held by third parties under a

[5] United Nations Development Programme, 'Papua New Guinea: Project Document – Forest Carbon Partnership Facility' (2015) ('UNDP Project Document'), www.pg.undp.org/content/papua_new_guinea/en/home/operations/projects/environment_and_energy/forest-carbon-partnership-facility-reddt-readiness-project.html, 5.

[6] Secretariat of the Pacific Island Community, above n 2, 6. The 'no regrets' approach to REDD+ is designed to keep 'options open to engage with possible future global instruments currently in development, whilst taking advantage of mechanisms already available'.

[7] See, e.g., The Government of Papua New Guinea, 'Readiness Preparation Proposal (R-PP) for Papua New Guinea (Final)' (9 December 2013) ('PNG R-PP') 33, www.unredd.net/documents/un-redd-partner-countries-181/asia-the-pacific-333/a-p-partner-countries/papua-new-guinea-188/r-pp-fcpf-1906.html; Office of Climate Change and Development, 'Guidelines on FPIC for REDD+ in Papua New Guinea' (Final Working Draft, Version 4, August 2014) ('PNG FPIC Guidelines') 8.

lease-hold arrangement with the state. The majority of the land is held under customary law.[8] Upon independence as a federation in 1980, the former Anglo-French condominium of New Hebrides, renamed 'Vanuatu' (meaning 'our land'), restored all land 'to the indigenous custom owners and their descendants'.[9] Ninety-nine per cent of land in Vanuatu is customary land. These circumstances in the Melanesian countries challenge assumptions about the basis of the rationale for safeguards for indigenous communities in REDD+[10] where the indigenous population is not a minority but where the communities may still be characterised as vulnerable and identify as 'customary' communities, rather than 'indigenous' people in public and political discourse. The draft PNG FPIC Guidelines, for example, 'use the term "customary landowners" to refer to indigenous peoples and communities who have customary rights over land and resources on the land', while 'recogniz[ing] that there may be communities who do not own land, but may be accorded use rights over land and resources on the land'.[11] Nonetheless, customary communities in each country are still captured by the indigenous peoples' policies and safeguards of United Nations development agencies and the World Bank.

In both countries, customary land rights are protected in the national constitution,[12] and these rights are inalienable except in accordance with customary law.[13] The constitutional position means that customary property can only be acquired by the government[14] or transferred to another indigenous community, consistent with established 'custom'. The circumstances of these Melanesian countries provide an opportunity to consider the complexities of integrating REDD+ with customary systems, especially the necessity of reconciling customary rights and formal title (i.e. derived from Westernised legal title) as well as securing FPIC in contexts where almost all tenure is held under communal customary law.

[8] H A Amankwah, J T Mugambwa and G Muroa, *Land Law in Papua New Guinea* (LBC Information Services, 2001) 62.

[9] Margaret Rodman, 'Breathing Spaces: Customary Land Tenure in Vanuatu' in R Gerard Ward and Elizabeth Kingdon (eds), *Land, Custom and Practice in the South Pacific* (Cambridge University Press, 1995) 65.

[10] Government of Vanuatu, 'Readiness Preparation Proposal (R-PP) for Vanuatu (Final)' (7 October 2013) ('Vanuatu R-PP') art 73, http://theredddesk.org/resources/readiness-preparation-proposal-r-pp-vanuatu.

[11] See, e.g., PNG FPIC Guidelines, above n 7, 8.

[12] Constitution of the Independent State of Papua New Guinea 1975.

[13] Land Act 1996 (PNG) s 132; Constitution of the Republic of Vanuatu 1980, art 74.

[14] Land Act 1996 (PNG) s 10: The government can only acquire customary land by agreement with the customary landholders.

Customary law in Melanesia is highly dynamic, complex and heterogeneous. This too presents challenges for both customary landholders and the third parties who may want to deal with the land to support resource development. Customary land is held communally at the tribal or clan level, inherited through patrilineal or matrilineal descent lines in accordance with locally specific custom, and 'has boundaries which are established by memory, transmitted through oral tradition, and often use natural features in their establishment'.[15] These features make it difficult to establish the scope and boundaries of customary land, and vastly complicate the process of identifying customary landholders for consultation and benefit-sharing purposes. In the past, in order to render customary interests more 'legible' to the state, resource developers and funding agencies, the governments in PNG and Vanuatu have designed various statutory schemes to 'translate' such interests into a form more easily administered by the state. These schemes typically have been the means of securing consent from communities and facilitating the distribution of benefits arising from project developments.[16] The various schemes have created legal processes and 'groups' through which governments and other entities interact with customary owners: in PNG, this interaction is accomplished via Incorporated Land Groups (ILGs), and in Vanuatu, through long-term leasing practices. These schemes enable dealings with the land that would otherwise breach protections for customary title by facilitating its alienation to third parties.

The schemes have reverberated across Melanesian society, with impacts including disenfranchising communities from decisions regarding use of their lands, and contributing to social destabilisation and disputes. These schemes predate REDD+ but often are used as a basis for REDD+ Readiness activities. The schemes may not conform to FAO Guidelines, including safeguards for recognising and protecting customary interests at the subgroup level.[17] Past experience in the two countries raises questions about the effectiveness of the guidelines, and whether they offer robust protections for forest-dependent communities to enable holders of customary interests to gain substantial benefits from REDD+. In this

[15] PNG R-PP, above n 7, 33.

[16] Dan Jorgensen, 'Clan-Finding, Clan-Making and the Politics of Identity in a Papua New Guinea Mining Project' in James F Weiner and Katie Glaskin (eds), *Customary Land Tenure and Registration in Australia and Papua New Guinea: Anthropological Perspectives* (ANU ePress, 2007) 57, 57.

[17] See Chapter 6.

respect, the challenges arising in REDD+ planning and implementation in Melanesia represent yet another instantiation of very long-standing debates about the intersection of custom and development, punctuated by controversial government interventions on tenure and local governance, which have been key issues for the World Bank and UN development agencies in their long history of engagement in Melanesia.

In PNG and Vanuatu, the UN-REDD Programme, the FCPF and other REDD+ funding and delivery organizations should seek to ensure that readiness activities and pilot projects allow for the complex patterns of tenure, resource rights and customary uses at the local community level, as well as the constraints posed by the diversity of custom and language groups.[18] Importantly, such an analysis requires attention to the wider social and political landscape and the governance mechanisms operative in each country in order to identify the barriers to successful REDD+ implementation. Here, issues such as the failure of governments to adequately resource or administer existing schemes – including giving effect to consent protocols – threaten to compromise the ability of communities to derive significant benefits from participation in REDD+.

### *1. The Likely Success of REDD+ in PNG and Vanuatu*

The Pacific Islands Regional Policy Framework for REDD+, while acknowledging the small scale of the Pacific forest sector in the global context, is optimistic about the potential for REDD+ in Melanesia to generate positive outcomes through carbon sales, community co-benefits, and climate change adaptation and mitigation outcomes.[19] In both PNG and Vanuatu, geography renders much of the countries' vast natural resources inaccessible, and contributes to the considerable cultural and language diversity found across the indigenous population.[20] PNG is comprised of more than 850 language groups, governed by highly fragmented and heterogeneous customary laws.[21] Eighty per cent of the

[18] Kanchana Kariyawasam, 'Protecting Biodiversity, Traditional Knowledge and Intellectual Property in the Pacific: Issues and Challenges' (2008) 16 *Asia Pacific Law Review* 73, 75–76.

[19] Secretariat of the Pacific Island Community, above n 2, 5.

[20] Rodman, above n 9, 69.

[21] Andrea Babon and Gae Yansom Gowae, 'The Context of REDD+ in Papua New Guinea: Drivers, Agents and Institutions' (Occasional Paper 89, CIFOR, 2013) 1, 17; Katherine Dixon, 'Working with Mixed Commons/Anticommons Property: Mobilizing Customary Land in Papua New Guinea the Melanesian Way' (2007) 31 *Harvard Environmental Law Review* 219, 226.

largely rural population remains dependent on semi-subsistence agriculture.[22] Customary land and associated arrangements provide the social safety nets and food security, of particular significance given the limited capacity of the central state.[23] This challenging environment is borne out in PNG's social indicators, which are amongst the worst in the Pacific and, indeed, the world.[24]

Vanuatu is a small island developing state located in the Southwest Pacific Ocean, comprised of an archipelago of more than eighty islands with a total land area of approximately 1.22 million hectares.[25] In 2016, Vanuatu's population was estimated to be around 271,000 people.[26] UN statistics for Vanuatu record that the country fares poorly on key development indicators and therefore is classified as a least-developed country. Landholdings in Vanuatu are dominated by two types of tenure: non-leased customary land (89.7 per cent) and leased customary land (9.3 per cent). A third category, public or government-owned land, accounts for a very small proportion of all land in Vanuatu (1 per cent).[27] In both countries, customary land ownership is protected in the Constitution and deeply woven into the fabric of land regulation.[28] In addition to their involvement in international climate change law and policy, both PNG and Vanuatu have undertaken international human rights obligations. Both countries have ratified the ICCPR,[29] and PNG has ratified CERD,

[22] Andrea Babon, 'Snapshot of REDD+ in Papua New Guinea' (CIFOR *Infobrief* No 40, August 2011) 1; Lawrence Kalinoe and Josepha Kiris, 'Socio-Economic Changes and Their Implications for Customary Land Tenure' in Charles Yala (ed), *The Genesis of the Papua New Guinea Land Reform Program: Selected Papers from the 2005 National Land Summit* (National Research Institute, 2010) 23, 23; Institute for Global Environmental Strategies (IGES), 'REDD+ Readiness in Papua New Guinea: State of Play (Draft)' (August 2012) 4.

[23] Kalinoe and Kiris, above n 22, 23.

[24] United Nations Development Programme, 'Human Development Report 2015' (2015), http://hdr.undp.org/sites/default/files/2015_human_development_report_1.pdf. According to the United Nations Development Programme's Human Development Index, PNG rated 158 out of 188 countries.

[25] UN Data, 'Country Profile – Vanuatu' (2016), http://data.un.org/CountryProfile.aspx?crName=Vanuatu#Environment.

[26] Ibid.

[27] World Bank, 'Leases in Vanuatu: Key Data from the World Bank Jastis Blong Evriwan Vanuatu National Leasing Profile' (December 2010), http://siteresources.worldbank.org/INTJUSFORPOOR/Resources/Vanuatu_Leasing_Data_Summary.pdf.

[28] Constitution of the Independent State of Papua New Guinea 1975, sch 2.1; Constitution of the Republic of Vanuatu 1980, arts 73–74.

[29] United Nations Human Rights Office of the High Commissioner, 'Status of Ratification Interactive Dashboard', http://indicators.ohchr.org.

leading some to argue that REDD+ has the potential to deliver significant benefits to indigenous communities.[30] On the other hand, unlike their Pacific neighbour Fiji, neither country has ratified ILO Convention 169 and both were absent during the General Assembly vote on the UNDRIP in 2007.

Cautious optimism about the potential for REDD+ implementation in PNG has been reinforced by the prominent role that PNG has played in promoting REDD+ internationally which, as we described in Chapter 2, included formally presenting REDD+ to the UNFCCC alongside Costa Rica.[31] However, despite early enthusiasm around 2008, domestic REDD+ implementation stalled soon after, undermined by a corruption scandal, political upheavals[32] and concerns expressed by many international organizations that PNG may not have the capacity to adequately implement and monitor REDD+.[33] Since 2011, PNG has received UN-REDD Programme and FCPF funding which, along with additional bilateral arrangements, has allowed the PNG government to move cautiously forward with readiness activities,[34] including by developing national climate change policies and legislative frameworks that will govern REDD+ planning and implementation.

Despite this modest progress, optimism about the political capacity to fully implement REDD+ in PNG should be tempered: the PNG economy remains highly dependent on natural resource extraction, particularly

[30] David Melick, 'Credibility of REDD and Experiences from Papua New Guinea' (2010) 24 *Conservation Biology* 359, 359; Nalau Bingeding, 'Reducing Emissions from Deforestation and Degradation of Forests in Papua New Guinea: Issues and Options' (National Research Institute Discussion Paper No 119, National Research Institute, Papua New Guinea, 2011) vi.

[31] Conference of the Parties, United Nations Framework Convention on Climate Change, *Reducing Emissions from Deforestation in Developing Countries: Approaches to Stimulate Action – Submission of Papua New Guinea and Costa Rica*, UN Doc FCCC/CP/2005/Misc.1 (11 November 2005) ('Reducing Emissions from Deforestation in Developing Countries').

[32] UN-REDD Programme, 'National Programme Semi-Annual Report: Papua New Guinea' (September 2013) 11, www.unredd.net/index.php?option=com_docman&task=doc_download&gid=11966&Itemid=53.

[33] See, for example, Greenpeace, 'PNG: Not Ready for REDD' (Media Release, 25 October 2010), www.greenpeace.org/australia/en/mediacentre/media-releases/forests-media-releases/png-not-ready-for-redd; Stephen Howes, 'Cheap But Not Easy: The Reduction of Greenhouse Gas Emissions from Deforestation and Forest Degradation in Papua New Guinea' (2009) 24 *Pacific Economic Bulletin* 130.

[34] Babon and Gowae, above n 21, provide a detailed account of institutional adjustments, consultation processes and policies developed in support of REDD+.

forestry and mining.[35] The UNDP reported in 2015 that land use, land use change and forestry account for 95 per cent of PNG's carbon emissions, but generate less than 20 per cent of the Gross Domestic Product.[36] The remaining 5 per cent of emissions derive from the 'mining, petroleum, manufacturing, transport and the energy sectors'.[37] These activities have an additional hold over the country: the government not only relies on export revenue from these extractive industries but also on associated investment to provide basic rural infrastructure, including healthcare facilities, schools and roads,[38] which remain largely absent or inadequate.[39] As such, despite apparent enthusiasm for REDD+, such entrenched dependence on foreign investment may compromise the government's commitment where REDD+ policies have an impact on industry. The government's capacity is further hampered by activities such as illegal logging.[40]

In Vanuatu, there is less pressure from commercial development on forests than in other Melanesian countries, although such pressure is increasing.[41] Forests still have relatively little protection in Vanuatu: in 2011, only 3 per cent of mid- to high-category forest and 0.7 per cent of low forest were in protected areas, (including as Community Conservation Areas under Vanuatu's Environmental Management and Conservation Act 2002).[42] As Vanuatu has relatively low rates of deforestation, REDD+ efforts are targeted towards enhancement of carbon stocks and related measures. Vanuatu is a participant in the FCPF and faces many challenges in developing a REDD+ policy framework and strengthening internal governance.

Vanuatu was an early participant in preparations for forest carbon mitigation schemes. The 1990–2000 remote sensing assessment programme advanced the understanding of the country's forestry stocks through satellite and other technologies to monitor deforestation,

[35] George Yapao, Lee Godden and Steven Pettigrove, 'Papua New Guinea: Conflicts, Customary Landholding and Resource Exploitation' in Marcia Langton and Judy Longbottom (eds), *Community Futures, Legal Architecture: Foundations for Indigenous Peoples in the Global Mining Boom* (Routledge, 2012) 77.

[36] UNDP Project Document, above n 5, 5; PNG R-PP, above n 7, 8.

[37] PNG FPIC Guidelines, above n 7, 9. [38] Babon, above n 22, 8.

[39] Institute for Global Environmental Strategies (IGES), above n 22, 3–4.

[40] Frédéric Mousseau and Peiley Lau, *The Great Timber Heist: The Logging Industry in Papua New Guinea* (The Oakland Institute, February 2016), www.oaklandinstitute.org/great-timber-heist-logging-industry-papua-new-guinea.

[41] Vanuatu R-PP, above n 10, 31.

[42] Department of Forests, 'Vanuatu Forest Policy: 2013–2023' (2011) ('Vanuatu Forest Policy') 10.

building on efforts to gauge national forestry levels under the Vanuatu National Forestry Inventory.[43] The national carbon monitoring and accounting system, institutional capacities and implementation efforts in Vanuatu remained more limited than in countries with high rates of current forest carbon loss.[44]

In 2006, the Vanuatu Carbon Credits Project was implemented as a pilot project to inform intergovernmental policy development on REDD. The project was to 'build capacity to assist Vanuatu to gain access to carbon and ecosystem services finance for climate change mitigation and sustainable development in the forest and energy sectors' and to 'inform international policy development as a consequence of demonstration/ pilot activities'.[45] Vanuatu subsequently released a National Forest Policy in 2011, which expressed an intention to develop a national policy for REDD+.[46] Since signing its R-PP in 2013, however, readiness progress in Vanuatu has been steady but slow due to finance and recruitment delays and a tropical cyclone in 2015.[47] Despite an intention to adopt a national approach to REDD+, Vanuatu still has no dedicated national REDD+ policies or legislation.

In summary, REDD+ has the potential to support climate change mitigation efforts in these two Melanesian countries, noting the particular susceptibility of PNG and Vanuatu to impacts from climate change in the form of rising sea levels and reduced water supplies.[48] The PNG government made a voluntary pledge to reduce its emissions by 50 per cent by 2030.[49] In a densely forested country where the vast majority of GHG

[43] P J Baldwin et al, 'Vanuatu National Forest Inventory' in G B Wood and B J Turner (eds), *Integrating Forest Information over Space and Time: IUFRO Conference 13–17 January 1992* (ANUTECH, 1992) 295.

[44] Sean Weaver and Martin Herald, 'The Forest Carbon Partnership Facility (FCPF) Readiness Plan Idea Note (R-PIN) Vanuatu' (Vanuatu National Advisory Committee on Climate Change, 29 July 2008) 2.

[45] Sean Weaver and Brian Phillips, 'Vanuatu Carbon Credits Project: Phase 2 Roadmap Workshop, 11–13 February, 2008' (Workshop Report, 2008).

[46] Vanuatu Forest Policy, above n 42.

[47] Forest Carbon Partnership, 'REDD+ Annual Country Progress Reporting: Vanuatu' (January 2015–August 2015), http://forestcarbonpartnership.org/sites/fcp/files/2015/September/FCPF_August2015_Vanuatu.pdf.

[48] International Bar Association, Climate Change Justice and Human Rights Task Force, *Achieving Justice and Human Rights in an Era of Climate Disruption* (2014).

[49] Ad Hoc Working Group on Long-Term Cooperative Action under the Convention, UNFCCC, *Compilation of Information on Nationally Appropriate Mitigation Actions to Be Implemented by Parties Not Included in Annex I to the Convention*, UN Doc FCCC/AWGLCA/2011/INF.1 (18 March 2011) 38.

emissions relate to land use and forestry,[50] the PNG government has identified forests as central to mitigation strategies.[51] Although the exact contribution of each primary driver of deforestation is contested,[52] it is generally agreed that commercial forestry is the largest (responsible for around fifty per cent), and dominated by a small number of Malaysian multinational logging companies.[53] It is widely accepted that, since the 1990s, there has been an acceleration in the rate of deforestation associated with (or under the guise of) commercial agriculture, a sector of increasing importance to the PNG economy.[54]

Vanuatu is increasingly interested in climate change mitigation due to its predominantly coastal population; it is one of the most vulnerable countries in the world to the impact of climate change.[55] Customary lands have come under increased pressure in recent years from foreign investment and development schemes, including logging.[56] These development projects, historically, have generated intra-community disputes.[57] Similar concerns arise around schemes to facilitate greater accessibility of third parties to customary land.[58]

Notably, given that almost all forests in PNG and Vanuatu are under some degree of customary ownership, REDD+ planning and

[50] In 2011, emissions from land-use change and forestry in PNG accounted for 86.83 per cent of national emissions: CAIT Climate Data Explorer: 'Papua New Guinea' (2013) World Resources Institute, http://cait2.wri.org/profile/Papua%20New%20Guinea.

[51] Government of Papua New Guinea, Office of Climate Change and Development, 'Interim Action Plan for Climate-Compatible Development' (Draft Document for Public Consultation)', (August 2010) 17, www.occd.gov.pg/images/stories/documents/PNG_Interim_Action_Plan.pdf.

[52] Colin Filer et al, 'Deforestation and Degradation in Papua New Guinea' (2009) 66 *Annals of Forest Science* 813; Phil L Shearman et al, *The State of the Forests in Papua New Guinea: Mapping the Extent and Condition of Forest Cover and Measuring the Drivers of Forest Change in the Period 1972–2002* (University of Papua New Guinea, 2008).

[53] Sam Lawson, 'Illegal Logging in Papua New Guinea' (Energy, Environment and Resources EER PP 2014/04, Chatham House, April 2014) 28.

[54] Ibid. 24.

[55] Government of the Republic of Vanuatu, 'Disaster Risk Reduction and Disaster Management National Action Plan 2006–2016' (2007).

[56] Joseph D Foukona, 'Management of Customary Land as a Form of Community Property in the Solomon Islands, Vanuatu and Fiji' in Lee Godden and Maureen Tehan (eds), *Comparative Perspectives on Communal Lands and Individual Ownership: Sustainable Futures* (Routledge, 2010) 263, 265.

[57] Anita Jowitt, 'Indigenous Land Grievances, Customary Land Disputes and Restorative Justice' (2004) 8(2) *Journal of South Pacific Law* 23.

[58] Vanuatu R-PP, above n 10, 59.

implementation will require activities on customary land.[59] Such a unique context presents an opportunity to examine the relationship of customary law, tenure and REDD+, and the complexities associated with superimposing standard tenure arrangements and associated consultation mechanisms over customary law. Further, as a large proportion of customary owners are dependent on traditional lands for subsistence needs, there is an additional challenge of finding a balance between the commercial opportunities presented by REDD+ and ensuring food security.

### 2. *Customary Land Tenure in Melanesia and Challenges for REDD+*

To understand the complex customary land arrangements, it is necessary to provide a detailed examination of these arrangements in PNG and Vanuatu. In PNG, customary land tenure was not extinguished by the British colonial powers.[60] Upon PNG gaining independence, customary landholding gained constitutional protection. The PNG Constitution also states that customary law forms part of the underlying law of PNG[61] to be adopted and applied, provided it is not inconsistent with statutory law.[62] A similar protection for customary landholding is found in Vanuatu's Constitution. The provision acknowledges that all land in Vanuatu belongs to the indigenous custom owners and their descendants;[63] rules of custom 'shall form the basis of ownership and land use';[64] and 'only indigenous citizens of Vanuatu who have acquired their land in accordance with a recognised system of land tenure have perpetual ownership of land'.[65]

In both countries, customary land rights are protected from government interference, except for a public purpose or interest, and with just compensation to be paid.[66] With a view to protecting customary

[59] Bingeding, above n 30, 2; Babon and Gowae, above n 21, 27.

[60] In *Gaya Nomgui & Others v The Administration of the Territory of Papua New Guinea* [1974] PNGLR 349.

[61] Constitution of the Independent State of Papua New Guinea 1975 sch 2.1 (repeated in section 4(2) of the Underlying Law Act 2000 (PNG)).

[62] Customary land ownership is also protected under the Land Act 1996 (PNG) s 4(1), which excludes customary land from land deemed to be the property of the state.

[63] Constitution of the Republic of Vanuatu 1980 arts 73–74. [64] Ibid. art 74.

[65] Ibid. art 75.

[66] Constitution of the Independent State of Papua New Guinea 1975 ss 53(1), 53(4), 38; Constitution of the Republic of Vanuatu 1980 art 80.

landowners from inequitable dealings, both countries prohibit certain land dealings with non-citizens. In Vanuatu, such dealings require consent from the government.[67] Consent may be withheld if the transaction is prejudicial to the interests of customary owners, the community in whose locality the land was situated, other indigenous citizens or the Republic of Vanuatu.[68] Customary landowners in PNG are unable 'to sell, lease or otherwise dispose of customary land or existing rights otherwise than to citizens in accordance with custom'.[69] Agreements that provide for customary land to 'be sold, leased, mortgaged, subdivided or otherwise disposed of', except in accordance with custom, are illegal and these instruments can be rendered legally void.[70]

As stated in PNG's 2013 REDD Readiness proposal, this may affect rural development:

> At the village level, the lack of investment opportunities is a more serious constraint. Poor infrastructure, remoteness from markets, the collapse of government extension services, and the high cost or lack of credit, impede the creation of business enterprise.[71]

This comment however, ignores the cultural values associated with land. In Melanesian understanding, land is of immense cultural as well as material value. Land ownership functions as an inherent social and spiritual connection, providing collective and individual identity. In Vanuatu, for example, in the words of one scholar, '[t]o ni-Vanuatu land is everything: it embodies their link to their past, present and future. It sustains everything they do in life – their beliefs and their daily interactions and activities'.[72] Rights to land and waters are determined by clan or familial relations expressed by connection with a specific area, although systems and laws of entitlement differ at both island and micro levels. Practices of intermarriage and exchange do result in the movement of persons and traditional resources over boundaries, and convey diverse interpretations of customary law. These movements contribute to the ongoing, conflictive process of evolution and retention of customary law amongst knowledge holders. While rules of custom (*kastom*) are locally situated, a distinctly Melanesian notion of communal ownership generally applies by which tribes and clans maintain

[67] Constitution of the Republic of Vanuatu 1980 art 79(1). [68] Ibid. art 79(2).
[69] Land Act 1996 (PNG) s 132. [70] PNG R-PP, above n 7, 34. [71] Ibid.
[72] Joel Simo, 'Land and the Traditional Economy: "Your Money, My Life" Hu I Kakae Long Basket Blong Laef' (Aid/Watch, April 2010) 40. 'Ni-Vanuatu' is the term used to denote indigenous peoples in Vanuatu.

custodianship of land but individuals within the group typically are granted holdings for subsistence use.[73]

Despite the significance of customary landholdings in Melanesia, certain areas of customary land effectively have been removed at a practical level from customary control in PNG and Vanuatu under leases and management agreements associated with development and resource extraction projects. Although customary owners still hold the underlying title to the land that is subject to lease and management arrangements, they are unable to possess and use the land for the duration of the arrangements (subject to any customary exemptions in such agreements, where they exist). In Vanuatu, while 99 per cent of land is held under customary ownership, almost 10 per cent of those lands are subject to registered leases (and the extent of unregistered leases is unclear).[74] Vanuatu's R-PP therefore notes that the complexity of existing tenure and leasing arrangements in Vanuatu, with inadequate checks and balances, leaves customary owners susceptible to corruption and illegal activities such as 'land grabbing' by local elites and foreign investors.[75] Indeed, customary owners in Vanuatu have expressed concerns in REDD+ consultation fora about potential formalisation of customary tenure as part of the REDD Readiness phase, for fears that this would intensify unequal power relations and land grabbing.[76]

By 2011, despite the retention of underlying customary ownership in PNG, use of the land in at least 30 per cent of the country effectively was not available to customary owners[77] due to the grant of leasehold interests to non-indigenous parties. Such leases are typically long term – often for periods of between fifty and ninety-nine years, and usually for the purposes of resource exploitation (generally agriculture or forestry). 'Forest management agreements' are the method by which the PNG government acquires rights to trees on customary land for the purpose of logging. 'Special agriculture and business leases' are a mechanism used

[73] Foukona, above n 56, 267.

[74] Siobhan McDonnell, 'Building a Pathway for Successful Land Reform in Solomon Islands' (2015) 31, www.academia.edu/15598962/Building_a_pathway_for_successful_land_reform_in_Solomon_Islands.

[75] Vanuatu R-PP, above n 10, 58. [76] Ibid. 59.

[77] The estimate was for FMA coverage of 10.5 million hectares out of a total 46.17 million (Shearman et al, above n 52, 9, 21) and the extent of land under SABL is estimated at 11 per cent of PNG landmass: Colin Filer, 'The New Land Grab in Papua New Guinea' (paper presented at the International Conference on Global Land Grabbing, Future Agricultures Consortium at the Institute of Development Studies, University of Sussex, 6–8 April 2011) 2.

by the PNG government to access customary land for extended periods for 'agricultural or development' projects.[78] Forestry-management agreements remove customary owners' rights to manage and harvest trees,[79] whereas special agriculture and business leases leave customary owners with only a reversionary interest during the term of the lease.[80] The extensive use of leases has led to claims that the government has facilitated customary land access for third parties, with suspicions that the leases act as a 'back door' logging concession to avoid cumbersome forestry-management agreement processes.[81] Concern about these arrangements prompted a Commission of Inquiry in 2011,[82] and condemnation from the CERD Committee as recently as 2016.[83] These schemes cover land either earmarked for logging or identified as having future potential for logging, areas which represent some of the 'lowest hanging fruit' with which the PNG nation might meet international REDD+ additionality requirements.[84]

An important consideration for REDD+ implementation in PNG and Vanuatu, with ramifications across Melanesia, is how to protect customary interests where the underpinning rights to customary lands are retained but the capacity of customary communities to control and manage customary land is diminished for the duration of third party leasehold interests.

### 3. *Customary Land and Challenges for REDD+*

The nature and 'ownership' of customary tenures, which often are largely undocumented, can be difficult to comprehend. Statutory schemes in

78 See Part C for more detail.

79 The 'timber rights' transferred include rights to fell or cut trees and any other vegetable growth, as well as the right to plant, grow, and manage trees and carry out regeneration and reforestation work (Forestry Act 1991 (PNG) s 2).

80 A reversionary interest is the underlying title interest that reverts to the customary holder at the end of the lease term.

81 Filer, 'The New Land Grab in Papua New Guinea', above n 77; Colin Filer, 'Why Green Grabs Don't Work in Papua New Guinea' (2012) 39 *The Journal of Peasant Studies* 599; Greenpeace, *Up for Grabs: Millions of Hectares of Customary Land in PNG Stolen for Logging* (Greenpeace Asia Pacific, 2012).

82 John Numapo, *Commission of Inquiry into the Special Agriculture and Business Lease (SABL): Final Report* (24 June 2013) 20 and 180–182, http://pngexposed.files.wordpress.com/2013/12/numapo-sabl-final-report.pdf.

83 José Francisco Calí Tzay, 'Letter from the Committee on the Elimination of Racial Discrimination to the Government of Papua New Guinea Re Special Agricultural Business Letters', 17 February 2016, http://tbinternet.ohchr.org/Treaties/CERD/Shared%20Documents/PNG/INT_CERD_ALE_PNG_7926_E.pdf.

84 See Chapter 2 for a discussion of additionality requirements.

Melanesia often inadequately reflect the intricate patterns of customary interests and cultural practices.[85] A review across Melanesia led Jorgensen to conclude 'that an untidy jumble of multiple overlapping claims is at least as common as clearly demarcated clan estates with similarly unambiguous lists of members'.[86] Customary law evolves and adapts in response to shifting relationships and influences within and outside customary communities,[87] including economic forces.[88] Customary landholding and rights within the group can be transfigured under land and resource legislation.[89] In PNG, this shift entrenched 'an ideology of land ownership' through which 'custom' has been reshaped as a source of rights to secure benefit flows from resource projects by way of compensation, rent or royalties.[90] In both PNG and Vanuatu, disputes about customary land, including the content of 'custom', draw on both customary and Western traditions.[91] In PNG, the resolution of disputes about the use, ownership and boundaries of customary land is generally governed by the Land Dispute Settlement Act 1975. The legislation created a system of land mediation designed to be 'close to the people' and a means by which traditional dispute measures can be implemented. Yet as Kalinoe explains, in PNG these laws and processes 'make up a dysfunctional and rather confusing system'.[92] Vanuatu initiated reforms to dispute settlement as part of wider land law reforms. The Customary Land Management Act introduced the concept of a '*nakamal*' – a customary organization that is the seat of governance for a particular area. Under the authority of that entity, Ni-Vanuatu people from the area meet to identify customary landowner groups and manage disputes about ownership. Determinations are recorded in writing and this

[85] Kenneth Brown, 'Customary Law in the Pacific: An Endangered Species?' (1999) 3 *Journal of South Pacific Law*, www.usp.ac.fj/index.php?id=13145.

[86] Jorgensen, above n 16, 67.

[87] John Wagner, 'Conservation as Development in Papua New Guinea: The View from Blue Mountain' (2007) 66 *Human Organization* 28, 35.

[88] See Kalinoe and Kiris, above n 22, 23.

[89] Emma Gilberthorp, 'Fasu Solidarity: A Case Study of Kin Networks, Land Tenure, and Oil Extraction in Kutubu, Papua New Guinea' (2007) 109 *American Anthropologist* 101, 104–105.

[90] Jorgensen, above n 16, 60; Colin Filer, 'The Double Movement of Immovable Property Rights in Papua New Guinea.' (2014) 49 *Journal of Pacific History* 76.

[91] Lawrence Kalinoe, 'Customary Land Dispute Settlement' in Charles Yala (ed), *The Genesis of the Papua New Guinea Land Reform Program: Selected Papers from the 2005 National Land Summit* (National Research Institute, 2010) 49, 52.

[92] Ibid. 49.

interest constitutes the basis for identifying traditional landowners who need to give their consent for development.[93]

In Vanuatu, as patterns of economic development have encouraged growth in areas such as tourism, real estate and construction, pressures toward alienable titles have been introduced, as well as the need for security for asset accumulation.[94] Land law reforms in 2014 sought to address a problems caused by development pressures.[95] In Vanuatu, during the period of the condominium by Britain and France, a system of land registration provided freehold title to colonists, enabling them to acquire large tracts of land.[96] After the reinstatement of customary ownership in Vanuatu's Constitution,[97] regulations enabled former colonists who previously held a freehold interest in land to remain in occupation until a lease was negotiated with the custom owners.[98] This practice evolved into a registration system for the leasing of land to non-customary owners under the Land Reform Act 1980 and subsequent legislation. The legislation gave the Minister for Lands extensive powers over lease transactions.[99] The 'interim' process became the de facto model for commercial land transactions in Vanuatu. From 2000, the unprecedented rate of leases to non-customary holders, a growing level of land disputes in the courts and the later unsuccessful delegation of disputes to the Land Tribunal built momentum for extensive land reform.[100] The Ministry of Lands convened a National Land Summit in 2006 to consider customary land management practices in Vanuatu.[101] The reforms, which came into effect in 2014, comprised four main changes: constitutional amendments to recognise customary institutions; advice functions for the Malvatumauri (National Council of Chiefs) in

[93] Siobhan McDonnell, Joseph Foukona and Alice Pollard, 'Building a Pathway for Successful Land Reform in Solomon Islands', http://bellschool.anu.edu.au/experts-publications/publications/5030/building-pathway-successful-land-reform-solomon-islands, 31–38, especially 33.

[94] Ibid. 265.

[95] Land Reform (Amendment) Act 2014, http://extwprlegs1.fao.org/docs/pdf/van148859.pdf; see also Vanuatu R-PP, above n 10, 56.

[96] Howard Van Trease, 'The History of Land Alienation' in Peter Larmour (ed), *Land Tenure in Vanuatu* (Institute of Pacific Studies, 1984) 17.

[97] Constitution of the Republic of Vanuatu 1980 art 73.

[98] Land Reform Regulation 1980 (Vanuatu).

[99] The leasing framework was subsequently expanded by the Land Leases Act 1983 (Cap 163) (Vanuatu).

[100] McDonnell, above n 93, 33.

[101] Sue Farran, 'Navigating Changing Land Use in Vanuatu' (2011) 34 *Pacific Studies* 250, 250.

relation to changes to land law; providing for customary institutions to make final determinations of customary ownership; and removing the power of the minister over customary land.[102] Central to the reforms to the land legislation was the introduction of a revised leasing process that adopts principles of FPIC for customary owner groups to any development on their land. The new leasing arrangements require environmental and planning safeguards to be met, and guarantees access to gardens, coastal and sea estates by customary owners.[103]

In both countries, the implementation of the REDD+ regime will have to successfully navigate these complexities of land law systems, shifting communal property patterns, evolving statutory frameworks and land reforms, and the patterns of third party leases over forests to ensure that safeguards are effectively implemented and co-benefits are equitably distributed within and between indigenous communities. It is in this dynamic context that REDD+ must operate.

## C. REDD Readiness in Melanesia

Despite being early participants in REDD+ programmes, both PNG and Vanuatu remain in the 'readiness' phase of REDD+. PNG joined the UN-REDD Programme as a participant state in 2008, although its National Programme Document was only approved in 2011. PNG joined the FCPF in 2011 and finalised its R-PP in 2013.[104] Vanuatu submitted its Readiness Plan Idea Note to the FCPF in 2008[105] and completed its R-PP in 2013.[106] Both countries are members of the Pacific Islands Regional Policy Framework for REDD+ (2012), along with Fiji and the Solomon Islands. Section 1 below explores how these funding bodies have framed readiness obligations for PNG and Vanuatu with respect to indigenous communities, in particular through FPIC and the recognition of legitimate interests in customary land tenure.

[102] For an excellent overview, see Siobhan McDonnell, 'Better Protection for Custom Owners: Key Changes in Vanuatu's New Land Legislation' (March 2014), http://pacificinstitute.anu.edu.au/outrigger/2014/03/04/better-protection-for-custom-owners-key-changes-in-vanuatus-new-land-legislation.

[103] Ibid.

[104] Government of Papua New Guinea et al, 'National Programme Document' (UN Collaborative Programme on Reducing Emissions from Deforestation and Forest Degradation in Developing Countries, 2011); PNG R-PP, above n 7.

[105] Weaver and Herald, above n 44, 2.

[106] Vanuatu R-PP, above n 10.

### 1. *Bilateral and Multilateral Funding*

Both PNG and Vanuatu have received funding for REDD+ activities from international donor organizations and other national governments, such as Germany.[107] Since 2013, Vanuatu has adopted a programmatic approach to REDD+ to fund land sector emissions-reduction activities through agricultural and development policies.[108] The necessary REDD Readiness funding for Vanuatu is estimated to be more than US$7 million, drawn in large part from a US$3.6 million FCPF grant committed in 2015 to support institutional requirements (including establishment of a Vanuatu National REDD+ Unit and provincial committees), decentralised stakeholder engagement and development of a national REDD+ strategy.[109] Most international funding for PNG's readiness activities, until recently, was directed towards technical elements of MRV.[110] The original National Programme Document for REDD+ in PNG focused on MRV and stakeholder consultation, although it expanded in 2011 to include references to tenure, FPIC and a greater emphasis on safeguards.[111]

The R-PPs of both countries identify tenure as a key issue to be addressed, following on from the Pacific Islands Regional Policy Framework for REDD+, which encouraged Pacific member states to 'clarify land tenure and forest carbon tenure arrangements as a key condition of REDD+ implementation'.[112] Vanuatu's R-PP provides:[113]

> Although customary land ownership in Vanuatu provides significant opportunities for maximizing certain equity components of REDD+, e.g. community engagement in REDD+ and benefits sharing, it also poses significant challenges. REDD+ activities will need to involve a large number of Ni-Vanuatu in decisions about the future management of their land and forest resources in order for landowners to understand the changes required by them.

[107] The REDD Desk, 'Papua New Guinea: Financing' (Global Canopy Programme), http://theredddesk.org/countries/papua-new-guinea/financing; Forest Carbon Partnership, 'REDD + Annual Country Progress Reporting: Vanuatu' (January 2015–August 2015), http://forestcarbonpartnership.org/sites/fcp/files/2015/September/FCPF_August2015_Vanuatu.pdf.

[108] Vanuatu R-PP, above n 10; Jennifer Corrin, 'The Root of the Problem: Carbon Rights and Natural Resources Issues in Solomon Islands and Vanuatu' (2014) 26 *New Zealand Universities Law Review* 1, 4.

[109] Vanuatu R-PP, above n 10, 120.

[110] Government of Papua New Guinea et al, above n 104, 48.

[111] Ibid. annex 1.

[112] Secretariat of the Pacific Island Community, above n 2, 4.6.3.

[113] Vanuatu R-PP, above n 10, 57.

The R-PP for PNG gives a similarly critical appraisal of the country's current and previous land tenure arrangements,[114] although it does little more than direct that tenure arrangements be reviewed to comply with safeguards.[115]

The UN-REDD Programme and the organizations endorsing the FCPF principles have yet to examine complex questions about how REDD+ will impact customary communities. Despite references in the R-PPs to the requirements contained in the policies and standards of the international funding body policies, REDD Readiness activities in both countries are directed towards informing policy development rather than locking in normative outcomes. They do not contain concrete proposals for tenure reform. Funding for pilots has not been tied to the resolution of safeguards and tenure arrangements or made conditional upon embedding multilateral donor standards into REDD Readiness activities. As a consequence, after eight years of REDD planning work, there has been limited progress on how the risks and opportunities entailed in REDD+ will play out in local communities. This approach bypasses or defers the need to resolve the complex trade-offs associated with ensuring FPIC or enabling customary institutions to actively engage with REDD+.

### 2. *UN-REDD Programme and FCPF REDD Readiness Activities in PNG*

PNG received REDD Readiness funding from the UN-REDD Programme in 2011, in furtherance of the PNG REDD+ National Programme[116] (although 'political instability and delays in the hiring of staff' meant that National Programme activities did not commence until early 2013).[117] Further support for PNG REDD+ Readiness came from the Australian government's International Forest Carbon Initiative for technical, scientific and analytical support for PNG's carbon monitoring and accounting systems.

National Programme activities were to conclude in December 2015, but were extended to the end of 2016.[118] The UN-REDD Programme–funded

[114] PNG R-PP, above n 7, 39. [115] Ibid. 49.

[116] US$6.378 million: 'Funding Support', Climate Change and Development Authority, www.ccda.gov.pg/Funding%20Support.html.

[117] UN-REDD Programme, 'National Programme Annual Report' (Draft, January–December 2015) 29.

[118] See Letter to Mario Boccucci, UN-REDD Programme Secretariat from Varigini Badira, UN-REDD Programme Papua New Guinea and Goodwill Amos, PNG Forest Activity, 'Request for No-Cost Extension of Papua New Guinea's UN-REDD National

National Programme thus overlaps with the FCPF's REDD+ Readiness Project initiated in 2015, which will run until 2018.[119] In 2016, PNG identified that the Technical Working Group on Social and Environmental Safeguards should lead REDD+ safeguards processes in that country. The establishment of the working group responds to the need for PNG to engage with the multiple international safeguards requirements (especially the UNFCCC and the FCPF Guidelines) through the adoption of a country-specific approach.

The UN-REDD Programme reports that 'National Programme activities are designed and implemented in consideration of continuation under FCPF Project'.[120] The FCPF Project will support the development of PNG's National REDD+ Strategy (expected to be completed in 2017)[121] and advance work on stakeholder engagement.[122] The project will be implemented by the UNDP as delivery partner in order to 'ensure continuity of the strategy and stakeholder engagement components of the PNG UN-REDD Programme',[123] and to ensure that 'funding . . . will be managed independently from government control by UNDP as its delivery partner'.[124] Work conducted under the UN-REDD Programme–funded activities included the completion of National FPIC Guidelines, and the drafting of the Climate Change (Management) Bill, enacted as the Climate Change (Management) Act in 2015. PNG reports that 'most of these [UN-REDD Programme] fund[s] have been and will continually be utilized to build and expand PNG's capacity in developing a National MRV system'.[125] The development of a framework for the design of a REDD+ benefit-sharing distribution system will continue under the auspices of the FCPF REDD+ Project,[126] along with work on a National Forest Inventory and Safeguards Information System.[127] The goal of the FCPF REDD+ Project is to 'establish the basis for emission reductions

Programme', 15 July 2015, www.unredd.net/documents/policy-board-86/intersessional-decisions-829/14750-no-cost-extension-request-and-justification-from-papua-new-guinea-national-programme.html Also UN-REDD Programme, 'National Programme Semi-Annual Report: Papua New Guinea (January–June 2015) 2.

119 US$3.8 million: Funding Support, above n 5; UNDP and Forest Carbon Partnership Facility (FCPF), 'Papua New Guinea REDD+ Readiness Project', www.pg.undp.org/content/dam/papua_new_guinea/docs

120 'National Programme Annual Report' (Draft, 2015), above n 117, 28.

121 The 'National REDD+ Strategy, funded by FCPF, is expected by end of 2016': 'National Programme Annual Report (Draft, 2015), above n 117, 17.

122 Ibid. 29. 123 'National Programme Semi-Annual Report' (2015), above n 118, 28.

124 PNG R-PP, above n 7, 7. 125 Ibid.

126 'National Programme Semi-Annual Report' (2015), above n 118, 13.

127 'National Programme Annual Report' (Draft, 2015), above n 117, 5–6.

and the receipt of REDD+ benefits' by 2018,[128] and to continue work that the UN-REDD Programme has insufficient funds to support.[129] Most recently, the PNG government has begun consultations on the design of a 'grievance redress mechanism' to receive complaints about the impact of REDD+ on affected communities.[130]

PNG is one of the few countries in the world that has passed national climate change legislation that is intended to govern the conduct of REDD+ projects.[131] Enacted in July 2015, the Climate Change (Management) Act 2015 establishes the Climate Change and Development Authority (CCDA), protects the rights of customary land holders, and specifies that climate-change related project agreements, including those initiated under REDD+, must be accompanied by the 'free prior and informed consent' of customary land holders. The CCDA's role is to 'promote and manage climate compatible development through climate change mitigation and adaptation activities'[132] and to serve as the 'designated National Authority' responsible for managing PNG's obligations and commitments under the UNFCCC.[133] It continues, with a legislative mandate, the role of the Office of Climate Change and Development established by an order of the National Executive Council (Cabinet Ministers and Heads of Department) in 2010.[134] CCDA's work is overseen by a nine-member National Climate Change Board comprised of seven officials directing the natural resource and development agencies of the PNG government, one representative of the National Women's Council and one representative of private sector interests.[135] NGOs are

[128] Forest Carbon Partnership Facility REDD+ Readiness Project, UNDP in Papua New Guinea, www.pg.undp.org/content/papua_new_guinea/en/home/operations/projects/

[129] 'However, the UN-REDD Programme has limited funding to continue its co-implementation role for information sharing and early dialogue with stakeholders, and this proposal will therefore be used to address that gap': PNG R-PP, above n 7, 24.

[130] CCDA Media Release, 'GoPNG Seeks to Strengthen the Community's Voice on Climate Change and Sustainable Management of Forests' (10 February 2017), www.ccda.gov.pg/images/REDD+%20GRM.pdf.

[131] Sylver Yagi, 'PNG First in the Region with Climate Change Law', *Loop*, 5 September 2015, www.looppng.com/content/png-first-region-climate-change-law.

[132] Climate Change (Management) Act 2015 (PNG) s 10(a).

[133] UNDP Project Document, above n 5, 9.

[134] Office of Climate Change and Development (OCCD), 'PNG REDD+ Readiness: Status and Funding Needs' (2011) 2. See also R J Cuthbert et al, *Analysis of National Circumstances in the Context of REDD+ and Identification of REDD+ Abatement Levers in Papua New Guinea* (report produced by the Wildlife Conservation Society for Papua New Guinea's UN-REDD National Programme, 2016) 24.

[135] Climate Change (Management) Act 2015 (PNG) s 16(2).

reported to have concerns about the lack of civil society representation on the Board.[136] In addition to overseeing and guiding the activities of the Authority,[137] the Board monitors the performance and reporting of REDD+ activities and administers REDD+ funding.[138]

As noted, the Climate Change (Management) Act 2015 provides that the consent of all customary landholders to proposed climate change–related agreements on customary land shall be obtained through FPIC, referencing regulations promulgated under the Act.[139] Customary land rights are to be vested in a land group incorporated as an ILG or otherwise registered, but where this is 'impracticable', the consent requirement can be satisfied if 87 per cent of the adult landholders resident on the land in question have given their consent to the proposed agreement.[140] The agreement must specify the monetary and other benefits 'if any' that are to be received by the landholders in exchange for the rights granted. The agreement must be accompanied by a map showing the boundaries of the land affected.[141] The Act further provides that all affected landholders 'shall participate and benefit from the incentives of a climate change related project' implemented on land or sea.[142]

The Act envisages that national FPIC guidelines will be implemented as regulations supporting the substantive consent requirement in the legislation. The Act specifies that the consent of customary landowners to climate change projects must be 'obtained through the "free, prior and informed consent" process prescribed in the Regulation'. The guidelines are expected to be finalised in late 2017, and these will then provide the substance of the regulatory process mentioned in the Act.[143] The efficacy of tenure arrangements and FPIC are tightly interconnected. The performance 'on the ground' of any FPIC guidelines or regulations in protecting the interests of local communities will depend not only upon resourcing and legal form but, as discussed, will critically rely on the relationship with other laws affecting customary lands, including ILGs and leases and agreements that facilitate resource development.

Despite these positive developments in PNG, many logistical and pragmatic challenges remain to the implementation of FPIC in Melanesia.

136 Liam Fox, 'NGO Raises Concern Over PNG's Climate Change Legislation', *ABC News* (9 December 2015), www.abc.net.au/news/2015-12-09/ngo-raises-concerns-over-pngs-climate-change/7014980.

137 Climate Change (Management) Act 2015 (PNG) s 13(1). 138 Ibid. s 13(2).

139 Ibid. ss 87, 88. 140 Ibid. s 89. 141 Ibid. s 90(1). 142 Ibid. s 93(2).

143 Ibid. s 87. Email from Dan Nekitel, OCCD, to Kirsty Gover, 20 February 2017.

Vanuatu, for example, is an extremely diverse country,[144] with numerous indigenous languages and differing customary land regimes on different islands.[145] This local context presents a number of challenges to implementing international frameworks for obtaining consent.[146] These factors may explain the tendency for countries to adopt a 'no regrets' approach to implementing REDD+ in the Pacific Region as demonstrated by the prominence of pilot projects.[147]

PNG, too, has experienced practical barriers to obtaining the consent of customary landowners for development projects. A review of the forestry industry in PNG concluded that the lack of informed consent for projects is unsurprising given the time frames, remoteness and number of communities covering vast areas.[148] Awareness raising is made more difficult given the complexity of the ILG agreements and low levels of education and literacy, which alongside the 'diverse population of owners having a spectrum of traditions of land ownership . . . and customs' would 'be a challenge to any forest administration in the world'.[149]

The complexity of PNG's customary and cultural make-up has raised issues concerning the progress of World Bank–funded projects outside of REDD+. Given the very high proportion of Papua New Guineans who are customary landholders, the World Bank's OP 4.10[150] conceivably could be triggered for every World Bank–funded project.[151] A survey of documents published on the Bank's PNG country-specific website shows that OP 4.10 has been engaged in at least eleven projects, with Indigenous Peoples Plans prepared by bank staff in each case.[152] This situation indicates that customary landowners are regarded by the Bank as falling

[144] UN Data, above n 25. [145] Foukona, above n 56, 267.

[146] Rodman, above n 8, 69. [147] See Chapter 9.

[148] Neil Bird et al, 'The Current Legal and Institutional Framework of the Forest Sector in Papua New Guinea: Papua New Guinea Forest Studies 2' (Overseas Development Institute, January 2007); Friends of the Earth Japan and Global Environment Forum, 'Evaluation of Social and Environmental Risks Accompanying the Procurement of Timber from Papua New Guinea' (June 2011) 11–12.

[149] 'Achieving the ITTO Objective 2000 and Sustainable Forest Management in Papua New Guinea: Report of the Diagnostic Mission' (ITTC(XLII)/7, International Tropical Timber Council, 4 April 2007) 14.

[150] See further in Chapter 4, Section B(a).

[151] World Bank Inspection Panel, 'Investigation Report, Independent State of Papua New Guinea: Small Agriculture Development Project (IDA Credit No 4374-PNG)' (Inspection Report No 64458-PG, World Bank Group, 19 September 2011), http://ewebapps.worldbank.org/apps/ip/PanelCases/62-Investigation%20Report%20%28English%29.pdf, xxvi.

[152] The World Bank, 'Papua New Guinea: Projects & Programs – All Projects', www.worldbank.org/en/country/png/projects/all.

within the OP.410 definition of 'Indigenous Peoples' in at least some circumstances, despite the fact that they constitute the vast majority of the population.

In addition, in 2010, customary landholders affected by a World Bank–funded oil palm production project relied on OP 4.10 (and other operational polices) to successfully request that the World Bank Inspection Panel investigate the Bank's non-compliance with its safeguards, including with regard to benefit-sharing[153] and consultation.[154] The Inspection Panel's report was issued in 2011.[155] In the report, the panel notes that 'projects in Papua New Guinea pose a unique challenge for the Bank in that the population is almost entirely composed of indigenous peoples belonging to more than eight hundred distinct ethnic groups',[156] and 'the diversity of languages presents a unique challenge for consultation and communication'.[157] On the Bank's alleged non-compliance with FPIC requirements in OP 4.10, the Inspection Panel made the following finding:

> [T]he analysis of the legal and institutional framework of customary law, leadership, decision-making and dispute-resolution processes, and the gathering of baseline information on indigenous communities, fell short of requirements, which may have affected consultation processes. Management also failed to provide relevant information prior to consultations in a culturally appropriate manner, form, and language. Further, the Panel was unable to find in Project documents ... any information documenting how broad community support was reached. The Panel finds that this is not in compliance with OP 4.10.[158]

On the issue of benefit-sharing, the panel further noted that:

> the Indigenous Peoples Policy ... requires the Bank 'to ensure that the Indigenous Peoples receive social and economic benefits that are culturally appropriate and gender and intergenerationally inclusive.' It is the

[153] World Bank Inspection Panel, 'Report and Recommendation, Papua New Guinea: Smallholder Agriculture Development Project (SADP) (IDA Credit No 43740-PNG)' (Inspection Report No 53280-PG, 10 March 2010), http://documents.worldbank.org/curated/en/456211468285074449/pdf/532800IPR0P079101Official0Use0Only1.pdf, 4 [15(i)], 8 [27] (re: poverty reduction and benefit-sharing, and management response on poverty reduction and benefit-sharing).

[154] World Bank Inspection Panel, 'Report and Recommendation', above n 153, 4 [15(ii)] (requesters on FPIC), 10 [34]–[35] (management response on FPIC).

[155] World Bank Inspection Panel, 'Investigation Report', above n 151. 19.

[156] Ibid. 139.

[157] World Bank Inspection Panel, 'Investigation Report', above n 151, 41 [147] (emphasis omitted).

[158] Ibid. vii.

> Panel's view that the Project design should have made provisions to respond to the significant differences among the Project areas. Further, Management did not conduct a review of the mill companies' institutional and financial viability, thus leaving the Project without a sound basis for revenue-sharing.

As discussed in Chapter 5, OP 4.10 is not triggered by REDD Readiness activities, unless there are exceptional circumstances. Once the FCPF begins to provide funding for emissions-reduction projects, as envisaged by its management of Carbon Fund 'results based payments', then OP 4.10 will likely apply to those projects. It is clear that the two Melanesian countries considered in this chapter present difficulties for 'top-down', standardised consent requirements that are not sufficiently attuned to local circumstances. Fundamentally, donor groups need to be aware of the demonstrated lack of capacity of governments to secure project consents in Melanesia, particularly where almost every project will involve indigenous communities to some extent. The sheer number of projects for which consent will be required, when combined with the diversity of customs and languages (and potentially inaccessible terrain) that must be navigated to secure it, adds considerable layers of complexity, time and effort to secure consent or consultation.[159]

## D. Third-Party Rights and Customary Tenure in Melanesia

The need to develop culturally appropriate means of aligning customary tenures to REDD+ technical requirements to enable participation in the REDD+ scheme continues to be emphasised in Melanesia. Corrin writes, '[i]n Melanesia, plural legal regimes governing natural resources present complex issues of "ownership" which need to be addressed before [REDD+] schemes can be put into effect with any certainty.'[160] The Pacific Islands Regional Policy Framework for REDD+ suggests that Pacific Island countries should use existing natural resources laws and regulations as a starting point for the development of the necessary laws to underpin REDD+ implementation.[161]

In 2013, the German government commissioned advice on customary land tenure and REDD+, providing the first fine-grained analysis

[159] See further in Chapter 5. [160] Corrin, 'The Root of the Problem', above n 108, 1.
[161] Secretariat of the Pacific Island Community, above n 2, 4.6.3.

of the complex trade-offs that might be required for REDD+ to become operational in PNG:[162]

> [C]ustomary land tenure, as presently structured, cannot legally support a market-based approach to REDD+ which involves site-specific forest carbon projects due to the legal restrictions which customary land tenure brings and the contractual obligations that underpin REDD+ projects.

Controversially, the advice concluded that:[163]

> The issue is resolved by removing land and the forest from customary ownership, preferably by tenure conversion to freehold or possibly by registering as Registered Clan Land [under recent changes to ILGs] ... thereby providing the best means of securing land and safely dealing in carbon rights and obligations.

The perceived need to convert customary tenures ignores two important factors. First, customary land is inalienable except in accordance with custom. Second, customary land is to be held in trust for future generations.[164] This is a central organizing principle in the PNG National Constitution, which identifies five 'National Goals', the fourth of which states:

> We declare our fourth goal to be for Papua New Guinea's natural resources and environment to be conserved and used for the collective benefit of us all, and to be replenished for the benefit of future generations.

This goal provides the mandate for much of PNG's conservation and climate change law and policy, and forms the basis of protections for customary land found in legislation.[165] Communities are consequently unable to enter into contractual arrangements to facilitate REDD+ projects on customary land if those agreements do not accord with custom. Unfortunately, the idea that customary tenures should be converted to Western-style rights continues to pervade many

162 Steven O'Brien, 'REDD+ and Forest Carbon Rights in Papua New Guinea: Background Legal Analysis' (SPC/GIZ Regional REDD+ Project: Climate Protection through Forest Conservation in Pacific Island Countries, November 2012) 3, http://theredddesk.org/resources/redd-and-forest-carbon-rights-papua-new-guinea-background-legal-analysis.

163 Ibid. 44.

164 Ibid. 49; see also John S Kawi and Wilfred W Golman, 'Project Financing in Papua New Guinea' in Eric L Kwa (ed), *Natural Resources Law of Papua New Guinea* (Law Book Company, 2001) 152.

165 See, e.g., s 53 of the Constitution of the Independent State of Papua New Guinea 1975 on customary land.

international analyses of land and resource development, but is an increasingly controversial view.[166]

Such 'conversion' advice is not new in the Melanesian context. Many resource development projects and supporting legislation have similarly sought to access otherwise inalienable customary land. These pressures to translate customary tenures to facilitate third party access provide a salutary warning about underestimating the challenges of ensuring adequate safeguards as a prerequisite for REDD+ implementation.[167]

### *1. PNG: Incorporated Land Groups*

Despite a constitutional commitment to 'achieve development primarily through the use of Papua New Guinean forms of social, political and economic organization',[168] PNG has struggled to reconcile the protection provided to customary land tenure with the needs of Western-style development projects that are premised on stable and clearly demarcated tenure arrangements. The difficulty in creating a workable interface between these two contrasting traditions has often led to laws governing resource management bypassing protections for customary land.[169]

A similar tension is evident with REDD+. The PNG government's R-PP indicates that: '[l]and registration processes [should] allow the formalization of land title and the opening of customary land for development and or/commercialization'.[170] Previous government attempts to formalise customary interests have faced strong opposition from customary owners.[171] In the 1990s, the PNG government encouraged the use of

166 Lee Godden and Maureen Tehan (eds), *Comparative Perspectives on Communal Lands and Individual Ownership: Sustainable Futures* (Routledge-Cavendish, 2010) 1–2.

167 See further in Chapter 5.

168 Constitution of the Independent State of Papua New Guinea 1975, Preamble, National Goal 5: 'Papua New Guinean Ways'.

169 PNG R-PP, above n 7, 34; more generally, see Dixon, above n 21.

170 PNG R-PP, above n 7, 34.

171 Almah Tararia and Lisa Ogle, 'Incorporated Land Groups and the Registration of Customary Lands: Recent Developments in PNG' in Tim Anderson and Gary Lee (eds), *In Defense of Melanesian Customary Land* (Aidwatch, 2010) 21, 23; Colin Filer, 'Local Custom and the Art of Land Group Boundary Maintenance in Papua New Guinea' in James Weiner and Katie Glaskin (eds), *Customary Land Tenure and Registration in Australia and Papua New Guinea: Anthropological Perspectives* (ANU ePress, 2007) 135, 143–144; Wycliffe Antonio, Max Wagi and Lewi Kari, 'Chapter 7: Incorporated Land Groups' in Charles Yala (ed), *The Genesis of the Papua New Guinea Land Reform Program: Selected Papers from the 2005 National Land Summit* (National Research Institute, 2010) 61, 61; PNG R-PP, above n 7, 34.

ILGs under the Land Groups Incorporation Act 1974 (PNG) to expedite customary landowner consent and enable benefit distribution in the mining and forestry sectors.[172] Some estimate that at least 18,000 ILGs are now registered.[173] The process was inadequate,[174] with no requirements for groups applying for incorporation to provide proof of rights or even to identify the customary land over which the ILG claimed an interest.[175] There were limited viable methods for dispute resolution.[176] Numerous reviews identified the absence of proper checks and balances as a significant contributor to the widespread abuse of the ILG system.[177] The Land Groups Incorporation (Amendment) Act 2007 sought to address systemic deficiencies while providing indigenous communities with a development option more compatible with customary land ownership.[178]

Although the reforms emerged from a longer process unrelated to REDD+,[179] the PNG government has been encouraging communities to form groups and register their land to facilitate their participation in REDD+ projects.[180] The reforms now allow for a mechanism through which ILGs can register their customary land.[181] Once the land is registered, 'derivative interests' can be granted.[182] The government has indicated that these 'derivative interests' are to provide the framework for REDD+ projects.[183] The new ILG incorporation process is considerably

[172] Lawrence Kalinoe, 'Incorporated Land Groups in Papua New Guinea' (paper presented at Australasian Law Teachers Association Conference, University of the South Pacific, Port Vila, Vanuatu, 1–4 July 2001) 6–7; David Lea, 'A Critical Review of the Recent Amendments to the Customary Land Tenure System in Papua New Guinea' (2013) 35 (3) *Social Development Issues* 63, 71.

[173] Filer, 'The Double Movement of Immovable Property Rights', above n 90, 84.

[174] PNG R-PP, above n 7, 36.

[175] Filer, 'Local Custom and the Art of Land Group Boundary Maintenance', above n 171.

[176] Land Groups Incorporation Act 1974 (PNG) s 23.

[177] Sarah N Stocks, 'Review of Forest and REDD+ Related Policy and Legislation in Papua New Guinea' (United States Agency for International Development (USAID) Lowering Emissions in Asia's Forests (LEAF) Program, 29 September 2014) 27–28; 'Achieving the ITTO Objective 2000 and Sustainable Forest Management in Papua New Guinea', above n 149, 23.

[178] Lea, above n 172, 79.

[179] See Charles Yala (ed), *The Genesis of the Papua New Guinea Land Reform Program: Selected Papers from the 2005 National Summit* (National Research Institute, 2010) 1.

[180] PNG R-PP, above n 7, 39.

[181] Land Registration (Customary Land) (Amendment) Act 2007 (PNG).

[182] Land Registration (Amendment) Act 2009 (PNG) s 34O.

[183] See PNG R-PP, above n 7, 35; for an alternative interpretation, see O'Brien, above n 162, 33.

more prescriptive and establishes tighter membership rules,[184] larger quorums and requirements for female representation.[185] It confers investigative powers on the Director of Customary Land Registration with respect to boundary disputes.[186]

Despite a greater emphasis on resolving disputes, the system still fails to identify and protect interests at the subgroup level.[187] Therefore, the creation of clan land as the mechanism to participate in REDD+ may override existing minority interests in the land. Allen and Monson found that in interviews they conducted there was a concern that this system would increase the frequency of land disputes, potentially exacerbating conflict and violence.[188] If superimposed government processes and structures fail to reflect community-specific forms of customary tenure, then benefit distribution and consent regimes may perpetuate historical injustices.

### 2. *Vanuatu: Leasing Practices*

In Vanuatu, leasing practices have been the dominant means for facilitating foreign investment and third party economic development.[189] The Land Leases Act 1984 (Vanuatu) remains the country's primary piece of legislation enabling the creation and transfer of leases over customary land for a maximum period of seventy-five years, resulting in registration of an indefeasible title.[190] This Act provided a 'clear and certain' Torrens-type title to facilitate land development, but failed to provide sufficient safeguards for customary owners.[191] There have been numerous cases of customary owners claiming to be unaware or opposed to

184 For example, Land Groups Incorporation (Amendment) Act 2009 (PNG) s 5(2)(c), sch 1.

185 Ibid. s 14B, sch 6.

186 Ibid. s 5A. The newly created registrar of titles has the power to refuse to incorporate the group when it suspects internal disputes regarding identity of representatives, officers or membership.

187 Lea, above n 172, 75.

188 Matthew Allen and Rebecca Monson, 'Land and Conflict in Papua New Guinea: The Role of Land Mediation' (2014) 10(2) *Security Challenges* 1, 4.

189 World Bank, 'Leases in Vanuatu', above n 27.

190 Land Leases Act 1984 (Vanuatu) ss 15, 17.

191 Reform may occur to the legislation. See report prepared by the Vanuatu Meteorological and Geo-Hazards Department (August 2016) which discusses possible amendments to the Land Leases Act 1984: see www.nab.vu/sites/default/files/documents/Urban%20Risk%20Management%20Strategy.pdf, 40–41.

leasing deals, and limited comprehension of the lease creation process, leasing conditions, or benefit-sharing options.[192] According to the World Bank, of the eighty-seven leases surveyed on Epi and Tanna islands, nearly 50 per cent had been subject to disputes.[193] The government came under increasing criticism in Ombudsmen's reports and commentaries,[194] as well as in court decisions.[195] The reforms under the Custom Land Management Act 2013[196] are being trialled in a number of places within Vanuatu. While the full impacts of the land reforms are yet to be evaluated, the Vanuatu experience highlights the difficulties that confront REDD+ project implementation.

Where indigenous custom owners have registered leasing arrangements, they can potentially acquire the right to benefit from forestry sequestration activities on the land. Vanuatu has a specific legislative framework for forestry rights, primarily under the Forestry Act (Cap 276) 2003, which makes provision for the management of forests, regulation of the forestry industry and establishment of a licensing system for commercial-forestry operations.[197] Customary owners may grant forestry leases under the Forestry Act 'and in accordance with the *Land Reform Act*', which are registerable under the Land Leases Act for a period of up to seventy-five years. The Forestry Act does not specifically refer to carbon sequestration rights, and it is unclear whether such rights (or emissions-reduction rights) would fall within the legislation.[198] If long-term leases are used for REDD+ projects, this would provide security for the period of carbon sequestration, exhibiting parallels with the lease conditions under Australia's Carbon Farming Act (Cth) discussed in Chapter 9.[199]

[192] Milena Stefanova, Raewyn Porter and Rod Nixon, 'Towards More Equitable Land Governance in Vanuatu: Ensuring Fair Land Dealings for Customary Groups' (Discussion Note, World Bank, May 2012) 10.

[193] Ibid. 13.

[194] Justin Haccius, 'The Interaction of Modern and Custom Land Tenure Systems in Vanuatu' (Discussion Paper 2011/1, School of International, Political and Strategic Studies, Australian National University, 2011); Ralph Regenvanu, 'Issues with Land Reform in Vanuatu' (2008) 12(1) *Journal of South Pacific Law*, www.paclii.org/journals/fJSPL/vol12no1/pdf/regenvanu.pdf.

[195] *Roqara v Takau* [2005] VUCA 5 (3 May 2005); *Ifira Trustees Ltd v Family Kalsakau* [2006] VUCA 23 (6 October 2006); *Solomon v Turquoise* [2008] VUSC 64 (8 August 2008).

[196] Custom Land Management Act (2013) (Vanuatu) Preamble.

[197] Forestry Act (Cap 276) (2003) (Vanuatu) pts 4, 5.

[198] See Chapter 2.

[199] See Chapter 9.

Notably, Vanuatu has a specific statutory framework providing for the grant of forest carbon rights on leased land, which prevails over the Forestry Act to the extent of any inconsistency. In fact, Vanuatu is the only Melanesian country that has a statutory framework for forest carbon rights.[200] The Forestry Rights Registration and Timber Harvest Guarantee Act 2000 allows for the grant of 'forestry rights', which may expressly include a 'carbon sequestration right'.[201] A carbon right in relation to land is 'a right conferred by agreement or otherwise to the legal, commercial or other benefit (whether present or future) of carbon sequestration by any existing or future tree of forest in the land', suggesting that carbon rights may be held as a distinct resource right, severable from the ownership of land. The proprietor of a lease may grant carbon forestry rights in a prescribed form, and in the absence of such a grant, the carbon rights remain with the lessor. The Act has not been widely used in Vanuatu.[202]

Aside from the Forestry Rights Registration and Timber Harvest Guarantee Act, there is little guidance on carbon rights in Vanuatu. Under the Constitution, 'all land in the Republic of Vanuatu belongs to the indigenous custom owners and their descendants',[203] but there is no express statement of rights to forests or soil, although it is widely accepted as a matter of practice, that forests belong to customary owners.[204] Thus, where unleased custom land is concerned, laws surrounding access, use and ownership of forests will be governed by customary law.

In summary, the future role for the leasing process in REDD+ projects is still unclear. Vanuatu's R-PP provides, in reference to REDD+ implementation and tenure arrangements, that '[a]ny process that requires leasing of customary land in Vanuatu should be avoided and will provide too great a challenge for REDD+ in Vanuatu unless dramatic reform occurs within the next 5–10 years'.[205] Accordingly, the Loru Forest Pilot Project has not adopted leasing practices to record customary tenure.

200 Vanuatu R-PP, above n 10, 58.

201 Forestry Rights Registration and Timber Harvest Guarantee Act (Cap 265) (2000) (Vanuatu) s 1.

202 Jennifer Corrin, 'REDD+ and Forest Carbon Rights in Vanuatu: Background Legal Analysis' (report prepared for Secretariat of the South Pacific Community and the Federal Ministry for the Environment, Nature Conservation and Nuclear Safety, November 2012) 42.

203 Constitution of the Republic of Vanuatu 1980, art 95(2).

204 Corrin, above n 202, 40.

205 Vanuatu R-PP, above n 10, 31.

It has relied instead on the registration of the land as a community conservation area under the Environmental Management and Conservation Act 2002. Amendments to that Act in 2010 allowed for community conservation areas for the specific purpose of providing ecosystem services, including climate mitigation.[206] This approach allows customary tenures and corresponding traditional management and representation arrangements to continue to operate in the REDD+ context without tenure amendments.

## E. FPIC in Melanesia

In Chapter 5, we discussed the requirement for FPIC within international law, noting that there is, as yet, no consensus on its content and scope.[207] Reflecting indeterminacy at the international level, REDD Readiness funding in the two Melanesian countries has made minimal provision for FPIC until recently. The Pacific Islands Regional Policy Framework for REDD+ includes a recommendation that 'Pacific Island countries and territories should recognise that the integrity and durability of REDD+ implementation will depend on definition and adherence to the principle of free, prior informed consent of resource owners', although there is no discussion of how this will be ensured.[208] While PNG has legislated the necessity of obtaining FPIC of customary landholders to climate change-related projects on their land,[209] Vanuatu does not have national FPIC legislation. Neither country has ratified ILO Convention 169, which would inform the content of FPIC at the national level.[210]

### 1. *National Guidelines*

The PNG government has produced draft national FPIC guidelines and, with funding from the FCPF, is working through a consultation process to further develop and finalise those guidelines.[211] The draft guidelines attempt to provide a step-by-step process for decision-making with

[206] Environmental Management and Conservation Act 2002 (Cap 283) (Vanuatu) s 35(ba).
[207] See Chapter 5. [208] Secretariat of the Pacific Island Community, above n 2.
[209] Climate Change (Management) Act 2015 (PNG) s 87.
[210] Convention (No 169) Concerning Indigenous and Tribal Peoples in Independent Countries, opened for signature 27 June 1989, 1650 UNTS 383 (entered into force 5 September 1991).
[211] Danny Nekitel, 'Draft Guidelines on FPIC for REDD+ in Papua New Guinea' (Office of Climate Change and Development, August 2014).

respect to REDD+ in PNG. While acknowledging the lack of consensus on the meaning of FPIC in international law, the guidelines specify that 'in the context of PNG', FPIC 'is the collective right of the people to give or withhold consent and applies to all activities, projects, legislative or administrative measures, and policies that take place in or impact the lands and resources or otherwise may affect the livelihoods of customary landowners and local communities'.[212] The guidelines further specify that '[c]onsent refers to the decision made by the customary landowners through their customary decision-making process' and that '[c]onsent must be sought and granted or withheld according to the unique formal or informal *political-administrative dynamic* of each community'.[213] They apply to a broad spectrum of activities requiring FPIC, including 'all activities, projects, legislative or administrative measures, and policies that take place in or impact the lands and resources or otherwise may affect the livelihoods of customary owners and local communities'.[214]

As explained in Chapter 5, ambiguity remains at the international level about whether, and how, FPIC is required to be sought in the case of indigenous groups as opposed to local or forest communities.[215] PNG's draft FPIC guidelines go beyond both of these international law categories to capture an even broader group. Unlike the World Bank's OP 4.10, the guidelines impose a consent requirement, given legislative force as section 87 of the Climate Change (Management) Act (2015). They seem to extend beyond the three situations requiring consent contained in the World Bank's new ESS 7, which will apply to Bank-funded projects beginning in 2018. Instead, the guidelines reference the constitutional protections extended to Papua New Guinean forms of social, political and economic organization, and rely on international law conceptions under the UNDRIP and ILO Convention 169 (despite the fact that PNG has not ratified this treaty).[216] The guidelines are to be 'strongly informed by the joint UN-REDD and FCPF Guidelines on Stakeholder Engagement in REDD+ Readiness',[217] and the current draft purports to 'build on' those joint guidelines.[218] It is clear that the guidelines and their accompanying legislative provisions accord more closely with the view of FPIC advanced by the UN-REDD Programme than the standard currently articulated in the World Bank's OP 4.10. In any case, since the FCPF has held that even in World Bank–funded REDD+ projects, where the recipient country or delivery partner adheres to FPIC

[212] Ibid. 10. [213] Ibid. 12. [214] Ibid. 10. [215] See Chapter 5.
[216] See Chapter 5. [217] PNG R-PP, above n 7, 27. [218] Nekitel, above n 211, 2.

standards that are more stringent than OP 4.10, those standards should be applied rather than OP 4.10. Given that the UNDP is the delivery partner for the current REDD+ project operating in PNG, either the UN-REDD Programme or PNG National Guidelines will provide a higher degree of protection for indigenous Papua New Guineans than OP 4.10 has in the past, by displacing the weaker 'free prior and informed consultation' requirement set out in that policy.

Applying these standards, the draft guidelines apply broadly to 'customary landowners and local communities', referring to 'clans and communities who have customary rights over the land and resources on the land'.[219] However, they also apply to 'other primary users of forests such as women, youth, marginalized and vulnerable groups who play either a direct or indirect role in forest management and use'.[220] The approach suggests that the key characteristic of communities entitled to FPIC protections is 'vulnerability', and reflects the large customary landholdings in PNG.

The draft guidelines effectively accord a power of veto to customary landowners and local communities for REDD+ related activity on the lands in which they have an interest, making it clear that any land or forest resources belonging to customary landowners or local communities within the vicinity of REDD+ activity which do not secure consent will not proceed.[221] In terms of dispute resolution, the draft guidelines encourage a preference for local decision-making processes. They refer specifically to ILGs being the unit for consent, but only when the ILG was formed with the FPIC of clan members, including women, and where there is no evidence of dispute about the authenticity of representation.[222] The draft guidelines take pains to encourage the inclusion of women in decision-making processes. PNG has also recently released a document titled, 'Situation Analysis and Recommendations for Improving Gender Inclusiveness and Participation in Papua New Guinea's National REDD+ Strategy and Policies',[223] which specifically considers the role of gender in the development and implementation of PNG's REDD+ framework. The approach accords with trends at the international level, including the requirement of the World Bank's ESS 7 for

[219] Ibid. 7. [220] Ibid. 7 (footnotes omitted). [221] Ibid. 17. [222] Ibid. 15.

[223] REDD+ Papua New Guinea, 'Situation Analysis and Recommendations for Improving Gender Inclusiveness and Participation in Papua New Guinea's National REDD+ Strategy and Policies' (Draft), www.pg.undp.org/content/dam/papua_new_guinea/FCPF/ROAR%20REports/Gender

'meaningful consultation, in a culturally appropriate and gender and inter-generationally inclusive manner'.[224]

The draft guidelines ultimately reference the 'norms, values and customs of the customary landowners and local communities involved in the consultation and decision-making methods they utilise'.[225] This reveals a tension between respecting customary decision-making processes and universal human rights protections, a theme explored in Chapter 4.[226]

Vanuatu has no national FPIC guidelines, and considerations of FPIC in its R-PP are cursory. The R-PP acknowledges that, as 95 per cent of land in Vanuatu is owned by indigenous people and subject to customary law, any decision about land use in relation to REDD+ is likely to trigger FPIC. The R-PP provides that, in the readiness phase, FPIC activities will focus on dispute resolution, although a need for a standalone FPIC process may arise in implementation. In the meantime, a Consultation and Participation Plan (under development) will provide a framework for the REDD+ process and for conflict resolution and redress of grievances. The R-PP identifies an immediate need for a conflict-resolution mechanism, as Vanuatu does not currently have a suitable grievance- or dispute-resolution mechanism apart from the land-management context.

FPIC policy is promoted under the UN-REDD Programme, of which Vanuatu is not a member. In the absence of a national policy other than in land law, REDD+-related activities in Vanuatu are not subject to the UN-REDD Programme's FPIC guidelines. Instead, OP 4.10, requiring broad community support based on 'free, prior and informed consultation', applies to REDD implementation (but not necessarily readiness). For projects commencing in or after 2018, ESS 7 applies.[227]

A future challenge will be ensuring robust implementation of FPIC principles in the forest sector in each country. In PNG, for example, the logging industry initially developed without adequate safeguards for obtaining the consent of customary owners. With assistance from the World Bank, PNG developed the Forestry Act 1991 to implement

224 'Environmental and Social Standard 7: Indigenous Peoples/Sub-Saharan African Historically Underserved Traditional Local Communities' in *Environmental and Social Framework: Setting Environmental and Social Standards for Investment Project Financing* (World Bank, 4 August 2016).

225 Nekitel, above n 211, 15.

226 See Chapter 4.

227 The two policies run in tandem until projects commenced under OP 4.10 are completed.

reforms.[228] A World Bank review in 2001 still found many landowners 'were not aware of the implications of belonging to an incorporated land group'.[229] Independent reviews into PNG forestry[230] have concluded that PNG's forestry laws are strong, with the weakness found in implementation and enforcement and the government's overestimation of its capacity to administer the new laws.[231] Critically, a Forest Trends' assessment revealed that 'widespread and demanding problems [in the forestry industry] are the failure to secure the informed consent of resource owners to timber harvesting ... and the consent that has been given for current and future forestry operations is clearly legally flawed'.[232] REDD+ project implementation will require strong and transparent implementation models if it is to overcome the history of relative neglect of FPIC in the operation of the PNG forestry sector.

### 2. *FPIC and National Tenure Reforms*

Recent national land-tenure reforms in PNG and Vanuatu seek to improve processes for consent to development on customary lands. Vanuatu's Custom Land Management Act 2013 sought to address issues where, in some cases, individual men (or groups of men) leased customary land without the knowledge or consent of other, often female, landowners.[233] The law reforms introduced notification periods for leasing, enhanced consent requirements and appeal rights, and referred specifically to women and children in the definition of customary institutions (*Nakamals*).[234] In 2015, the Ministry of Lands began a pilot of

[228] Frances J Seymour and Navroz K Dubash, *The Right Conditions: The World Bank, Structural Adjustment, and Forest Policy Reform* (World Resources Institute, 2000) 32–33, 36.

[229] Tony Power, 'Reconciling Customary Ownership and Development: Incorporated Land Groups' in *Making Land Work* (Australian Agency for International Development (AusAID), 2008) vol. 2 (Case Studies on Customary Land and Development in the Pacific) 3, 9, http://dfat.gov.au/about-us/publications/Pages/making-land-work.aspx.

[230] Bird et al, above n 148; Friends of the Earth (Japan) and Global Environment Forum, above n 148.

[231] Jim Fingleton, *Legal Papers Online No 73: Pacific Land Tenures: New Ideas for Reform* (FAO, 2008) 27.

[232] Forest Trends, *Logging, Legality and Livelihoods in Papua New Guinea: Synthesis of Official Assessments of the Large-Scale Logging Industry, Volume 1* (2006) 6, www.forest-trends.org/publication_details.php?publicationID=105.

[233] McDonnell, above n 74, 31. [234] Ibid. 35.

the new legal arrangements, but given the instability characterising Vanuatu's recent political history, uncertainty remains about their long-term viability.

PNG's ILG framework now has prescriptive requirements regarding group decision-making processes, with tight quorum requirements to allow for some gender equity. A quorum for meetings is at least 60 per cent of group members (with at least 10 per cent being women); 60 per cent approval at such a meeting is required for group decisions.[235] However, 40 per cent of group members still are excluded from decisions regarding the development of land.[236]

Given that the PNG government has indicated that national FPIC guidelines are to be implemented through ILG decision-making processes, these voting requirements influence the character of 'consent' that can be secured. The 60 per cent requirement may fail to safeguard or even account for 'interests' at the sublevel. In a country where such a significant proportion of the population remains dependent on customary lands for subsistence needs, ensuring adequate consent or consultation requirements at the subgroup level takes on heightened significance.

A considerable amount of customary land in PNG is subject to a forestry-management agreement or special agriculture and business lease (see Section A), which allow third parties to develop resource projects on customary land. Given the documented prevalence of fraud and abuse in the use of such mechanisms, additional processes may be needed to ensure the effective FPIC of customary owners.[237]

How REDD+ integrates with other consent requirements has broader implications. If REDD+ catalyses the alignment of consent provisions associated with other forms of resource development with international FPIC standards, REDD+ may well improve the practices applied across the forestry, mining and agricultural industries for consulting and engaging with customary landowners – a substantial benefit for indigenous communities. If harmonisation fails to occur, and the potentially expensive and time-consuming FPIC requirements are not universally applied, this double standard will increase the relative disadvantage of REDD+ projects[238] under circumstances where the opportunity costs of REDD+

[235] Land Groups Incorporation (Amendment) Act (2009) (PNG) ss 14D(3)–(4).
[236] O'Brien, above n 162, 33. [237] Stocks, above n 177, 29–30.
[238] Lisa Ogle and Celina Yong, 'What Does It Take to Make a Local Consultation a Success?' Input Paper III: Asia Pacific Region', (BMZ, FCPF, UN-REDD Joint Expert Workshop, 1 October 2013) 1, 24.

may already pose a formidable barrier.[239] Past experience in Melanesian countries shows that, even with the most rigorous obligations, ensuring tenure reforms are adequately implemented and resourced is a real challenge, due to practical and logistical difficulties associated with securing consent.

## F. Pilot Programmes

As part of REDD+ programing, the PNG and Vanuatu governments have developed pilot emissions-reduction projects, implemented by local and international NGOs, including The Nature Conservancy, Conservation International, Live & Learn, and the Wildlife Conservation Society.[240] In both countries, the pilot projects have faced challenges in identifying customary owners and facilitating robust community consultation and consent. While the examples below are not subject to FCPF and UN-REDD Programme processes, and do not depend on funding from those bodies, they provide insights into the complexities of integrating REDD+ with customary land ownership.

The April Salumei project, a PNG REDD+ style pilot,[241] was developed by Pacific Forest Alliance with cooperation from PNG government departments, research institutions and NGOs.[242] The project exemplifies the vexed historical arrangements for obtaining consent and reflects a history of disputation.[243] The project is located in one of the largest undisturbed tracts of lowland forest in PNG, where the customary owners remain highly dependent on a subsistence economy based on shifting cultivation, and have limited access to the cash economy.[244] The aim was for the landowners to sell credits on the voluntary carbon market rather than participate in logging.[245]

239 At a local community level, where the competing uses include commercial agriculture, and at the national level due to the reliance on logging revenue (Bingeding, above n 30, 6); Colin Hunt et al, 'Asking the Question: Is Papua New Guinea REDD-Ready?' (2012) 4, www.colinhunt.com.au/index.php/download_file/view/59/139.

240 PNG R-PP, above n 7, 46. 241 Ibid. 65.

242 A full list is provided in Pacific Forest Alliance Limited, 'April Salumei Sustainable Forest Management Project' (VCS Feasibility Assessment, 4 May 2012) 3, http://theredddesk.org/countries/initiatives/april-salumei-sustainable-forest-management-project-east-sepik.

243 See Chapter 9.

244 Matthew Leggett and Heather Lovell, 'Community Perceptions of REDD+: A Case Study from Papua New Guinea' (2012) 12 *Climate Policy* 115, 121.

245 Pacific Forest Alliance Limited, above n 242, 1.

Concerns about tenure arrangements and FPIC were raised during planning for the pilot project given a history of land disputes. In the late 1990s, a forestry-management agreement sought to rectify consent defects in the previous timber-rights agreement. At the same time, an NGO was negotiating a Wildlife Management Agreement over the site. The customary landowners claimed that they had not agreed to the forestry-management agreement,[246] resulting in a long-running dispute. The consultation process for April Salumei ran through 2009 and 2010,[247] but the project received validation under the Climate, Community and Biodiversity Alliance Standards in 2011.[248] The project submitted a separate application for approval under the Verified Carbon Standard. It is required to provide verification reports every five years throughout the project.[249] The Climate, Community and Biodiversity Alliance Standards are the 'gold standard' of forest carbon certification, given their potential to provide exceptional biodiversity benefits. The standards include requirements to obtain informed consent,[250] requiring project developers to 'document and defend' the process used to identify communities potentially affected and involve them in project design.[251] Once affected communities are identified, the project developer must seek FPIC from those whose property interests will be affected,[252] and must resolve any tenure disputes before the project commences, although the affected community does not have a power to veto proposed projects.[253] The standards expressly refer to OP 4.10[254] as a tool to assist project developers to assess the background tenure conditions.[255]

246 Filer, above n 81, 612.

247 Scientific Certification Systems, 'Final CCBA Project Validation Report, April Salumei, East Sepik, Papua New Guinea' (Validation Report, 30 May 2011) 31, https://s3.amazonaws.com/CCBA/Projects/April_Salumei_Sustainable_Forest_Management_Project/

248 Ibid. 5.

249 Pacific Forest Alliance Limited, above n 242. However, since this application, the Climate, Community and Biodiversity Alliance has been taken over by the Vulnerability and Capacity Assessment.

250 The Climate, Community and Biodiversity Alliance, 'Climate, Community and Biodiversity Project Design Standards (Second Edition)' (December 2008) 7, http://s3.amazonaws.com/CCBA/Upload/ccb_standards_second_edition_december_2008+%281%29.pdf.

251 Ibid. G3.8. This includes requiring project developers to establish a grievance procedure (G3.10).

252 Ibid. G5.3. 253 Ibid. G5. 254 See Chapter 5, Section B(a).

255 The Climate, Community and Biodiversity Alliance, above n 250, 36.

As discussed in Chapter 5, the standard requires only that 'free prior and informed *consultation*' be undertaken with indigenous peoples.

Pilot programmes have been used in Vanuatu to 'test a community-owned model of REDD+', which provides benefits for forest communities while reducing emissions and conserving and enhancing forests.[256] The project in Vanuatu is part of a regional REDD+ pilot, the Nakau Programme; implemented at project sites in Fiji, Solomon Islands and Vanuatu by NGO Live & Learn.

The Nakau Programme is a Payment for Ecosystem Services Programme intended to finance the costs of forest protection and enhancement, including opportunity costs to landowners who give up the right to log or deforest.[257] In Vanuatu, the pilot site for the Nakau Programme is the Loru Forest Project, an avoided deforestation programme on 166 hectares of tropical rainforest on eastern Espiritu Santo; on land degraded by commercial logging operations in the 1990s.[258] It was initiated in 2013 with funding from the European Union as a joint venture between the Nakau Programme, Live & Learn Vanuatu, and the Serakar Clan landowners, combining coastal rainforest protection with agroforestry.[259] The project is certified to the 'Plan Vivo' standard, developed with elements of the Verified Carbon Standard and the Climate, Community and Biodiversity Alliance Standards.[260]

The project is piloting 'community-owned and community-led' implementation of REDD+ in Melanesia to demonstrate the potential for carbon-management projects to generate wealth for customary landholders.[261] It resulted in the sale of 3,711 credits (3,711 tonnes avoided $CO_2e$) on the voluntary market to buyers overseas and within Vanuatu during

256 Vanuatu R-PP, above n 10, 27.

257 Sean Weaver, Robbie Henderson and Anjali Nelson, *Nakau Methodology Framework: General Methodology for the Nakau Programme: An Indigenous Forest Conservation Programme through Payments for Ecosystem Services* (Live and Learn International, April 2014), http://theredddesk.org/sites/default/files/nakau_methodology_framework_d2.1_v1.0_20140428_0.pdf.

258 Anjali Nelson et al, *Loru Forest Project – Project Description (PD), Part A: General Description: An Avoided Deforestation Project at Loru, Santo, Vanuatu* (Live & Learn Environmental Education, 2015). www.nakau.org/uploads/5/2/2/5/52251233/loru_pd_part_a_d3.2a_v1.0_20151009b.pdf

259 Sean Weaver, Anjali Nelson and Robbie Henderson, *Loru Forest Project Monitoring Report 1* (Live and Learn Environmental Education, October 2015), www.nakau.org/uploads/5/2/2/5/52251233/loru_monitoring_report_1_d3.3__1__v1.0_20151009b.pdf

260 See Chapter 2 for more on voluntary carbon market standards.

261 The REDD Desk, 'Pilot Effective Models for Governance and Implementation of REDD+ in Small Island Developing States to Provide Equitable Benefits for Forest-Dependent

2016.[262] This is the first Pacific project to general carbon credit sales. Landowners have elected to sell rainforest carbon offsets (credits) to generate revenue for local economic development instead of clearing land for coconut plantations.

While the project has accomplished a number of goals for the community, tensions around tenure reforms and cultural governance are apparent. For example, the Loru Forest Project's Idea Note suggests that the future viability of the project is undermined by the customary status of land tenure in the project area, because the task of identifying the correct landowners and their interests is difficult.[263] This analysis represents another instance of resource and environmental schemes, and now REDD+ style projects, failing to understand and respond to the complexity of customary tenures – with ramifications for the future equitable distribution of benefits within the community.[264] The project is further looking into new laws recently developed to allow for land registration by local customary tribunals. The project documentation notes that 'this law can provide further clarification of ownership and certainty over carbon and land tenure rights. Recent changes will take some time to be feasibly implemented within the project due to slow reform processes and current attitudes towards registration of land.'[265]

The continuing tensions around attempts to 'formalise' customary land holdings are evident in this statement. In light of past history of dealings with customary lands, customary landowners in Vanuatu are wary of externally imposed processes. The project may have been more effective if it had used the mechanism of a Community Conservation Area under the Environmental Management and Conservation Act 2002. This approach, used for other projects, may avoid the difficulties inherent

Local and Indigenous People', http://theredddesk.org/countries/initiatives/pilot-effective-models-governance-and-implementation-redd-small-island.

[262] 'First Sales of Rainforest Carbon Credits Sold in the Pacific – Vanuatu Leading the Charge', *Vanuatu Daily Post*, 29 November 2016.

[263] Anjali Nelson et al, *Project Idea Note: Loru Forest Project* (Live & Learn International, 9 October 2014) 27–28 (Land Tenure Context), see also 12 (Secure Tenure), www.ekos.org.nz/uploads/8/2/1/0/8210062/loru_pin_d3.3_v1.0_20140606.pdf; see further, Lisa Ogle, *Vanuatu: Legal Framework for REDD+* (SPC/GIZ Regional Project 'Climate Protection through Forest Conservation in Pacific Island Countries', August 2014, 11–13 (on the complexity of land tenure in Vanuatu), https://lrd.spc.int/spcgiz-qclimate-protection-through-forest-conservation-in-the-pacific-islandsq/studies-and-reports/cat_view

[264] See Chapter 6.

[265] Nelson et al, *Loru Forest Project – Project Description (PD)*, above n 258, 27–28.

to the practice of leases in Vanuatu. This approach to communal property, together with allowing autonomy for the customary group to retain its own internal governance structures, may be more beneficial than an insistence on tenure 'conversion'. Such an approach gives greater autonomy to customary owners, which in turn allows customary forms of tenure and accompanying traditional authority structures to endure.[266] The Loru Forest Project experience also confronts many assumptions that underpin the international policy framework for REDD+, where customary tenures must be 'formal' to be sufficiently certain to enable participation in projects.[267]

## G. Conclusion: Future Prospects for REDD+ Implementation

REDD+ has the potential to play a substantial role in Melanesia, in allowing alternative mechanisms for community development, as well as enabling PNG and Vanuatu to meet international climate change mitigation targets. At the same time, it will require interventions into ongoing, systemic and endemic problems regarding the use of forest resources on customary land. Previous efforts to create an interface between customary land tenure and Western-style resource projects have been hampered not only due to the complexity of customary relations, whose diversity defies simple categorisation as 'communal' or 'individual', but also because customary land tenure arrangements have been realigned in response to the impacts of resource projects. In the past, realignment around 'clear and certain tenures' has engendered significant intra-group disputes over the allocation of benefits. These effects can be exacerbated by ambitious international policy agendas, which national governments struggle to fully implement. Systemic problems with capacity and resourcing extend to consent determination and dispute-resolution processes.

The systemic issues evident in both Melanesian countries have obvious implications for REDD+ implementation, and for the articulation of policies in the readiness phase that can successfully incorporate the unique features of customary law and tenure arrangements in the two countries. These factors will determine whether REDD+ achieves its promised co-benefits and protects customary interests.[268] In the PNG

[266] See discussion in Section C and Chapter 6.

[267] See further in Chapter 5.

[268] See Chapter 4.

context, 'there has been little work to examine how the impacts and influences of locally situated institutions, relationships and cultural contexts have obfuscated the translation of high-level REDD policy statements into grass-roots reality'.[269] Through detailed examination of localised, contextual issues, together with the rapidly expanding REDD+ Readiness framework in each country, this chapter has demonstrated the complex trade-offs that REDD+ must navigate if readiness is to move to implementation.

In this context, a prescriptive adherence to the standards of the UN-REDD Programme and the FCPF in respect of safeguards, tenure arrangements and co-benefit arrangements may need to be adjusted to more fully acknowledge the potential for distributive inequalities in PNG and Vanuatu, and to anticipate the diverse customary environment and unique geographical challenges that Melanesia presents. Importantly, such standards must also be cognisant of the limited ability of national governments to resource and administer complex regimes. A perceived lack of governmental capacity may explain the measured pace toward REDD+ implementation in these countries. Undoubtedly, ongoing international support will be required to administer existing legal processes, develop new frameworks and carry out the demanding platform for FPIC, and to ensure adequate dispute-resolution procedures. Such support in these (and other) Melanesian countries is essential for providing REDD+ with firm foundations for ensuring that it can incorporate the interests of the majority of the population – the customary owners – who hold the land and forest on which REDD+ outcomes depend.

[269] Leggett and Lovell, above n 244, 118.

# 9

# Indigenous Land Tenures and Carbon Mitigation Schemes

## Lessons from Northern Australia

MAUREEN TEHAN*

### A. Australia Overview

Australia is not a recipient of REDD+ funds, but the country has been a prominent supporter of the REDD+ scheme in international fora. Moreover, at a national level, Australia has developed sophisticated emissions-reduction models in the forestry and land-use sectors and bio-sequestration schemes of general application, ranging from trading to direct funding grant models.[1] The Australian states (provincial governments) have enacted statutory schemes that provide an independent legal status for sequestered carbon and emissions reductions that form a distinct right, severable from the surrounding land and vegetation. Statutory carbon 'rights' of this type,[2] initially were designed to interact with national cap and trade credit schemes for mitigation of emissions through the promotion of bio-sequestration. Current models for carbon abatement and emissions reductions in the land sector focus on funding specific, voluntary emissions-reduction projects, and include unique

* The author gratefully acknowledges the research contributions of Carly Godden and Dr Elizabeth Macpherson to this chapter.

1 These arrangements were implemented through a related package of nineteen pieces of legislation known as the Clean Energy Future Package, including the Carbon Credits (Carbon Farming Initiative) Act 2011 (Cth); Clean Energy Act 2011 (Cth); Climate Change Authority Act 2011 (Cth); and a suite of consequential legislation. See Cathy J Robinson et al, 'Australia's Indigenous Carbon Economy: A National Snapshot' (2014) 52 *Geographical Research* 123. The trading model was replaced in 2014 by the Emissions Reduction Fund: Clean Energy Legislation (Carbon Tax Repeal) Act 2014 (Cth), and the Carbon Farming Initiative Amendment Act 2014 (Cth), which amended aspects of the 2011 Act.

2 The status of carbon has been characterised in different ways across jurisdictions, ranging from a proprietary right to a sui generis right.

programmes to reduce emissions from the burning of savanna lands. Some Aboriginal and Torres Strait Islander peoples (described generally as 'indigenous' in this chapter) have participated in these REDD+ style schemes, including savanna-burning projects, that utilise traditional knowledge, but other groups have had minimal opportunities for involvement. In this chapter, we discuss how the various models that underpin indigenous peoples' tenures and rights in Australia[3] determine the opportunities for Aboriginal and Torres Strait Islander peoples to participate in emissions-reduction and carbon-sequestration schemes. This detailed analysis offers rich insights to inform REDD+ regimes and associated programmes of ways to reduce and avoid emissions in other contexts and countries. Significantly, the Australian experience underlines the importance of recognising the complexity of legal regimes providing for indigenous tenure, and therefore the variable impact of carbon abatement legislation and emissions-reduction governance frameworks on them. An acknowledgement of the diversity of types of tenures and associated rights is critical to providing an effective and equitable basis for all indigenous peoples to participate in carbon sequestration and emissions-reduction schemes designed to avoid forest degradation.

REDD+ safeguards around tenure (see Chapter 5) implicitly tend to assume that there is a generic, indigenous tenure model that operates across countries.[4] Australian experience, by contrast, reveals that there are multiple forms of indigenous tenure and rights that intersect with carbon-sequestration and emissions-reduction schemes. The variety of tenures and rights not only leads to a constellation of differential impacts upon indigenous communities, but also critically influences how readily indigenous groups are included within a 'carbon economy'. It is necessary also to acknowledge that not all communities may seek to be involved. Accordingly, this chapter provides information upon which a critical analysis of international law and REDD+ safeguards for dealing with indigenous and forest-dwelling communities may draw.

A critical factor in determining the access of particular indigenous communities (and groups within communities) to REDD+-style projects, and any attendant benefits, is the coverage of the legal models and

[3] Models cover recognition, as discussed in Chapter 5, as well as other legal forms of transferring control and ownership to indigenous occupants or those with customary associations to land. While all these models might come under a broad 'recognition' rubric, some are not strictly recognition forms.

[4] See Chapter 5.

classificatory systems for emissions reductions that are developed. Where Aboriginal and Torres Strait Islander peoples have not had their claims to land and waters recognised by the state, the adoption of a REDD+-style scheme – even with 'safeguards' in place – is unlikely to occur.[5] Such situations are analogous to the international context where the formalisation of tenure is regarded as a prerequisite to the adoption of a REDD+ model. Yet Australian experience also demonstrates that, even where tenure and rights are formally recognised by the nation state, there remain difficult issues around robust consultation and participation measures, and uncertainty as to exactly how indigenous communities are to share in the value and benefits generated by emissions-reduction schemes.

### *1. Lessons for REDD+ from Australia*

Three key factors demonstrate why the Australian experience offers a strongly analogous case study for examining the interaction between indigenous peoples' rights that can be extrapolated to international REDD+ models and the national implementation of REDD projects. First, Australia is highly dependent on its land sector – economically and culturally. This dependency is evident in the prominence of land sector responses in its nationally determined contributions under the Paris Agreement.[6] Further, the importance of the land sector to Australia has generated extensive experimentation with statutory schemes for the management of carbon and emissions reduction in forest and savanna lands. While there has been a recent rollback of these schemes, Australia has developed a number of emissions-reduction and avoided deforestation projects, many of which have been implemented and continue to operate. Indeed, Australia is one of few jurisdictions with an established land-use framework for carbon management and emissions reductions with links for offsets – as set out in the Commonwealth's Carbon Credits (Carbon Farming Initiative) Act 2011 (Cth) as amended by the Carbon Farming Initiative Amendment Act 2014 (Cth).

Secondly, Australia, albeit belatedly, has engaged extensively with according statutory land rights and recognising indigenous peoples' tenures and rights at a federal (i.e. national) government and state

[5] We note that some political 'agreements' may establish a basis for indigenous participation, even in the absence of fully formalised tenure rights.

[6] See Chapter 2.

(provincial) level over the past forty years. The two main land tenure schemes are native title and statutory land rights legislation. These schemes provide for institutions and frameworks for indigenous communities to exercise decision-making on land management, administration, governance and benefits distribution on indigenous lands.

The third significant factor is the development of models which directly incorporate indigenous traditional knowledge and practices in the savanna-burning projects. Australia developed a 'methodology' for determining emission credits that combines traditional knowledge and scientific measures.[7] There is potential to consider how these schemes, which blend traditional practices and ecosystem-services models, could be adapted to other regions where the customary burning of vegetation takes place.[8] Australian experience with savanna burning also provides well-established models that offer useful guidance on how indigenous governance, which is associated with particular tenure forms, interfaces with REDD+ style schemes when evaluating the implementation of projects at national and provincial scales.

More generally, the experience of indigenous peoples in dealing with the 'carbon economy' is that, while it has delivered benefits for some, it has not been uniformly inclusive. It highlights that Australia, even as a first-world economy with well-functioning institutions of government, land and resource administration, together with a history of recognition of indigenous land interests, has still not managed a smooth transformation. Interestingly, this chapter identifies some similar challenges to the other country case studies in this volume, including: conflicts arising from multiple jurisdictions; inconsistent schemes for managing land and resources; irregularity in the recognition of indigenous peoples' land tenure; access to resources[9] and consent requirements;[10] and in the regimes for managing carbon rights and co-benefits.[11] Thus, it seems that formalising indigenous tenures within carbon-sequestration and land sector emissions-reduction regimes, while significant, does not

[7] The Carbon Credits (Carbon Farming Initiative – Emissions Abatement through Savanna Fire Management) Methodology Determination (2015) (Cth) made 25 March 2015. See Australian Government, 'Savanna Fire Management', Department of the Environment and Energy, www.environment.gov.au/climate-change/emissions-reduction-fund/methods/savanna-burning.

[8] Stephanie Niall et al, 'Climate Change and REDD+: Integrating Customary Fire-Management Schemes in East Malaysia and Northern Australia' (2013) 28 *Sojourn: Journal of Social Issues in Southeast Asia* 538, 553–534, 562–563.

[9] See Chapter 5. [10] See Chapter 4. [11] See Chapter 6.

necessarily resolve the myriad challenges involved in integrating indigenous peoples into the carbon economy.

### 2. *Challenges to Indigenous Peoples' Participation in Carbon-Management Activities*

Australia, under successive governments, has implemented several different legislative models to promote emissions reduction and carbon-sequestration schemes that may operate with respect to indigenous land tenure. The major initiative to incorporate traditional knowledge in national level emissions-reduction schemes is now based on a direct funding model known as the Emissions Reduction Fund.[12] An indigenous carbon economy is slowly emerging in Australia, especially in northern regions.[13] Yet, integrating such schemes into the complex legal setting of the multiple jurisdictions comprising Australia's federal system, alongside the layered and varied forms of indigenous land tenure, poses significant challenges. Across the Australian jurisdictions, there are a range of indigenous land tenure models, and land management, administration and consent processes. These variously provide for elements of FPIC[14] as well as schemes for benefit-sharing when third party activities, such as mining, impact indigenous communities. Within this general structure, many unresolved issues impinge upon the interaction between emissions-reduction schemes and indigenous peoples' tenure regimes. These include whether indigenous land tenure extends to a right to benefit from carbon-management activities; how indigenous rights are affected by third-party interests; whether a native title resource right in respect of emissions reductions can be recognised; and the overarching relationship between these various rights and regulatory schemes for emissions reduction and carbon sequestration. The analysis provides insights for the development of institutional and process responses as

[12] The Direct Action Package: Clean Energy Legislation (Carbon Tax Repeal) Act 2014 (Cth); Carbon Farming Initiative Amendment Act 2014 (Cth).

[13] Robinson et al, above n 1; Catherine J Robinson et al, 'Indigenous Benefits and Carbon Offset Schemes: An Australian Case Study' (2016) 56 *Environmental Science and Policy* 129.

[14] Australian law does not have a direct correlate of international principles around FPIC, although various statutes such as the Native Title Act 1993 (Cth) and the Aboriginal Land Rights (Northern Territory) Act 1976 (Cth) provide for designated aspects of those principles. Such provisions are discussed below.

they emerge in many other countries that are adopting REDD+.[15] Such analysis contributes to evaluating the contribution of REDD+ schemes in different legal contexts where competing local, state and international regimes are at play. It highlights key considerations for appropriate dealings with indigenous peoples and local communities in relation to tenure, FPIC and co-benefits.

### (a) Indigenous Tenure and 'Ownership'

A key challenge in the Australian context is the complex nature of indigenous peoples' tenure, which is not adequately accounted for within carbon and emissions-reduction regimes. Within Australia, indigenous 'tenure' can range from non-exclusive rights that coexist with third party rights (non-exclusive native title) to exclusive possession native title (a collection of rights, including the right to exclude others from the land, but still subject to government regulation), to exclusive possession, such as the fee simple (freehold) titles under land rights legislation. The nature of the rights that are held by indigenous peoples will directly influence the scope of the opportunities for engagement with emissions-reduction and carbon-sequestration projects under the Emissions Reduction Fund.

Where indigenous communities hold exclusive possession native title, they may be deemed eligible to participate in federal carbon management initiatives because the legislation *specifically provides* for eligibility. Where indigenous communities have not established exclusive-possession native title, including where they have been subject to historical dispossession, there is limited potential for indigenous participation in REDD+ style schemes. Participation may be dependent upon the carbon legislative schemes specifically carving out a place for indigenous landholders. Similar concerns apply to REDD+ (as discussed in Chapter 5), where the emphasis on securing tenure formalisation may advantage groups in current occupation, while disadvantaging those communities claiming rights that are based on an earlier historic possession.[16] The Australian experience demonstrates that often there are protracted time frames for many indigenous communities in securing tenure recognition (over two decades in many instances). These time frames do not sit well with the pressures to implement avoided deforestation and sequestration regimes under international law. REDD+ style schemes thus may exacerbate

[15] See Chapters 7 and 8. [16] See Chapter 5.

existing inequalities between those groups whose claims are recognised, and those communities yet to achieve formal state substantiation of their claims.

### (b) Property in Carbon

Another key theme explored is whether indigenous 'ownership' is necessary to gaining co-benefits under emissions mitigation schemes. In Chapter 5, we highlighted uncertainty in the REDD+ context as to the degree of indigenous 'ownership' of land and carbon that was necessary for co-benefits to flow to forest-dependent communities under national REDD+ activities.[17] While robust models of indigenous land tenure should be supported where possible, the viability of an indigenous 'right to carbon' as a resource right requires examination. Australian jurisprudence on the recognition of native title rights to resources for commercial purposes[18] may provide a viable route for other indigenous communities to participate in REDD+ schemes. In this context, the severability of 'carbon' as an independent legal entity, as well as the legal status of the 'severed' resource, requires a highly efficient and transparent system, such as registration, for managing all the myriad layers of tenure and resource rights to ensure that indigenous interests are not overridden.

Given the novelty of the concept of a carbon right in many legal systems, this chapter briefly outlines the Australian national and state government schemes providing for carbon mitigation and sequestration, focusing on one state scheme that disaggregates carbon rights from land. The chapter considers the complex institutional and legal arrangements for indigenous land tenure, management and governance regimes, and concludes with an assessment of the risks, safeguards and benefits emerging from the connection of recognition, land tenure arrangements and carbon.

## B. Mitigation and Sequestration of Carbon in Australia

In Australia, the regulation of 'forestry' and 'carbon sequestration rights' falls under the jurisdiction of Australian states.[19] All Australian states have enacted carbon rights legislation that identifies and distributes rights in respect of emissions-reduction capacity and carbon

[17] See Chapter 5. [18] *Akiba v Commonwealth* (2013) 250 CLR 209.

[19] The federal government has developed both trading mechanisms for managing carbon credits and direct-funding models that work in conjunction with the state legislation.

sequestration in the land-use sector. The Northern Territory and the Australian Capital Territory have not done so, but may access federal carbon mitigation schemes. The characterisation of carbon varies widely across states.[20]

State statutory schemes separate rights to carbon in vegetation (and, in some instances, in soil) from landholding. Despite attempts to harmonise the legislation, there remain many inconsistencies between different state regimes for forestry and carbon rights. States have significantly different terminology and, in certain instances, different legal constructs in recognising carbon rights. In turn, each of these regulatory frameworks interacts with national legislation.[21] To gain some sense of the nature of these statutory carbon rights, and their interaction with national legislation and current policy initiatives promoting land sector emissions-abatement projects, it is useful to consider the arrangements in Western Australia. Section 1 discusses the regulation of carbon management activities in Western Australia; a jurisdiction with a relatively high indigenous population and extensive areas of native title determinations.

## 1. *State Schemes for Carbon Management: Western Australia*

State government carbon management schemes are a key component of the overall operation of carbon mitigation within Australia and reflect reliance on comprehensive statutory models and land management practices. The Carbon Rights Act 2003 (WA) provides for a carbon right in land where an approved carbon rights form is registered over freehold or Crown land.[22] The legislation treats the carbon right as a separate, statutory-based proprietary interest. The owner of the carbon right need not be the proprietor of the land. Once the carbon right is registered, its owner acquires all the accruing legal and commercial benefits, as well as

[20] In some states, rights to carbon are regarded as a common law form of property. In New South Wales and Tasmania, they are a type of profit à prendre (a property interest in land), while in South Australia such rights are defined as a chose in action (a personal property right rather than an interest in land). In Western Australia, Queensland and Victoria, carbon rights are deemed a sui generis form of statutory property interest: Conveyancing Act 1919 (NSW) ss 87A, 88AB; Forestry Rights Registration Act 1990 (Tas) ss 3, 5; Forest Property Act 2000 (SA) s 3A; Carbon Rights Act 2003 (WA) ss 5, 6, 8, 10; Land Act 1994 (Qld) s 373R; Land Title Act 1994 (Qld) s 97N; Forestry Act 1959 (Qld) s 61K; Climate Change Act 2010 (Vic) ss 3A, 3B, 26.

[21] This is an inevitable consequence of the diverse histories of the states and the federal system and is not only limited to this field of activity.

[22] Carbon Rights Act 2003 (WA) ss 5, 6.

responsibility for risks, such as destruction by fire.[23] Only one carbon right can be registered over a specified part of land.[24] In Australia's federal system, the legal treatment of 'severable carbon' can result in myriad regulatory systems as well as layers of potential rights and interests which, in turn, exacerbate the complexities of the multiple forms of indigenous land arrangements.

A separate mechanism, a carbon covenant, is used to ensure commercial certainty and utilisation.[25] The owner of a carbon right may enter into a carbon covenant with the other persons who have an interest in the land (i.e. the 'owner') through registration of the carbon covenant on the title.[26] A carbon covenant in effect is a contractual agreement setting out the terms and conditions between the landholder and carbon right holder. For instance, a carbon covenant may outline any management responsibilities over the forest vegetation to which the carbon relates.[27] The same person (or legal entity) holds the carbon right and carbon covenant.[28]

The legal character of the registered carbon right and carbon covenant is not directly defined in the legislation, but it has attributes of a property right.[29] The legislation does identify the carbon right and carbon covenant as being enforceable against future purchasers of the interest in the land. Importantly also, the scheme does not prescribe the type of land holding that the carbon right can attach to, which means there is nothing to prevent any person, 'from applying to register the carbon right irrespective of land ownership'.[30] Potentially this may include holders of non-exclusive native title.[31] Ironically, a non-indigenous person conceivably may hold carbon rights in land subject to native title.

[23] See Pamela O'Connor et al, 'From Rights to Responsibilities: Reconceptualising Carbon Sequestration Rights in Australia' (2013) 30 *Environmental and Planning Law Journal* 403.

[24] Carbon Rights Act 2003 (WA) s 7(2).

[25] Sandra Eckert and Richard McKellar, 'Securing Rights to Carbon Sequestration: The Western Australian Experience' (2008) 8(2) *Sustainable Development Law & Policy* 30, 31.

[26] Carbon Rights Act 2003 (WA) s 12.

[27] Ibid. s 10(2); Samantha Hepburn, 'Carbon Rights as New Property: The Benefits of Statutory Verification' (2009) 31 *Sydney Law Review* 239, 251.

[28] Carbon Rights Act 2003 (WA) s 11(2)(g).

[29] It is described as both an encumbrance and a hereditament – suggesting a proprietary character.

[30] Hepburn, above n 27, 250; Jeremy Dore et al, 'Carbon Projects and Indigenous Land in Northern Australia' (2014) 36 *The Rangeland Journal* 389, 395.

[31] Although this does not flow from their status as a native title-holder.

Steps for registration of the carbon rights and carbon covenants are set out under the Transfer of Land Act 1893 (WA).[32] For freehold land and government (Crown) land, the written consent of each person having a registered interest in the specified land is required.[33] The minister may apply for the state to become the registered proprietor of a carbon covenant, enter into a carbon covenant on government (Crown) land, and deal with carbon rights and carbon covenants.[34]

### 2. *The Direct Action Plan – Emissions Reduction Fund*

The national scheme for carbon mitigation and sequestration activities in the land sector is now a direct action model that operates as a voluntary Emissions Reduction Fund for emissions-reduction projects.[35] This programme subsumed the Carbon Farming Initiative (CFI).[36] The CFI Act established a statutory-based bio-sequestration and carbon offsets scheme that sought to deliver sequestration and avoidance of emissions over and above 'business-as-usual' practices in the land sector. The Act addressed 'additionality' requirements[37] by providing statutory recognition for eligible CFI projects as determined against particular scientific and technical rules and standards (methodologies).[38] The 'additionality' requirement is slightly altered by the Carbon Farming Initiative Amendment Act 2014 (Cth).[39] The amendments now require that '[p]rojects must be new, not required by regulations and not funded by another

[32] Transfer of Land Act 1893 (WA) pt VI div 2A. [33] Ibid. s 104B.

[34] Land Administration Act 1997 (WA) s 18A; Dore et al, above n 30.

[35] These schemes are considered further below.

[36] Established under the Carbon Credits (Carbon Farming Initiative) Act 2011 (Cth) (CFI Act).

[37] 'Additionality' (now called 'newness') is generally achieved through requirements that the project is an eligible activity as set out in the regulations; it has not begun to be implemented; and is not already required to be carried out by or under a Commonwealth, state or territory law or is unlikely to be carried out by a government programme or scheme: Ibid. Criteria for the declaration of an eligible offsets project are provided in ibid. s 27(4).

[38] The changes to the Emissions Reduction Fund have resulted in transitional provisions for reaccreditation of methodologies: Australian Government, 'Carbon Farming Initiative Project Transition into the Emissions Reduction Fund', Department of the Environment and Energy, www.environment.gov.au/climate-change/emissions-reduction-fund/carbon-farming-initiative-project-transition.

[39] Carbon Farming Initiative Amendment Act 2014 (Cth) s 107; amending the Carbon Credits (Carbon Farming Initiative) Act 2011 (Cth) with the insertion of s 27(4A), which allows for other additionality measurements in lieu of 'newness'.

Government programme'.[40] Eligible projects fall into two distinct categories: emissions-avoidance offsets projects and sequestration offsets projects.

The CFI initially rested upon a unique hybrid instrument: a carbon pricing mechanism, which was to transition into an Emissions Trading Scheme in 2015. Under the CFI, project proponents were issued a credit unit for every tonne of emissions avoided or sequestered as a result of CFI activities. These credits could then be sold on the domestic carbon market to entities with obligations to account for their emissions or on the international voluntary carbon market, enabling entities to 'offset' their emissions.[41]

Following a change of government in 2014, the carbon pricing mechanism was abolished and replaced by the Emissions Reduction Fund,[42] which modified and extended the application of the CFI model to other types of emissions-reduction activities beyond the land sector.[43] Under the Emissions Reduction Fund, project proponents bid for government funds from the Emissions Reduction Fund via a reverse auction process where the lowest-priced bids are awarded 'carbon abatement contracts'. Payment then flows in exchange for delivery of emissions reductions in accordance with contractual arrangements entered into between the Clean Energy Regulator and the successful bidder.[44] The Australian Government now expects to meet its emissions-reduction commitment (NDC) under the Paris Agreement via the Direct Action Plan.[45]

The requirement for additionality or newness for emissions-reduction projects has specific ramifications for indigenous peoples whose land is

40 Explanatory Memorandum, Carbon Farming Initiative Amendment Bill 2014 (Cth) 22–3; see also 29–30 [1.51]–[1.55].

41 For an overview, see Rodney J Keenan et al, 'Science and the Governance of Australia's Climate Regime' (2012) 2 *Nature Climate Change* 477, 477–478.

42 Clean Energy Legislation (Carbon Tax Repeal) Act 2014 (Cth).

43 Carbon Farming Initiative Amendment Act 2014 (Cth).

44 Ibid. Currently $2.55 billion has been allocated to the Emissions Reduction Fund; see Australian Government, 'About the Emissions Reduction Fund', Department of the Environment and Energy, www.environment.gov.au/climate-change/emissions-reduction-fund/about.

45 Some have questioned the ability of the scheme to deliver the reduction required as NDCs are meant to increase with ambition every five years: Australian Government, Climate Change Authority, 'Australia's Climate Policy Options' (Special Review, Second Draft Report, November 2015) 8; Frank Jotzo, 'Australia's 2030 Climate Target Puts Us in the Race, but at the Back', *The Conversation* (12 August 2015), http://theconversation.com/australias-2030-climate-target-puts-us-in-the-race-but-at-the-back-45931.

part of the conservation estate, as the existing conservation-management requirements in respect of the land may mean that such land will not satisfy the newness requirement for the CFI.[46] This analogous situation highlights the need for the REDD+ to ensure that its abatement schemes adequately account for indigenous customary and conservation practices as part of additionality requirements.

An offsets project must be conducted under a government-approved method established under a 'methodology determination'.[47] The integrity standards require that the emissions abatement by the project is capable of measurement and verification.[48] The Australian government has approved methodologies for projects including soil carbon sequestration, reforestation and afforestation, environmental plantings, avoided deforestation, and fire management amongst others.[49] Once a project has been approved, ongoing monitoring and reporting requirements apply, aimed at ensuring verification of credits.[50] The crediting period – that is, the duration of contractual arrangements between the government and a project proponent – for an offsets project is generally a maximum of seven years for an emissions-avoidance project and twenty-five years for a sequestration offsets project, although this period can be varied by the methodology.[51] Proponents of new sequestration projects can nominate a permanence period of twenty-five to one hundred years when they apply to register their project.[52]

Several provisions address permanence and the enforceability of offset credits generated under the scheme. The proponent of an eligible offsets project must be the person responsible for the project and hold the legal right to carry out the project.[53] In addition, all persons

[46] As at June 2014 there was one abatement project in an Indigenous Protected Area: Australian Government, 'Indigenous Protected Areas: Balanggarra', Department of the Environment, www.environment.gov.au/indigenous/ipa/declared/balangarra.html.

[47] Carbon Credits (Carbon Farming Initiative) Act 2011 (Cth) ss 106(1)–(3).

[48] Ibid. s 133(1)(b). See also, Australian Government, Department of the Environment, 'Emissions Reduction Fund' (Methodology Determination Guidelines, January 2015), www.environment.gov.au/climate-change/emissions-reduction-fund/publications/methodology-determination-guidelines.

[49] Australian Government, 'Emissions Reduction Fund Methods', Department of the Environment and Energy, www.environment.gov.au/climate-change/emissions-reduction-fund/methods.

[50] Carbon Credits (Carbon Farming Initiative) Act 2011 (Cth) s 76.

[51] Ibid. pt 5.

[52] Ibid. s 23(1)(g).

[53] Ibid. ss 5, 27(4)(e). Where there are multiple parties, each party will be a project proponent and have responsibility for carrying out the offsets project: Ibid. s 135.

holding an 'eligible interest in the land' to be covered by the offsets project must have consented to the making of the application, although the offsets project can be registered conditional on consent being obtained.[54]

An 'eligible interest in land' includes freehold land[55] and leasehold interests.[56] The definition also includes interests in Crown (government) land, 'freehold land rights land'[57] and determined native title rights (including non-exclusive native title), exercised by the relevant Registered Native Title Body Corporate, but only the holder of exclusive possession native title can be a deemed project proponent.[58]

A number of identified difficulties emerge from this scheme of relevance to REDD+ legislative, policy and project design. These matters include the effect of drafting and definitions in legislation that may or may not encompass indigenous land interests (or rights) as well as the specificity of the interests covered. For example, the state scheme makes no provision for native title-holders, whereas the national scheme deals specifically with native title but ultimately distinguishes between the rights of exclusive possession and non-exclusive possession holders.[59] Thus, protection of indigenous interests is dependent upon the capacity of legislative schemes to adequately capture the full range of such land and resource interests.[60] Consideration of the application of these Australian laws to the indigenous estate raises further questions. Accordingly, this chapter now examines the various complex regimes providing for indigenous tenure and rights.

## C. Indigenous Tenure and Recognition: Common Law and Statute

Determining the extent to which Aboriginal peoples and Torres Strait Islanders may benefit from and participate in the carbon economy is complicated by the diversity of tenure arrangements in Australia.

54 Carbon Credits (Carbon Farming Initiative) Act 2011 (Cth) s 28A.
55 Technically, this is an estate in fee simple in Torrens Title Land.
56 Leasehold land is leased for a term which can be any length, e.g. ninety-nine years.
57 Similar to indigenous statutory land-rights land in this chapter. See Carbon Credits (Carbon Farming Initiative) Act 2011 (Cth) s 5.
58 Ibid. ss 44, 45, 45A, 46.
59 Non-exclusive interests may be conceived as akin to 'resource rights'.
60 Chapter 5 considers the difficulties in achieving such an outcome.

Whether indigenous communities hold a 'clear and secure' title (i.e. formalised title) is significant as it is a major determinant of whether these groups can participate in the carbon mitigation schemes. Yet across Australia the rights exercised by various indigenous communities vary depending on the form of legal recognition, the nature of the title, and the terms of the settlement or grant, as do the management and governance arrangements. The Aboriginal Land Rights (Northern Territory) Act 1976 (Cth) offers the most robust model with a fee simple title and associated decision-making rights, but native title is the most widespread and common form of title across all jurisdictions. These two regimes reflect a formalisation of indigenous land relationships that are cognisable to the settler legal system. The desirability of other forms of formalisation, such as individual titles, is an ongoing issue.[61]

### 1. *The Historical Context*

The legal landscape surrounding indigenous tenure in Australia has, from inception, been characterised by its multi-layered complexity. The Australian continent was colonised by the British from 1788 onwards, resulting in six separate colonies that formed the federal Commonwealth of Australia in 1901.[62] Each colony developed its own land administration and resource management laws. Powers to make laws in relation to indigenous people passed to the federal parliament in 1967. This historical and constitutional matrix results in multiple layers of interaction – state laws governing land administration and management of environmental and mineral resources may be subject to federal laws in relation to indigenous peoples (such as the native title regime) or to laws made pursuant to Australia's international obligations.[63]

[61] See Lee Godden and Maureen Tehan (eds), *Comparative Perspectives on Communal Lands and Individual Ownership: Sustainable Futures* (Routledge, 2010); Lee Godden and Maureen Tehan, 'Translating Native Title to Individual "Title" in Australia: Are Real Property Forms and Indigenous Interests Reconcilable?' in Elizabeth Cooke (ed), *Modern Studies in Property Law* (Hart Publishing, 2007) vol. 4, 263; Leon Terrill, *Beyond Communal and Individual Ownership: Indigenous Land Reform in Australia* (Routledge, 2016).

[62] Commonwealth of Australia Constitution Act 1900 (Imp) 63 & 64 Vict, c 12.

[63] *Commonwealth v Tasmania* (1983) 158 CLR 1 ('*Tasmanian Dam Case*'), 218–219, 220–221 (Brennan J), 126–9 (Mason J), 168–170 (Murphy J), 254–256, 259–260, 261–262 (Deane J).

At the time of colonisation, Australia was legally considered terra nullius and all land was Crown land.[64] There were no negotiated settlements or treaties nor was there formal legal recognition of pre-sovereignty indigenous rights and interests until the 1992 High Court decision in *Mabo v Queensland*.[65] In response to that decision, the Federal Parliament passed the Native Title Act 1993 (Cth) as a statutory basis for the recognition of pre-sovereignty rights. In the recent native title era, a 'culture of agreement making' over land access and resource use has developed.[66] To date however, there has been no constitutional recognition of prior indigenous occupation, and no constitutional entrenchment of indigenous rights.[67] The recognition of indigenous tenure and rights must be 'established' within the context of existing property law and land and environmental management regimes.[68]

From the 1970s onwards, federal and state governments began to recognise indigenous tenures in Australia, including through statutory land rights legislation. The indigenous estate[69] now covers approximately 43 per cent of Australia, comprising 903,222 $km^2$ under statutory land rights schemes and reserves, 865,574 $km^2$ of exclusive-possession native title and 1,521,579 $km^2$ of non-exclusive possession native title.[70] Further land is still under claim.

Aside from native title and statutory land rights, there are numerous other indigenous tenures in Australia, as well as a range of other laws that

[64] *Cooper v Stuart* (1889) 14 App Cas 286; *Attorney-General v Brown* (1847) 1 Legge 312; *Milirrpum v Nabalco Pty Ltd* [1972–73] ALR 65.

[65] *Mabo v Queensland (No 2)* (1992) 175 CLR 1. An earlier claim for native title in the Northern Territory Supreme Court had been unsuccessful and was not appealed: *Milirrpum v Nabalco Pty Ltd* [1972–1973] ALR 65.

[66] Marcia Langton et al (eds), *Settling with Indigenous People: Modern Treaty and Agreement-Making* (Federation Press, 2006); Marcia Langton et al (eds), *Honour Among Nations?: Treaties and Agreements with Indigenous People* (Melbourne University Press, 2004).

[67] The Racial Discrimination Act 1975 (Cth) provides limited protections against racially discriminatory actions by governments and third parties.

[68] See Chapter 3 for an examination of regime interaction.

[69] This term is used to describe the full extent of indigenous interests in land under multiple schemes and tenures.

[70] Jon Altman and Francis Markham, 'Burgeoning Indigenous Land Ownership: Diverse Values and Strategic Potentialities' in Sean Brennan et al (eds), *Native Title from Mabo to Akiba: A Vehicle for Change and Empowerment?* (Federation Press, 2015) 126, 134–135 (2016 figures on file with author); see also Francis Markham and Jon Altman, 'Values Mapping Indigenous Lands: An Exploration of Development Possibilities' (National Native Title Conference 2013, Alice Springs, Australia, 3 June 2013) (copy on file with author).

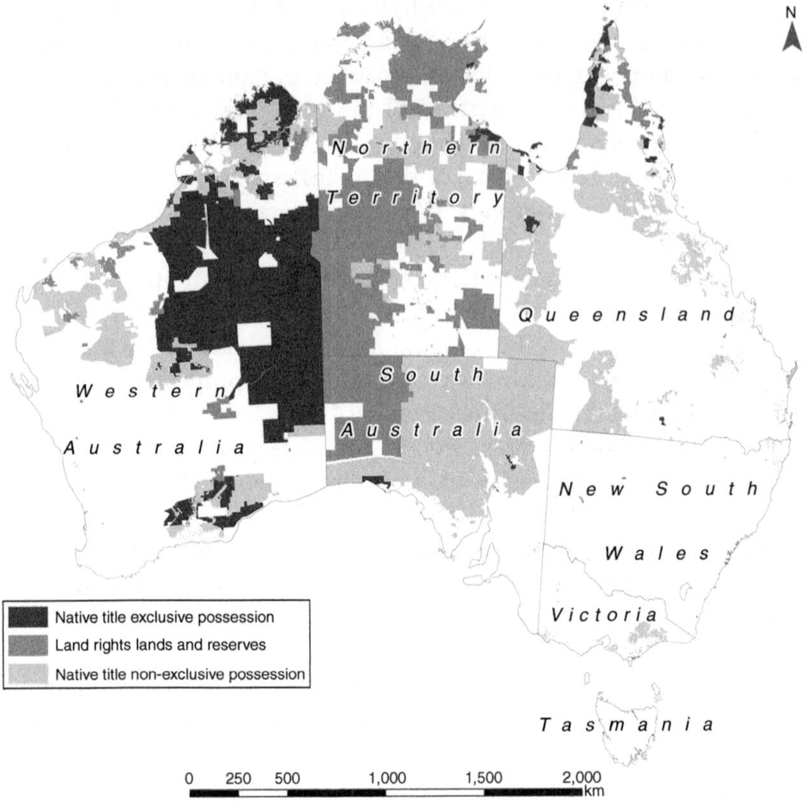


**Figure 9.1.** Indigenous-held land in Australia[71]

affect indigenous rights to land and resources such as heritage protection regimes.[72] The Indigenous Land Corporation funds land purchases for indigenous communities through normal commercial land transactions.[73] Indigenous Protected Areas (that are implemented in respect of various forms of indigenous tenure) are established by land management

[71] Ibid., 126–142 (Figure 9.1 updated for this volume).

[72] Cultural heritage legislation provides procedural rights only. While heritage provisions might provide a legal shadow for some negotiation about the terms of access to and activities on land (such as for carbon-related purposes), they are of limited effect in providing meaningful control by indigenous peoples of activities on their traditional land.

[73] Land Fund and Indigenous Land Corporation (ATSIC Amendment) Act 1995 (Cth); Australian Government, 'About Us', Indigenous Land Corporation, www.ilc.gov.au/Home/About-Us.

agreements between the Commonwealth government and indigenous groups for the purposes of biodiversity protection and conservation.[74] Land owned by indigenous communities may also be subject to joint management agreements for conservation and state-owned conservation land can be the subject to co-management arrangements with indigenous peoples.[75] Across these various indigenous tenure and land management arrangements, the basis for indigenous peoples' consent to projects and activities can differ widely.

In Chapter 4, we examined how the policy framework for REDD+ includes safeguards around the FPIC of indigenous communities. Although Australian law does not have a direct correlate of international principles around FPIC, the various statutory frameworks for land grants and native title set the parameters for decision-making and draw on aspects of FPIC in respect of activities on land. The Australian schemes establish a three-stage process for *internal* decision-making: advising the indigenous owners of a proposal, confirming that they understand the proposal and ensuring that they consent to it.[76] Decisions made in this internal process are then placed within a broader context of state regulatory provisions for consents, grants and licensing in relation to the development proposal. This process, *external* to the group, then determines whether, and on what conditions a proposal may proceed. In many instances this occurs notwithstanding a lack of consent or on terms different from those consented to by the indigenous group. In Section 2, we describe the main tenure forms and rights, including the consent provisions attaching to them.

## 2. *Native Title: What Type of Tenure and Rights?*

The Native Title Act 1993 plays a key role in providing for indigenous tenure in Australia. The legislation was designed to protect native title and establish processes for claims and, through the 'future acts regime', decision-making by all governments in relation to proposed

[74] Lee Godden and Stuart Cowell, 'Conservation Planning and Indigenous Governance in Australia's Indigenous Protected Areas' (2016) 24 *Restoration Ecology* 692.

[75] Altman and Markham, above n 70.

[76] Aboriginal Land Rights (Northern Territory) Act 1976 (Cth) s 23(3); Anangu Pitjantjatjara Yankunytjatjara Land Rights Act 1981 (SA) s 7; Native Title (Prescribed Bodies Corporate) Regulations 1999 (Cth).

activities on land where native title exists (as determined by a court) or is subject to a claim.[77]

Specific legal 'recognition'[78] extends to indigenous groups that were in occupation of land at the time sovereignty was claimed by the British and have, as a group, continued to acknowledge and observe traditional laws and customs until the present day.[79] The rights and interests, called native title, are subject to extinguishment by valid government actions at any time since sovereignty was claimed, generally without compensation.[80] Often, they are diminished rights to land and resources, as they are subject to, or coexist with the rights of government and third parties in the same land. Areas of exclusive-possession native title exist but these tend to be in remote areas. Further, exclusive possession under the Native Title Act is construed as one element in a collection of rights rather than a right to ownership of land.[81] Native title did extend recognition to indigenous peoples[82] across the country and most significantly in many parts of the country not covered by statutory land rights, particularly in Western Australia and Queensland.[83]

While the resources and land administration regimes of state governments (arguably including those covering carbon sequestration and emissions reduction) continue to apply, the future acts provisions of the Native Title Act overlay them and bind the states in terms of *how* they must make proposed grants of third party rights or government actions on native title land.[84]

[77] For background, see Maureen Tehan, 'A Hope Disillusioned, An Opportunity Lost? Reflections on Common Law Native Title and Ten Years of the *Native Title Act*' (2003) 27 *Melbourne University Law Review* 523.

[78] This strict recognition model contrasts with other forms of indigenous land tenure extended by governments such as inalienable freehold titles which may or may not be based on prior occupation of specific lands and continuing acknowledgement and observance of traditional laws.

[79] *Mabo (No 2)* (1992) 175 CLR 1, 57 and 59 (Brennan J).

[80] Ibid. 15 (Mason CJ and McHugh J).

[81] *Western Australia v Ward* (2002) 213 CLR 1, 89–90 [76]–[79] (Gleeson CJ, Gaudron, Gummow and Hayne JJ).

[82] Often called traditional owners or native title-holders. Other jurisdictions might use terms such as 'customary owners'.

[83] Although Queensland had land-title schemes under its Deed of Grant in Trust system and subsequently under the Aboriginal Land Act 1991 (Qld) and the Torres Strait Islander Land Act 1991 (Qld), native title dramatically extended the amount of land available for claim by allowing claims over pastoral leases.

[84] Native Title Act 1993 (Cth) Div 3.

The details of the rights and interests comprising native title[85] derive from 'the law acknowledged and customs observed' by the relevant group.[86] Groups must prove that, as a vibrant group, the acknowledgement and observance of traditional laws and customs has continued and been passed down to each generation substantially uninterrupted since sovereignty was acquired over the land.[87] This 'continuity' is required regardless of any government actions resulting in dispossession or removal of groups from their traditional lands.[88] Many groups are unable to meet this standard. If a native title applicant can satisfy these requirements, native title might include a right to exclusive possession.[89] At its narrowest conception, native title comprises rights to use the land for limited social or ceremonial purposes or to forage for food. Further, the extinguishment of native title doctrine has significantly reduced the amount of land, as well as the extent of the rights that groups may obtain pursuant to a determination of native title, even if they can meet the continuity requirements.[90]

Native title has been characterised as a collection or bundle of rights, as opposed to a right to the land itself. The components of the bundle can be separately extinguished by different acts of government.[91] Native title rights have been substantially eroded by the gradual grant of a range of rights to third parties over government land, extinguishing or diminishing native title. Large areas of land in Australia are now subject to multiple titles – native title, government title and third party rights. Where there are multiple interests in land, it is not always clear who has the right to access and manage resources on that land, including in the case of carbon abatement and sequestration. This is the case where native title is held in land subject to non-exclusive pastoral leases,[92]

[85] Ibid. s 225. [86] Ibid. s 223; *Western Australia v Ward* (2002) 213 CLR 1.

[87] *Members of the Yorta Yorta Aboriginal Community v Victoria* (2002) 214 CLR 422, [46], [47], [87]–[89].

[88] *Mabo [No 2]* (1992) 175 CLR 1, 59 (Brennan J).

[89] This does not equate directly to western property 'ownership'.

[90] Native title is extinguished by valid inconsistent government actions such as the grant of inconsistent interests in land to third parties or, in some instances, reservation and vesting of land for its own use: *Western Australia v Ward* (2002) 213 CLR 1, 146–147 [249] (Gleeson CJ, Gaudron, Gummow and Hayne JJ).

[91] Ibid. 89 [76] (Gleeson CJ, Gaudron, Gummow and Hayne JJ).

[92] Pastoral leases are a peculiarly Australian form of landholding created by statute in the nineteenth century to permit and regulate sheep and cattle grazing: Daniel Walker and Allan Dale, 'Land Tenure in Northern Australia: Opportunities and Challenges for Investment' (CSIRO, 2013).

which cover approximately fifty per cent of northern Australia. Pastoral leases are not grants of exclusive possession. Therefore, while they extinguish the right to control access, they do not necessarily extinguish all native title rights.[93] Thus, land covered by a pastoral lease may be subject to native title rights, with respective rights determined relative to one another.[94] Pastoral leases are particularly important for sequestration and savanna burning, as they cover large parts of northern Australia where these activities are likely to occur. Together, the continuity requirement and extinguishment regime disadvantage indigenous communities claiming tenure rights based on prior possession, a theme explored in Chapter 5.[95]

### (a) Carbon Rights and Native Title

In the REDD+ context, Fisher and Lyster suggest that the concept of a 'resource' tenure for indigenous peoples and forest-dwelling communities should be considered, even though the rights may not necessarily accord with ownership constructs as a 'full bundle of rights'.[96] There is potential for Australian native title rights to operate within a resource-rights model for managing carbon and avoided emissions. There are some limitations, however, with any resource-rights model that simply replicates the 'bundle of rights' construct in native title with all its constraints.[97] As a set of 'use rights', including use of flora and plant resources, native title rights might encompass sequestered carbon or carbon in other forms. Carbon-related projects are not specifically included in the activities in the future acts scheme of the Native Title Act,[98] although it is possible that some carbon-related activities might attract some aspect of the scheme.[99] Some commentators have

93 *Western Australia v Ward* (2002) 213 CLR 1, 126–31 [177]–[195] (Gleeson CJ, Gaudron, Gummow and Hayne JJ).

94 Ibid. 126–131 [177]–[195] (Gleeson CJ, Gaudron, Gummow and Hayne JJ).

95 See Chapter 5.

96 Robert Fisher and Rosemary Lyster, 'Land and Resource Tenure: The Rights of Indigenous Peoples and Forest Dwellers' in Rosemary Lyster, Catherine MacKenzie and Constance McDermott (eds), *Law, Tropical Forests and Carbon: The Case of REDD+* (Cambridge University Press, 2013) 187, 197.

97 Katy Barnett, 'One Step Forward and Two Steps Back: Native Title and the Bundle of Rights Analysis' (2000) 24 *Melbourne University Law Review* 462.

98 Native Title Act 1993 (Cth) pt 2 Div 3.

99 Michael O'Donnell, 'Native Title – A Right to Burn and Fire the Land? Savanna Burning and the Carbon Farming Initiative in Northern Australia' (2013) 30 *Environmental and Planning Law Journal* 553; Dore et al, above n 30; Jeremy Dore, 'Is There a Native Title Right to Burn Country?' (27 August 2014) Aboriginal Carbon Fund, http://aboriginalcarbonfund.com.au/blog/2014/8/21/is-there-a-native-title-right-to-burn-country.

suggested that a carbon right (whether in the context of sequestration or as general mitigation) might itself be a native title right.[100] O'Donnell and Dore et al. suggest that a right to manage country, by way of controlled burning of savanna lands, may constitute an exclusive-possession native title right.[101] It may be encompassed by a general right to use and manage resources in accordance with traditional law and custom. Indeed, subject to the evidence brought to establish each native title claim, the Australian High Court decision in *Akiba on behalf of the Torres Strait Regional Seas Claim Group v Commonwealth of Australia (Akiba)*,[102] where the majority held that native title could comprise a broadly-framed right to use resources, conceivably could support some broader right to carbon resources.[103] Sheehan, earlier had also suggested that a right to stored carbon might be eventually recognised as a native title right,[104] especially where there is no other carbon right holder.[105]

The extent to which native title rights might encompass commercial rights and thereby, potentially, allow indigenous communities to benefit economically and culturally from the carbon economy is also relevant. The Australian High Court in *Akiba* also confirmed that native title rights might include use of marine resources for any purpose; which would include commercial purposes where the evidence supports this finding.[106] In this regard, the Australian Federal Court in making a determination of exclusive possession native title for the Pilki Peoples applied the concept of 'the right to access and take for any purpose the resources of the land and waters'.[107] The determination relied on an

[100] See Section D.
[101] O'Donnell, above n 99; Dore et al, above n 30; Dore, above n 99.
[102] *Akiba v Commonwealth of Australia* (2013) 250 CLR 209 [21], [28].
[103] For an early analysis on the scope of native title rights see Greg McIntyre, 'Native Title Rights after Yorta Yorta' (paper presented at Native Title Conference 2003: Native Title on the Ground, Alice Springs, Australia, 3–5 June 2003); Dore et al, above n 30.
[104] John Sheehan, 'Indigenous Carbon Property Rights' (paper presented at the 16th Annual Pacific Rim Real Estate Society (PRRES) Conference, Wellington, New Zealand, 25 January 2010), www.prres.net/papers/Sheehan_Indigenous_Carbon_Property_Rights.pdf.
[105] Dore et al, above n 30.
[106] *Akiba v The Commonwealth* (2013) 250 CLR 209 [66], [67]. See also *Akiba v Queensland [No 2]* (2010) 204 FCR 1.
[107] *Willis on behalf of the Pilki People v State of Western Australia [No 2]* [2014] FCA 1293 (2 December 2014) Determination, (McKerracher J). See also *Western Australia v BP (deceased)* (2014) 223 FCR 488.

earlier Federal Court decision related to the recognition and determination of the native title right to take and use resources of the land and waters for commercial purposes. In that instance Justice North held that the evidence supported the existence of patterns of trading, which in turn, encompassed the commercial use of resources, such as ochre.[108] In this sense, native title might be conceived as a set of rights to use resources including for commercial purposes, similar to the resource-rights focus for forest communities.[109]

Whether a carbon right is ultimately established as a native title right, its utility outside of the specific carbon management system is likely to be limited. This is due to the narrow protection for native title rights and interests offered by the future acts management regime in the Native Title Act. Further, any such right would still be subject to the general regulatory regime for resource management. Based on the Australian experience of incorporating carbon rights into the land claims process, alternative forms of providing a legal basis for indigenous and local communities to gain benefits from carbon sequestration and emissions-reduction activities are preferable.

### (b) Gaining Indigenous Peoples' Consent under the Native TitleAct

The Native Title Act prescribes a regime for facilitating indigenous consent for developments on native title lands. The requirement to consult with native title-holders about carbon and emissions-reduction activities on native title lands however does not fully accord with international standards of FPIC. Further, this consent regime operates within the overall approvals process for 'future acts' (activities and projects) on native title land.[110]

The requirements include notice, consultation or negotiation, depending on the activity proposed, and provide a legislative code for decision-making about activities on native title land. Compensation will generally be payable if native title is affected by an activity but consent is not required. Even a failure to comply with some procedures will not result

[108] *Willis on behalf of the Pilki People v State of Western Australia [No 1]* [2014] FCA 714 (4 July 2014) [57] (North J). This decision was based on evidence of a pre-sovereignty and post-sovereignty trade in commodities, such as 'ochre, shell, grindstones, ground stone axes, stone knives, wooden implements and tobacco'.

[109] Rosemary Lyster, 'Reducing Emissions from Deforestation and Degradation: The Road to Copenhagen' in Rosemary Lyster (ed), *In the Wilds of Climate Law* (Australian Academic Press, 2010) 97.

[110] Native Title Act 1993 (Cth) pt 2 Div 3.

in the invalidity of any grant or act.[111] As pointed out above, carbon sequestration and mitigation are not listed,[112] but some carbon-related activities might attract the future acts regime.

While the native title regime requires consent internal to the indigenous group, there is no requirement for that consent to be mirrored under the future acts regime for land access or resource use on native title land. At most, the native title regime requires negotiation subject to arbitration (for example, regarding mineral and petroleum extraction proposals),[113] consultation or simply notice of some proposed activities.

### 3. *Agreement-Making for Indigenous Tenure as an Alternative to Native Title*

One consequence of more than twenty years of operation of the native title scheme is the changing culture amongst governments, resource companies, other land users and native title-holders. An emerging culture of agreement-making[114] has seen negotiation, whether mandated under the Act or otherwise, as the dominant mode of interaction between native title-holders and others.[115] This mode of decision-making about land access and resource use operates in parallel to and in the shadow of the statutory frameworks. Agreements are a feature of the savanna-burning examples referred to below.

Negotiation and agreement-making might encourage stronger indigenous engagement in carbon management projects. Yet, while agreement-making has allowed native title-holders to be involved in decision-making, it has been criticised as reinforcing inequities in bargaining power and resources, resulting in variable and often unfavourable outcomes for native title-holders.[116] This experience suggests that power imbalances are a key consideration in agreement-based regimes.

111 *Harris v Great Barrier Reef Marine Park Authority* (2000) 98 FCR 60.

112 This absence is discussed further in Section D.

113 Native Title Act 1993 (Cth) ss 26–44.

114 Langton et al, *Settling with Indigenous People*, above n 66; Langton et al, *Honour among Nations*, above n 66.

115 Ibid.

116 David Ritter, *The Native Title Market* (UWA Publishing, 2009); Ciaran O'Faircheallaigh, 'Native Title and Mining Negotiations: A Seat at the Table, But No Guarantee of Success' (2007) 6(26) *Indigenous Law Bulletin* 18; Ciaran O'Faircheallaigh, 'Native Title and Agreement Making in the Mining Industry: Focusing on Outcomes for Indigenous Peoples'

Many negotiations are embodied in an Indigenous Land Use Agreement, a key instrument under the future acts regime in the Native Title Act.[117] Once registered, an agreement authorises the relevant activity, meaning that, subject to terms of the agreement, activities can be undertaken without any further consents being necessary. The agreements bind all native title-holders, including their successors in title,[118] but not the successors in title of non-native title parties. It has taken twenty years of the native title regime for these agreement-based processes to become stable and functional, an important lesson about the time it can take to establish clear and certain tenure regimes and processes in the context of REDD+.

The tendency towards agreement-making is evident in the Traditional Owner Settlement Act 2010 (Vic) which arose in response to deficiencies in the native title processes.[119] This regime enables the state (provincial) government to negotiate outcomes with groups of traditional owners (including those that may not be able to prove native title), and results in an Indigenous Land Use Agreement. This umbrella agreement can subsume other agreements relating to fee simple title to land for economic or cultural purposes, governance mechanisms, land use and natural resources.[120] The Act makes direct provision for traditional owners to enter into carbon agreements with the state.[121] The legislation provides a trigger of procedural rights for traditional owners to be engaged when a

(Issues Paper No 25, Native Title Research Unit, Australian Institute of Aboriginal and Torres Strait Islander Studies, 2004). The authors agree that this is an issue in many cases.

117 Native Title Act 1993 (Cth) ss 24AA–24EC.

118 Native Title Act 1993 (Cth) s 24EA.

119 Reflected in the denial of native title in *Members of the Yorta Yorta Aboriginal Community v Victoria* (2002) 214 CLR 422. Following this decision, a Traditional Owner Land Management Agreement was negotiated with the Yorta Yorta under the Conservation Forests and Lands Act 1987 (Vic). Two agreements have been reached: the Gunaikurnai agreement and the Dja Dja Wurrung agreement.

120 State Government Victoria, 'Agreements with Traditional Owners', Environment, Land, Water and Planning, www.depi.vic.gov.au/forestry-and-land-use/managing-land/indigenous-land-management/agreements-with-traditional-owners#Yorta_Yorta_Agreements.

121 Traditional Owner Settlement Act 2010 (Vic) s 27(1)(i). For example, under the Dja Dja Wurrung Land Use Activity Agreement, a Carbon Sequestration Right is prescribed as an Agreement Activity. Section 40(4) requires that a person holding a carbon sequestration agreement must reach agreement with the Dja Dja Wurrung group, including the provision of community benefits, if any: Land Use Activity Agreement between Dja Dja Wurrung Clans Aboriginal Corporation (Indigenous Corporation No 4421) and the State of Victoria, Being Part of the Recognition and Settlement Agreement under Section 4 of the Traditional Owner Settlement Act 2010 (Vic) (24 October 2013) sch 3, s 5.

carbon sequestration right is created,[122] and a Land Use Activity Agreement is to be negotiated with the group. While the legislation creates a robust example of an agreement-based approach to involving indigenous communities in carbon management, it is strongly dependent on the specific scheme that is cognisant of indigenous 'carbon' interests and a government willing to negotiate in good faith with indigenous landholders.

### 4. *Statutory Land Rights*

Australia has adopted statutory schemes for indigenous land rights outside of the native title process. These models offer highly relevant lessons for considering statutory tenure reform as part of the REDD+ process. The secure nature of the title granted under statutory schemes, including the right to exclude third parties, means that the indigenous title-holders have broader scope than in native title to enter into carbon management activities. Moreover, the rights and powers that accompany title provide stronger potential for indigenous landholders to undertake emissions-reduction activities. A number of schemes have been established by states,[123] but the best practice model for indigenous land tenure in Australia is the Aboriginal Land Rights (Northern Territory) Act 1976 (Cth).[124] It provides for grants of inalienable freehold title to indigenous

[122] Traditional Owner Settlement Act 2010 (Vic) pt 4.

[123] Pitjantjatjara Land Rights Act 1981 (SA); renamed Anangu Pitjantjatjara Yankunytjatjara Land Rights Act 1981 (SA). (Similar legislation was passed for another indigenous group: Maralinga-Tjarutja Land Rights Act 1984 (SA).) The South Australian parliament passed special legislation under which a single, inalienable freehold title was granted to Anangu Pitjantjaraku (now known as Anangu Pitjantjatjara Yankunytjatjara (APY)), a body corporate comprising all Pitjantjatjara, Yankunytjatjara and Ngaanyatjarra people who are traditional owners of the land granted: Anangu Pitjantjatjara Yankunytjatjara Land Rights Act 1981 (SA) ss 4, 5. The land grant covers 103,000 square kilometres. Governance is a mixture of the whole membership and the executive of the body corporate, the executive being required to ensure that traditional owners understand and consent to proposals on their land: Ibid. s 7. The Anangu Pitjantjatjara Yankunytjatjara Land Rights Act 1981 (SA) reflected a political settlement after an extended period of negotiation. The title (the fee simple estate) provides a high level of tenure security for land-management activities such as carbon abatement and allows the indigenous owners to control third-party access to the land. See also Aboriginal Land Rights Act 1983 (NSW); Aboriginal Land Act 1991 (Qld); Torres Strait Islander Land Act 1991 (Qld).

[124] The statute recognises indigenous land rights held under Aboriginal customary (traditional) law. The statute has parallels in the other common law jurisdictions examined, i.e. Malaysia, Papua New Guinea and Vanuatu.

peoples as 'traditional owners'.[125] This federal legislation, although only applicable in the Northern Territory, establishes a claims process under which traditional owners can claim unallocated Crown land (and some indigenous-owned pastoral lease land).[126] If the claimants satisfy statutory requirements, the Land Commissioner recommends to the federal minister that a grant of an inalienable freehold title be made to a Land Trust.

Land Trusts comprise elected traditional owners who are required to manage that trust's land for the benefit of all its traditional owners. The trust acts on the direction of the relevant Land Council.[127] Land Councils play a key governance role for the relevant indigenous peoples, including providing advice on land and resource proposals and ensuring that the traditional owners understand and consent to proposals.[128] These councils, statutory bodies comprising elected representatives from all land trusts in the region, are relatively well resourced, with institutional capacity, backed by expert professionals in law, anthropology and land management. They play a crucial part in the relationships between traditional owners and governments and third parties.[129]

Under the Act, in order to manage third party interests, cross claims and intra-indigenous disputes about land (a key issue in indigenous and tenure resolution, as described in Chapter 5), there is a process for third parties who claim 'detriment by a grant of title' (i.e. that their interests would be adversely affected) to be heard before a land grant is made.[130] Once granted, the title is registered in the Northern Territory Land Registry, a mainstream, Torrens Title land registry. Registration provides

[125] Traditional owners are defined as 'a local descent group ... who (a) have common spiritual affiliations to a site on the land, being affiliations that place the group under a primary spiritual responsibility for that site and for the land; and (b) are entitled by Aboriginal tradition to forage as of right over that land': Aboriginal Land Rights (Northern Territory) Act 1976 (Cth) s 3.

[126] Ibid. s 50(1)(a).

[127] Ibid. s 5. The five land councils each comprise a board of elected traditional owners from the relevant region.

[128] Ibid. s 23(3); this aspect is discussed further below.

[129] Land councils' operations and powers have been controversial and the subject of review: see John Reeves, *Building on Land Rights for the Next Generation: The Review of the Aboriginal Land Rights (Northern Territory) Act 1976* (Commonwealth of Australia, 1998); Amos Aikman, 'Aboriginal Land Rights Out of Step?' *The Australian* (31 July 2014).

[130] Aboriginal Land Rights (Northern Territory) Act 1976 (Cth) s 50(3)(c). Detriment might be claimed by third-party commercial interests or by indigenous peoples who have had some use of the land claimed.

a 'clear and secure' title for land-management activities, such as carbon management, and includes the right for indigenous holders to exclude third parties.[131]

### 5. *Inalienable Title, Powers and Consents*

Inalienability of title is a key feature of these statutory schemes, sometimes criticised for its damaging effect on the economic development potential of the land.[132] However, once the grant of inalienable title is made, there are few limitations on the use of the land by its owners, including for carbon abatement or sequestration.

Third parties must comply with statutory governance requirements before undertaking carbon management activities on this land. These provisions ensure that the indigenous owners' consent to activities on their land, either by other owners or third parties, encompassing elements of free, prior and informed consent or consultation. In the Northern Territory, as there is no carbon legislation, robust contractual arrangements between the indigenous peoples who hold title and third parties are required to make clear who holds the rights to emissions reductions.

In most statutory regimes, subject to the consent provisions, the land may be leased or licenced by its owners for varying terms to traditional owners, governments or third parties,[133] and it is foreseeable that they could lease lands to third parties for carbon management projects.[134] Schemes have additional requirements for mineral and petroleum projects, although some enable a decision to withhold consent by a traditional owner group to be overridden.[135] Thus, there is a framework for

[131] As a fee simple title, it can be encompassed by statutory definitions of 'land' unrelated to its indigenous elements.

[132] Although it is not without controversy: see, e.g., Maureen Tehan, 'Customary Land Tenure, Communal Titles and Sustainability: The Allure of Individual Title and Property Rights in Australia' in Godden and Tehan (eds), above n 61, 354; Terrill, above n 61; Leon Terrill, 'The Days of the Failed Collective: Communal Ownership, Individual Ownership and Township Leasing in Aboriginal Communities in the Northern Territory' (2009) 32 *University of New South Wales Law Journal* 814.

[133] Aboriginal Land Rights (Northern Territory) Act 1976 (Cth) ss 19, 19A; Anangu Pitjantjatjara Yankunytjatjara Land Rights Act 1981 (SA) s 6(2).

[134] Tehan, above n 132; Terrill, above n 132.

[135] Under the Aboriginal Land Rights (Northern Territory) Act 1976 (Cth), the landowners effectively have a veto over such projects, as a rejection of a proposal means that the project cannot proceed unless the federal minister determines that it should proceed in the national interest. This has never been done. The Anangu Pitjantjatjara Yankunytjatjara Land Rights Act 1981 (SA) adopts an arbitration model in relation to mineral and

consultation and informed consent in relation to most land-based activities with special provisions for mineral and petroleum projects.

In summary, in Australia there are a wide range of land titling and tenure arrangements that deliver varying levels of rights and interests in land and resources to indigenous peoples with attendant mechanisms for management, access and governance. Each scheme has implications for the management of emissions reductions in land-based activities and access to carbon as a resource under various schemes.[136] In Australia, there is considerable experience of benefit-sharing with indigenous communities in respect of many forms of resource extraction and land use activities that can provide useful comparative insights for REDD+. Much of the Australian experience comes from the long history of benefits flowing to indigenous communities from mineral and petroleum resource extraction.[137] More recently, there are examples of benefit-sharing and co-benefits arising out of ecosystem services and payments for land management (Indigenous Protected Areas are a pertinent example),[138] or rents for leasing of indigenous lands for conservation purposes.[139] Indigenous peoples within Australia increasingly have become more active participants in managing their own 'country' and in deriving cultural, economic and social benefits from a growing range of land-based activities. These activities include land uses and value-creation systems that have developed as part of Australia's response to climate change and the emissions-reduction measures in the land sector. Importantly, it is the rights and powers that accompany title or tenure arrangements, rather than the title itself, that are the central

petroleum projects. If the parties cannot agree to a project proceeding or the terms on which it might proceed, either party may seek to have the dispute arbitrated. The arbitrator's decision is binding. In respect of other land-based activities, such as leasing land for particular purposes, the decision of the relevant Land Trust or Anangu Pitjantjatjara Yankunytjatjara body corporate is binding: Anangu Pitjantjatjara Yankunytjatjara Land Rights Act 1981 (SA) ss 7, 20.

[136] Issues surrounding community benefit-sharing in the REDD context are considered in Chapter 6.

[137] Marcia Langton and Odette Mazel, 'The Resource Curse Compared: Australian Aboriginal Participation in the Resource Extraction Industry and Distribution of Impacts' in Marcia Langton and Judy Longbottom (eds), *Community Futures, Legal Architecture: Foundations for Indigenous Peoples in the Global Mining Boom* (Routledge, 2012) 23.

[138] Godden and Cowell, above n 74.

[139] Jon Altman and Seán Kerins (eds), *People on Country: Vital Landscapes, Indigenous Futures* (Federation Press, 2012).

consideration in determining the potential for indigenous landholders to participate in REDD+ style activities.

## D. Interaction of Tenure, Carbon Regulation and Mitigation Schemes

Section D examines the complexities and opportunities that emerge when schemes designed to reduce emissions and avoid deforestation and degradation intersect with indigenous land tenure systems. It considers national (federal) and state (provincial) schemes for carbon management, and highlights the difficulty experienced in Western Australia in attempting to account for indigenous tenure within the state scheme. It concludes with three case studies to reflect on the challenges and possibilities of the scheme for indigenous peoples, particularly in relation to land and resources tenure.

At the outset, it is important to reaffirm the two key ways in which to conceptualise the rights that may be realised in relation to carbon mitigation/emissions reductions. First, there are rights in the resource and the land in which the carbon is stored or on which abatement activities occur. The second method is the credits or units that such activities generate, and the consequent financial benefits.[140] Credit trading schemes are one source of potential benefits but contractual purchase of credits or direct-funding of schemes such as sequestration offsets also produce benefits.[141] The right to control access to the land or commodity for the purpose of carbon management activities in itself may be a valuable benefit for indigenous title-holders.

As noted, schemes for attributing and distributing 'property' in carbon and the extent of its severability from other interests in land are found within the Australian legal system. The legal frameworks in Australia provide useful insights into the ramifications of a 'resource right' approach.[142] Where indigenous interests are construed as resource rights, it is important to ensure that there is a regulated system for managing the resource rights. Significantly, the Australian experience underlines the importance of ensuring coherence across jurisdictions in defining the statutory scope of the requisite rights and the distribution of rights in

[140] Leo Peskett and Gernot Brodnig, 'Carbon Rights in REDD+: Exploring the Implications for Poor and Vulnerable People' (Working Paper 65864, World Bank and REDD-net, 2011) 3. See the discussion of co-benefits in Chapter 6.

[141] Pesket and Brodnig, above n 140, 5.

[142] See Section C above.

carbon, in accordance with clear and transparent values that comprehend the need for the system to interface with other complex forest and land management systems,[143] in this instance, various forms of indigenous land tenures and embedded safeguards, including 'consent'.

At one level, severing 'carbon' from land ownership may expand options for indigenous peoples in their engagement with REDD+.[144] The Australian native title experience, which has led to 'vulnerable' usufructuary or resource use rights vis-à-vis the relatively more robust protections of the statutory land-title schemes, should sound a note of warning. Nonetheless, a severance model may avoid some of the disadvantages flowing from the perceived need for land tenure changes to 'align' customary land title with REDD+ requirements, which has occurred in other countries in the context of REDD+ and related project implementation.[145]

The potential for a resource-style approach for REDD+ was considered by the World Bank[146] to address questions about 'which actors have the right to exploit the benefits of [greenhouse gas emissions] and removals in REDD+'.[147] At that time (2011), carbon rights were seen as 'legislative and contractual arrangements that allow the recognition of separate benefits arising from the sequestration of carbon in the biomass'.[148] Bank policy staff regarded carbon as 'a new form of property in forest ecosystems that has potential value because of the creation of new markets and funds'.[149] Operating in a similar time frame and mind set, state (provincial) governments in Australia from 2003 developed carbon 'rights' statutes in anticipation of booming national and international carbon markets – that, in many instances, have yet to materialise.

### 1. *Carbon Rights and Tenure*

Carbon interests, as noted above, have been characterised in different ways amongst the jurisdictions in the state legislative frameworks that

143 See Chapter 2. 144 Lyster, above n 109. 145 See Chapter 5 and Chapter 8.
146 Peskett and Brodnig, above n 140. 147 Ibid. 1.
148 Terrestrial Carbon Group Project, 'Legal and Institutional Foundations for the National Implementation of REDD: Lessons from Early Experience in Developing and Developed Countries' (Policy Brief 4, Terrestrial Carbon Group and UN-REDD Programme, 2009); Charlotte Streck and Robert O'Sullivan, 'Legal Tools for the ENCOFOR Programme' (Joanneum Research, 2007); cited in Peskett and Brodnig, above n 140, 3.
149 Peskett and Brodnig, above n 140, 2.

developed. The exact nature of the legal characterisation of the rights in each state may affect the ability of native title-holders and statutory land-rights holders to engage in carbon management activities on their land. The outcome is dependent upon how the interaction between the carbon right and indigenous peoples' tenure and/or resource rights is legally construed. In turn, the overarching Emissions Reduction Fund legislative requirements will be influential.[150] Further, in some situations this interaction may be the subject of agreement-making.[151] Importantly, the carbon rights will be subject to general land-management practices such as land registration – again we use the example of Western Australia.

Delineating and registering carbon interests under legislation provides advantages of clarity, certainty, comprehensiveness and stability. Registration creates the property interest: '[u]ntil registration, the carbon right and carbon agreement have no proprietary [effect] ... registration validates not only the new title of the holder, but the very existence of the right'.[152] Yet, these statutory processes, while promoting clarity for the 'right', can introduce limitations where those statutory processes are not sufficiently inclusive of all possible right holders. This is the case in Western Australian legislation, which is uncertain and disengaged with respect to native title-holders and, to a lesser extent, in respect of other land in which there may be indigenous interests, such as land held by the Aboriginal Land Trust. The Western Australian legislation does not indicate whether native title-holders as of right can hold carbon rights on native title land. The Carbon Rights Act 2003 (WA) is silent on how it applies to native title-holders. Moreover, even if carbon rights were in the future to be determined by the courts as a native title right, it is unclear whether native title-holders would require the consent of the Minister for Lands for transactions, such as entering into a carbon covenant.[153] The Western Australian experience points to the dangers of focusing primarily on indigenous tenure without also giving attention to the broader legal context of land administration and carbon regulatory regimes.

The relationship between carbon management schemes and indigenous land tenure is particularly complex and unsettled in the case of non-exclusive native title rights.[154] For example, generally, pastoral leases

150 Alexander Zahar, *International Climate Change Law and State Compliance* (Taylor and Francis, 2014).

151 See further on agreement discussion in Section C 3 above.

152 Hepburn, above n 27, 269.

153 See Land Administration Act 1997 (WA) s 18.

154 This complexity is exacerbated when the carbon farming entitlements are overlaid.

do not completely extinguish native title, so that non-exclusive native title rights that are not inconsistent with the pastoral purposes granted over the land remain intact.[155] Further, a lessee who sequesters carbon needs to be granted carbon rights under the lease and requires ministerial consent for carbon activities.[156]

The legislative techniques used to extend statutory property rights to carbon in Western Australia highlight the challenges inherent in designing legislative schemes to incorporate the interests of indigenous communities. Despite the recognition of indigenous rights in land and waters, and a substantial and growing number of native title claims determined in Western Australia, the statutory scheme for carbon management in Western Australia has largely failed to incorporate the interests of native title-holders. The Western Australian case underlines the importance of specific legislation for dealing with indigenous interests, especially if severable carbon resource rights are created. The Western Australian model of statutory carbon rights is narrowly conceived in respect of indigenous interests and it should be contrasted with the national 'carbon farming' legislation, as well as the specific savanna-burning projects where indigenous interests have been more directly acknowledged.

### 2. *The Emissions Reduction Fund and Indigenous Land*

Broadly, the CFI treats indigenous statutory 'land-rights lands' as akin to freehold land.[157] Where the land rights land is a lease, is held by the Australian government or a statutory authority, or is less than freehold under state or territory legislation, the relevant minister is taken to hold an eligible interest to participate in the CFI and must provide their consent.[158] Where land-rights land is equivalent to freehold (such as

[155] See pastoral lease discussion, Section C 2 above.

[156] Land Administration Act 1997 (WA) s 18. In 2015, all eligible pastoral leases were renewed, granting tenure for a further fifty-year period without express consideration of carbon rights or indigenous interests: Guy Greer and Claudia Henfry, 'Roe 8 Delay: The Decision of *Save Beeliar Wetlands v Jacob*,' *Gilbert & Tobin: Insights – WA Resources Update* (23 March 2016), www.gtlaw.com.au/wa-resources-update-march-2016.

[157] The definition of freehold land-rights land basically includes the grant of freehold estate or vesting of land under a law of a state or a territory, or the Commonwealth where the grant (or vesting) is for the benefit of Aboriginal peoples or Torres Strait Islanders: Carbon Credits (Carbon Farming Initiative) Act 2011 (Cth) s 5.

[158] Ibid. s 44(6)–(7).

Aboriginal Land Rights (Northern Territory) Act land) and Crown (government) land is involved, there is a requirement to notify the relevant Crown lands minister.[159] In an acknowledgement of the extensive range of specific indigenous land legislation across Australian jurisdictions, the Act makes provision for regulations to ensure that any arrangement that is not otherwise captured can be accommodated.[160]

The treatment of native title land under the CFI is more complex. The first aspect is the requirement that the project proponent hold the relevant rights in the land and be responsible for the project.[161] For native title land, the registered native title body corporate[162] will be taken to be the project proponent.[163] The second requirement is that, to apply for a project, native title-holders must hold the legal right to carry out the project and, in the case of sequestration projects, the carbon sequestration right.[164] In development of the scheme, a statutory solution that avoided the need for native title-holders to seek a court determination was adopted. The Act deems the registered native title body corporate to be the project proponent where the project area is determined exclusive-possession native title land,[165] and no other person (except for the government) holds the carbon right or the legal right to carry out the project.[166] The Act does not require the government to provide their consent to the project.[167] As no rights are actually conferred under the Carbon Farming Initiative Act, this approach is not a future act under the Native Title Act 1993.[168] Finally, the legislation recognises that a carbon right may be exercised by another party under an Indigenous Land Use Agreement.[169] This allows a third party to undertake a project on native

[159] Ibid. s 47.

[160] Replacement Explanatory Memorandum, Carbon Credits (Carbon Farming Initiative) Bill 2011 (Cth) ch 4, 37 [4.8]; Carbon Credits (Carbon Farming Initiative) Act 2011 (Cth) ss 44(5), 45(5).

[161] Carbon Credits (Carbon Farming Initiative) Act 2011 (Cth) s 5.

[162] The native title-holding body incorporated under the Corporations (Aboriginal and Torres Strait Islander) Act 2006 (Cth).

[163] Carbon Credits (Carbon Farming Initiative) Act 2011 (Cth) s 46.

[164] Ibid. s 5.

[165] Subject to the proviso that there is a registered native title body corporate for the project area.

[166] Carbon Credits (Carbon Farming Initiative) Act 2011 (Cth) s 46(1).

[167] Ibid. s 46 (assuming the body corporate meets the requirements of this section, that is, has the legal right to carry out the project and holds applicable carbon sequestration right.)

[168] See 'future acts' Section C 2 above.

[169] Carbon Credits (Carbon Farming Initiative) Act 2011 (Cth) s 43(10).

title land effectively with the consent of native title-holders expressed through the agreement.

The Act does not provide any special treatment for non-exclusive native title (in many parts of Australia this will constitute the majority form in which native title rights are held).[170] This situation reflects the view that such non-exclusive native title access or rights of usage would be less likely to include carbon sequestration rights. The removal of a requirement that the project proponent must hold a carbon sequestration right for all other carbon sequestration projects, while maintaining it for native title land, entrenches a discriminatory characterisation of non-exclusive native title.[171]

### 3. *Emissions Reduction and Carbon Mitigation Projects in Northern Australia*

There are currently no carbon sequestration projects in Australia on indigenous land or involving indigenous peoples. Thus, the elaborate statutory infrastructure designed to facilitate the participation of indigenous peoples in REDD+ style activities has to date proved inadequate to the task of actual implementation of the projects. The difficulty rests not with the legislation, but with a raft of entrenched and systemic factors within Australia that work against the effective realisation of indigenous economic development based in natural resources.[172] However, there are some encouraging signs in respect of other REDD+ style emissions-reduction activities and ecosystem-service projects.[173] Experience in remote northern Australia indicates that the strategic reintroduction

170 See the maps on the National Native Title Tribunal - www.nntt.gov.au/.

171 The Carbon Farming Initiative Amendment Act 2014 (Cth) removed the former requirement. This change was made to encourage greater flexibility on the basis that 'some landholders may be unwilling to transfer [carbon sequestration] property rights to the project proponent. Landholders may be more willing to grant the project proponent a legal right to undertake a project on their land'.

172 Markham and Altman, above n 70; Altman and Kerins, above n 139.

173 In November 2016, eighteen burning projects had been funded by the Emissions Reduction Fund or its predecessors, Jeremy Dore, 'ERF and Indigenous Savanna Entwined' (29 November 2016) Aboriginal Carbon Fund, http://aboriginalcarbonfund.com.au/blog/2016/11/29/erf-and-indigenous-savanna-entwined. See generally Robinson et al, above n 13; Robinson et al, above n 1; Linda Hansen, 'Carbon in Country: Legal Pathways and Barriers to Indigenous Participation in Australia's Carbon Market through Savanna Fire Management under the Carbon Farming Initiative' (Working Paper No 3, Centre for Resources, Energy and Environmental Law, Melbourne Law School, September 2015). See further on co-benefits in Chapter 6 and Chapter 3.

of fire-management practices based on traditional indigenous knowledge across savanna rangelands can produce scientifically quantifiable emissions reductions. As Walton and Fitzsimons note, 'In Australia, savannas occur in the wet-dry tropics and feature both trees, dominated by eucalypts, and herbaceous plants, principally grasses.'[174] The methodology for traditional fire management, accepted under the CFI for 'newness' purposes, provides for a planned series of controlled burns undertaken early in the dry season, when fires produce lower levels of greenhouse gases.[175] This traditional practice of 'mosaic' burning lessens the intensity and scope of late dry-season fires to reduce overall emissions. In accordance with the methodology, abatement levels are calculated through the use of vegetation and fire maps.[176] These schemes are illustrative of the types of indigenous land use that could generate benefits or payments for indigenous peoples in accordance with sequestration and emissions-reduction schemes.

The eighteen savanna-burning projects involving indigenous peoples involve quite extensive land and resource management in association with the emissions reductions achieved through traditional burning practices.[177] The current eighteen projects are contracted to deliver

[174] Nerissa Walton and James Fitzsimons, 'Payment for Ecosystem Services in Practice: Savanna Burning and Carbon Abatement at Fish River, Northern Australia' in Penelope Figgis et al (eds), *Valuing Nature: Protected Areas and Ecosystem Services* (Australian Committee for IUCN Inc, 2015) 78, 78: 'A defining feature of savannas is the existence of a dry season lasting up to nine months of the year. Grasses in the savanna have short, intense growing periods during the wet season, then cure rapidly during the dry'.

[175] Carbon Credits (Carbon Farming Initiative – Emissions Abatement through Savanna Fire Management) Methodology Determination 2015 (Cth); Australian Government, 'Savanna Fire Management', above n 7.

[176] Walton and Fitzsimons, above n 174. For earlier work on the development of this practice, see Tropical Savannas CRC, 'West Arnhem Land Fire Abatement Project – What's Been Achieved?' (2007), http://savanna.cdu.edu.au/information/walfa_achievements.html; Aboriginal and Torres Strait Islander Social Justice Commissioner, *Native Title Report 2007* (Australian Human Commission, 2008), www.humanrights.gov.au/our-work/aboriginal-and-torres-strait-islander-social-justice/publications/native-title-report-2007; Australian Government, *Carbon Farming Initiative (CFI): Savanna Burning* (2014) Department of the Environment and Energy, www.environment.gov.au/climate-change/emissions-reduction-fund/cfi/publications/factsheet-savanna-burning. For a detailed discussion of this activity, see Jeremy Russell-Smith, Peter Whitehead and Peter Cooke, (eds) *Culture, Ecology and Economy of Fire Management in North Australian Savannas: Rekindling the Wurrk Tradition* (CSIRO, 2009).

[177] For an example of a project see Australian Government, 'Fish River Fire Project: Northern Territory' (2015) Indigenous Land Corporation, www.ilc.gov.au/Home/What-We-Do/Project-Profiles/Fish-River-Fire-Project.

8.8 million tonnes of carbon over the next ten years. Three of these projects, discussed below, illustrate the complexities and possibilities of emissions-reduction projects in savanna lands under the Australian legal framework.

### (a) West Arnhem Fire Abatement Project

Traditional indigenous burning practices have been used in West Arnhem land (remote northern Australia) from the mid-1990s as part of projects designed to measure the reduction in carbon emissions.[178] Individual projects in the area are now funded under the Emissions Reduction Scheme.[179] An initial carbon abatement project, known as the West Arnhem Fire Abatement Project, arose as the resource company Conoco-Phillips – as a condition of obtaining project development approval – was required to fund carbon 'offsets' for its Liquefied Natural Gas Facility in Darwin. In 2006, the West Arnhem Fire Management Project Agreements were concluded between the Northern Territory Government, Conoco-Phillips, the Northern Land Council and traditional owners from areas in West Arnhem land adjacent to Kakadu National Park.[180] Under the agreements, a minimum of 100,000 tonnes of carbon abatement, as measured under a methodology approved by the National Greenhouse Gas Inventory,[181] was to be produced by the Abatement Project. Conoco-Phillips was required to pay $1 million per year[182] for seventeen years.[183] Funds have been expended on activities such as monitoring and auditing and the employment of 'on country' rangers to carry out the required fire work.[184] Set-up costs were shared between a research organization, the 'Tropical Savannas CRC', and traditional owner organizations.

[178] 'Fire Brings Back Country' (2009) (January–July) *Savanna Links*, www.environorth.org.au/environorth/teach/downloads/Fire_brings_back_mosaic.pdf.

[179] Australian Government, 'West Arnhem Land Fire Abatement (WALFA) Project' (27 January 2017) Clean Energy Regulator, www.cleanenergyregulator.gov.au.

[180] This was a voluntary agreement, ie traditional owners sought to be involved: Jeremy Russell-Smith, 'Fire Agreement to Strengthen Communities' (1 November 2013) North Australian Land Manager, http://savanna.cdu.edu.au/view/250363/fire-agreement-to-strengthen-communities.html.

[181] Aboriginal and Torres Strait Islander Social Justice Commissioner, above n 176, 267.

[182] In the five years to 2010, the project actually abated 707,000 tonnes: North Australian Land and Sea Management Alliance (NAILSMA), 'WALFA West Arnhem Land Fire Abatement Project' (2012), www.nailsma.org.au/walfa-west-arnhem-land-fire-abatement-projecthtml.html.

[183] Ibid.

[184] Aboriginal and Torres Strait Islander Social Justice Commissioner, above n 176, 267.

The burning is carried out on land held by Aboriginal peoples under the Aboriginal Land Rights (Northern Territory) Act (fee simple title).[185] No issue has arisen as to who holds the right to participate in carbon abatement practices on the land.[186] There are five traditional owner/ranger groups in the partnership: the Jawoyn Association Aboriginal Corporation, the Bawinanga Aboriginal Corporation, Warddeken Land Management, Mimal Rangers and Adjumarlarl Rangers (hosted by the Northern Land Council).[187] As the project is carried out by traditional owners, or corporations connected with traditional owners, issues in relation to access are obviated but consents for the fire work are required from traditional owners.[188] The project employs local rangers to carry out on-the-ground burning and to assist with the larger-scale helicopter burning.[189] This early project has now been expanded and has received funds from the Emissions Reduction Fund auctions, thus involving both the voluntary market and direct funding.[190]

### (b) Fish River Fire Project

The Fish River Fire Project was the first eligible 'offsets' savanna fire-management project to operate and receive carbon credits under the Carbon Farming Initiative.[191] The project was later funded in the 2015 auction under the Emissions Reduction Fund.[192] Located in the Daly River Catchment area in the Northern Territory, the 178,000-hectare pastoral lease property is owned and managed by the Indigenous Land Corporation,[193] having been purchased with contributions from the Australian government's Caring for Our Country programme, The Nature Conservancy and Pew Environment Group, with a view to

[185] See ALRA title discussion in Section C 4 above.
[186] The Northern Territory does not sever carbon 'rights' from underlying land tenure.
[187] Northern Land Council, 'Fire Management', www.nlc.org.au/articles/info/fire-management.
[188] This emphasises the control exercised by traditional owners.
[189] Aboriginal and Torres Strait Islander Social Justice Commissioner, above n 176, 263.
[190] Australian Government, above n 179.
[191] Carmen Brown, 'Credits Roll in for Territory Carbon Project' (23 April 2013) *ABC Rural News*, www.abc.net.au/site-archive/rural/news/content/201304/s3743095.htm.
[192] Australian Government, 'Fish River Fire Project', Clean Energy Regulator, www.cleanenergyregulator.gov.au.
[193] The Indigenous Land Corporation funds land purchases for indigenous communities through normal commercial transactions. It is a statutory authority that holds land on behalf of indigenous groups or eventually transfers ownership to them.

establishing a carbon demonstration project and ultimately transferring the land to the traditional owners.[194] The pastoral property is the subject of a yet-to-be determined native title claim. In 2012, the project reduced the area burnt in the late dry season from an average of 36 per cent during the baseline period (2000–2009) to approximately 1 per cent. It generated 25,884 Australian carbon credits during the 2011–2012 reporting period.[195] Since approval by the Clean Energy Regulator, the project has generated 44,114 carbon credits.[196] The project has now received funding from the Emissions Reduction Fund.[197]

The land tenure arrangements for the Fish River project were relatively less complex as the pastoral lease on which the project is implemented was held by the Indigenous Land Corporation as a perpetual Crown lease interest, although the effect of any determination of native title may add a complicating layer. The current situation is simplified also as legislation in the Northern Territory does not sever carbon 'rights' from underlying land tenure. The land is Crown (government) land and it is covered by a perpetual lease for grazing and ancillary purposes. Under the Crown Lands Act 1992 (NT), leased land can only be used for this purpose. Dore et al.[198] suggest that it has been accepted that burning is an activity ancillary to grazing and therefore consistent with the lease.[199] The resolution of the issue of whether traditional burning practices constitute requisite activities under the terms of the various pastoral leases that operate across much of northern Australia is an important one for the longer-term viability of these schemes. If there was to be more intensive savanna burning or a sequestration project on a pastoral lease in the future, which cannot be characterised as ancillary to existing tenure arrangements, then consent from the Northern Territory government (i.e. the landowner) would be required. The creation of new rights such as sequestration could also engage the future acts provisions of the Native Title Act 1993 (Cth), adding additional requirements for decision-making on native title land.[200]

194 Australian Government, 'Fish River Fire Project: Northern Territory', above n 177.
195 Ibid.
196 At Australian Government, 'Emissions Reduction Fund Project Register' (10 February 2017) Clean Energy Regulator, www.cleanenergyregulator.gov.au/ERF/project-and-contracts-registers/project-register.
197 Walton and Fitzsimons, above n 174.
198 Dore et al, above n 30, 396–397, 399.
199 Ibid. see also ibid. 398.
200 Dore et al, above n 30, 392–394.

### (c) Kimberley Projects

Fire-abatement projects also have been conducted on exclusive-possession native title land in the Kimberley region in north Western Australia.[201] The Kimberley Land Council (a representative political body with statutory obligations under the Native Title Act 1993 (Cth)) strongly advocated the co-benefits of such savanna-burning projects to COP21 in Paris, in 2015, including conservation and economic benefits to the indigenous community and its traditional country.[202] The Kimberley Land Council reports that between 2012 and 2014 North Kimberley Traditional Owners generated about 230,000 Kyoto Australia carbon credits through undertaking fire-management practices.[203] As with the other projects, selling credits into the voluntary market is important here. The Kimberley experience, however, also highlights that, even when indigenous landholders have 'clear and certain' title, the engagement of Aboriginal peoples in the carbon economy is still vulnerable to policy change. For example, most Kimberley Land Council applications for funding for carbon abatement projects on community-held native title lands from the Emissions Reduction Fund have been unsuccessful.[204] In addition, there have been difficulties for the Kimberley Land Council in obtaining permits for Aboriginal landowners to legally burn on their lands as part of a carbon offset project.[205] Other potential barriers to the

[201] See, e.g., 'Qantas Backs North Kimberley Carbon Project' (2014) (September) *Kimberley Land Council Newsletter* 10.

[202] Natalie Jones, 'WA Group Represents Indigenous Australia at UN Climate Change Conference', *ABC News* (25 November 2015), www.abc.net.au/news/2015-11-24/wa-groups-heads-to-climate-change-conference/6971072.

[203] Kimberley Land Council, 'Traditional Owners to Lose Millions of Dollars under Emission Reduction Fund', Media Release (9 July 2014), www.klc.org.au/docs/default-source/Media-Releases/traditional-owners-to-lose-millions-of-dollars-under-emission-reduction-fund?sfvrsn=0.

[204] Emma Fagan, 'New Methods and Greater Price Certainty Makes the Second ERF Reverse Auction More Attractive' (1 May 2015) Energetics, www.energetics.com.au/resources/latest-news/climate-change-matters/new-methods-and-greater-price-certainty-makes-the; Sarina Locke, 'Red Meat Producers Biggest Winners from Government's Multi-Billion Dollar Emission Reduction Fund', *ABC News* (1 December 2016), www.abc.net.au/news/2016-12-01/carbon-auctions-deliver-funds-to-beef-cattle-producers/8081904; see also Hansen, above n 175, 12.

[205] Land Rights Now, Kimberley Land Council and Oxfam, 'Making Rights a Reality: The Ongoing Struggle for Land Justice in the Kimberley Region, Australia' (Briefing Note, 26 September 2016) 7, www.oxfam.org.au/wp-content/uploads/2016/09/bn-australia-kimberley-land-rights-260916-en.pdf; Kimberley Land Council, 'KLC Puts Aboriginal Fire Management on the World Stage' (28 October 2016), www.klc.org.au/news-media/newsroom/news-detail/2016/10/28/klc-puts-aboriginal-fire-management-on-the-world-stage.

traditional burning projects have been mooted, including the creation of laws that require state government consent for carbon offset projects on native title land (because the Crown holds underlying sovereign title), and requiring native title-holders, unlike non-indigenous landholders, to enter into agreements with the state in relation to the projects.[206] Such barriers to economic and cultural sustainability can occur at many points for indigenous peoples within the Australian legal system, particularly in the treatment of native title as a lesser form of tenure than 'mainstream' landholding.[207] Nonetheless, there remains significant potential for a wide range of land sector activities involving indigenous peoples to come within the umbrella of nationally determined contributions for Australia under the Paris Agreement.[208]

Thus, indigenous participation in carbon mitigation projects, principally the traditional burning projects in northern Australia, are instructive for REDD+ in a number of ways. The projects under the current, direct-funding model or the earlier emissions-trading frameworks provide clear examples of how indigenous communities utilising traditional knowledge in forest/savanna vegetation contexts can produce emissions reduction and the avoidance of further degradation of ecologically significant tropical forest and rangeland vegetation. Significant environmental co-benefits have been demonstrated in these projects and the measurements of emissions reductions have been scientifically evidenced. The projects have engaged in relatively successful market-based and voluntary carbon market schemes, though the longer-term viability of carbon credit trading, especially in voluntary markets, is yet to be substantially proven. The projects have captured a wide range of community co-benefits. Aside from providing employment and income opportunities for indigenous Australians, these benefits include positive, demonstrable impacts on the reinvigoration of indigenous culture through an

[206] Ibid.

[207] Maureen Tehan and Lee Godden, 'Legal Forms and Their Implications for Long-Term Relationships and Economic, Cultural and Social Empowerment: Structuring Agreements in Australia' in Marcia Langton and Judy Longbottom (eds), *Community Futures, Legal Architecture: Foundations for Indigenous Peoples in the Global Mining Boom* (Routledge, 2012) 111.

[208] See, e.g., Conference of the Parties, United Nations Framework Convention on Climate Change, *Report of the Conference of the Parties on Its Twenty-First Session, Held in Paris from 30 November to 13 December 2015 – Addendum – Part Two: Action Taken by the Conference of the Parties at Its Twenty-First Session* (UN Doc FCCC/CP/2015/10/Add.1 (29 January 2016) Decision 1/CP.21), Annex ('Paris Agreement') preamble. See Chapter 3.

increased local ability to care actively for traditional country. Other outcomes are an increased level of intergenerational transfer of traditional knowledge, and the fostering of cross-cultural skills that are required for other economic activities, such as tourism.[209] In this manner, the projects have positive attributes that mark them as paradigmatic UN-REDD Programme models.

We acknowledge the relatively more robust position of the burning projects that operate on 'freehold equivalent land' under statutory land-rights arrangements. The projects utilising these indigenous lands, and the potential for carbon 'resource' rights, indicate that well-structured legislative and regulatory arrangements can provide a basis for indigenous participation in reducing emissions and avoiding deforestation and degradation of significant vegetation complexes, such as tropical savannas, through market-based and voluntary market regimes. The variety of indigenous land-tenure arrangements, such as in Australia, complicate, but should not be the basis for, the exclusion of indigenous participation in projects which seek to achieve reductions in emissions and climate mitigation objectives, while providing a basis for culturally appropriate economic activities.

It is clear that successful savanna-burning projects remain heavily dependent on a facilitative socio-economic context that transcends the adoption of safeguards of FPIC and tenure security, to embrace constitutive governance of the projects by indigenous organizations. Further, the projects operate in a socio-economic context where project implementation is anchored in a decades-long transition in the understanding and acceptance of indigenous-led fire burning and co-management of traditional country across Australian tropical savannas, combined with the progressive integration and valuing of indigenous traditional knowledge in these activities.[210]

Such a normative transition in recipient countries is vital to the viable extrapolation of this style of project. The biophysical conditions required for fire-management emissions-reduction programmes are not confined to Australia. Similar savanna landscapes and traditional

[209] See, e.g., Cooperative Research Centre for Tropical Savannas Management (Australia) and Ecological Society of Australia, *Savanna Burning* (Blackwell Science Asia, 2012).

[210] Toni Bauman, Chris Haynes and Gabrielle Lauder, 'Pathways to the Co-Management of Protected Areas and Native Title in Australia' (AIATSIS Research Discussion Paper No 32, May 2013), http://aiatsis.gov.au/sites/default/files/products/discussion_paper/pathways-to-comanagement-of-protected-areas-and-native-title.pdf.

land-management practices exist in parts of Asia, sub-Saharan Africa and southern Africa – highlighting the importance of information sharing processes in the REDD+ regime. In many of these regions, the exercise of traditional swidden land-use practices has been disrupted or severely limited, resulting in cultural disassociation from the land and unmanaged, high-intensity and damaging fire regimes.[211] The Australian model of savanna burning may yet gain cogency and enhanced traction for funding within the UN-REDD Programme.[212]

## E. Lessons from Australia

While the narrative in northern Australia focuses on successful indigenous engagement at a project level in emissions reductions from 'forest' management, the participation of Aboriginal and Torres Strait Islander communities is much less robust at a national, systemic level. The enabling legal architecture for carbon sequestration and reducing emissions from avoided deforestation and degradation in Australia is detailed and prescriptive, covering statutory land laws and registration systems, climate mitigation law and policy, and state (provincial) government 'carbon' resource rights; typically overlain by contractual requirements. In turn, these frameworks interface with indigenous tenures under the two central tenure models within Australia – statutory land rights and native title. Within this framework, non-exclusive native title is generally ignored – thus excluding a relatively large percentage of native title-holders from participation in emissions-reduction activities. The complicated nature of indigenous tenures (that bring together both underpinning possession and/or resource or use rights) is not adequately captured by the REDD+ style regime and its current tenure models. In these instances, however, it is not simply the lack of formal recognition of tenure per se that disadvantages indigenous communities; rather, the problem lies in a legislative design that is not flexible enough to deal with the fine-grained, holistic nature of indigenous tenures. Further, the perceived 'precariousness' of native title in the carbon regime is due to an endemic vulnerability inherent to the manner in which the title is

[211] Jefferson Fox, Jean-Christophe Castella and Alan D Ziegler, 'Swidden, Rubber and Carbon: Can REDD+ Work for People and the Environment in Montane Mainland Southeast Asia?' (2014) 29 *Global Environmental Change* 318.

[212] Australian indigenous groups have been active promoters of the benefits of these schemes in international fora.

recognised by the Australian legal system,[213] rather than the absence of formalisation procedures per se.[214] This highlights that formalisation of tenure in itself is not determinative of 'justice' for indigenous communities. Rather, attention should be directed as much to the nature of the rights that accompany recognition or formalisation of tenure – the way that recognition is effected, the strength of consent principles, and the degree to which the rights accord to indigenous groups' robust measures to manage land and exercise autonomous decision-making. In short, what is the capacity for governance that the formalisation of the rights confers on indigenous communities?[215] Fisher and Lyster[216] argue that governance is perhaps more important to robust indigenous outcomes under the UN–REDD Programme than land tenure.[217] Notably, even where recognition under the state-based system occurs in Australia, the respective indigenous rights may be overridden by administrative discretions, and generally relevant projects and activities will be subject to complex land and resource management, planning and environmental regimes.[218]

Clearly, though, formalisation of tenure within a legal system is a necessary but not sufficient 'first step'. Using tenure formalisation as the basis for inclusion within sequestration and avoided-deforestation/degradation schemes risks dealing inadequately with groups who have been historically dispossessed. In Australia, the position of indigenous communities who are still to have native title claims determined, or who are precluded from claims due to the stringency of the legal requirements for native title and the impacts of historical dispossession, are more clearly disadvantaged in terms of potential participation in a carbon economy. This situation has clear parallels with that of forest communities in many tropical nations who seek state recognition of their customary rights.[219]

The degree of formalisation also will typically operate synergistically with a 'consent' process. The model, in Australia, in respect of third party access to indigenous lands (i.e. a passive model) has a different basis to the international FPIC model, as it is embedded in the statutory

[213] Lee Godden and Maureen Tehan, above n 61. [214] See Chapter 5.
[215] Lyster, above n 109, 123. [216] Fisher and Lyster, above n 96. [217] Ibid. 188.
[218] Ibid. 197.
[219] For Malaysia, see, e.g., Amity Doolittle, 'Native Customary Land Rights in Sabah, Malaysia 1881 – 2010' in Marcus Colchester and Sophie Chao (eds), *Divers Paths to Justice: Legal Pluralism and the Rights of Indigenous Peoples in South east Asia* (Forest Peoples Programme and Asia Indigenous Peoples Pact, 2011) 81.

requirements under land rights and native title legislation. Moreover, an important theme to emerge from Australia is that indigenous peoples as a group are often 'the doers' – and accordingly that FPIC and benefit-sharing models may need to encompass inter-group and intra-group dimensions.

In conclusion, even where the tenure and rights of indigenous peoples are relatively secure, there remain strongly contested positions in respect of how indigenous communities are to be enabled to share in the value and benefits generated by forest and land-sector mitigation schemes. Australia has relatively 'robust' rights for indigenous communities; it is a first-world economy with resources that allow it to fund well-functioning land and resource administration systems to implement REDD type programmes, yet its indigenous communities continue to be marginalised. Indeed, there are many parallels between Australia and the challenges identified in the other case study chapters including: conflicts arising from multiple jurisdictions with inconsistent schemes for managing land and resources; ambivalence and delay in granting recognition of land tenure for some groups; and uncertainty in the coverage of regimes for managing carbon rights. Australia offers a cautionary tale; it is possible to instigate viable REDD type regimes, but the time frames will be protracted, particularly where they depend on national legal systems for tenure clarity. Australia also demonstrates that tenure formalisation may be partial or incomplete and remain so for many indigenous communities. In the interim, providing indigenous communities with various, 'carbon resource' rights has merit if the accompanying safeguards are robust, while acknowledging that single measures cannot solve the myriad challenges that confront indigenous peoples and local communities in continuing to inhabit and manage traditional forest lands, while meeting the imperatives of global climate mitigation.

# PART IV

# Implications and Conclusions

# 10

## Interacting Regimes and Experimentalism

MARGARET YOUNG

This book has shown how REDD+ is shaped by local, domestic and international laws and practices. Given legal pluralism and the multi-layered structures of shared authority, the impacts and implications for indigenous peoples and forest-dwelling communities are diffuse and difficult to predict. In this context, one central question to pose is whether the key actors in regime interaction at the international level – which include but are not limited to states – are subject to rights or accountability frameworks.[1] To answer this, we must move beyond governments to consider governance. Merging models of global experimentalist governance seek to account for localised, contextualised implementation of rules and standards combined with peer-review and oversight.[2] From a related research agenda, ideas of polycentric governance also support the pursuit of local, collective goals at different levels of government or private associations; indeed, this literature suggests that monitoring and rule enforcement is best devolved to a local level where trust and reciprocity can operate to achieve compliance.[3] The approaches in these diverse literatures allow for deeper theoretical reflection on the different state and nonstate entities that may hold correlative duties on global issues, and point to further research questions on the emerging REDD+ regime's interaction with other regimes.

This chapter is in three sections. In Section A, we reshape duties and rights in the REDD+ context, and in the main we address our findings and future research agenda to scholarship on public international law.

[1] Margaret A Young, 'Fragmentation, Regime Interaction and Sovereignty' in Christine Chinkin and Freya Baetens (eds), *Sovereignty, Statehood and State Responsibility* (Cambridge University Press, 2015) 71.

[2] Gráinne de Búrca, Robert O Keohane and Charles Sabel, 'New Modes of Pluralist Global Governance' (2013) 45 *New York University Journal of International Law and Politics* 723, 781.

[3] Harini Nagendra and Elinor Ostrom, 'Polycentric Governance of Multifunctional Forested Landscapes' (2012) 6 *International Journal of the Commons* 104.

In Section B, we draw on such models of 'experimentalist governance' and ask how they apply to the REDD+ context, with implications beyond public international law to more general theories of governance. In Section C, we offer conclusions on this theme. We then move to more general conclusions in our final chapter of the book.

## A. REDD+ and Regime Interaction: Implications for Indigenous Peoples and Forest Communities

This book has called attention to the variety of existing international regimes and bodies of law that coexist with REDD+, and we have shown how this implicates international, domestic and local legal processes. An international objective to preserve forests for climate change mitigation overlaps or conflicts with a number of other existing legal and institutional arrangements. For example, forestry regimes (including SFM) and domestic and transnational forest-certification schemes will intersect with any climate mitigation policy involving forest carbon sequestration.[4] The international regime governing biodiversity is implicated not only by the danger that REDD+ incentivises the planting of monocultural forests, but also because the legal framework for the exploitation of genetic resources has a more established system of benefit-sharing. There are also conscious intersections made between REDD+ and human rights and indigenous rights, especially by guidance notes from the organizations implementing REDD+ on the ground, such as the UN-REDD Programme. In many instances, the intersecting norms do not of themselves fall under the umbrella of an international regime: they emerge through domestic property law, including indigenous property rights or tenure arrangements or through international aspirations that develop into binding norms of customary international law.

While regime interaction has occurred largely in the negotiation of REDD+ (and especially in the COP decisions, such as the Cancun Safeguards), many of the problems and opportunities surrounding REDD+'s further legal development and implementation will be concentrated in sites of interaction with existing international regimes. These 'sites of interaction' are themselves evolving and are apparent at different international, domestic and local law levels. They may be readily visible, such as through published

[4] See, e.g., Chapter 7; see also Jessica Rae, Mahala Gunther and Lee Godden, 'Governing Tropical Forests: REDD+, Certification and Local Forest Outcomes in Malaysia' (2011) 7(2) *Macquarie Journal of International and Comparative Environmental Law* 40.

works and activities of the UN-REDD Programme and the FCPF, which have reproduced guidance from treaties such as ILO Convention 169 even without ratification by the particular REDD+ recipient.[5] They may be more subtle, such as the advice given by the CBD's executive secretary to the UNFCCC COP about safeguards for biodiversity and adequate benefit-sharing.[6] As such, there is an ongoing need to account for REDD+'s influences on, and by, other forest-focused and forest-related regimes, and indeed other regimes in international law. In light of this, we point to the role of states, international organizations and nonstate actors in REDD+ below.

### 1. *Protections: Recipients and Donors*

The book has shown that regime interaction flows from states' existing international legal obligations as well as from the Cancun Safeguards and other practices. The safeguards may yet be regarded as 'subsequent practice' in the interpretation of UNFCCC obligations.[7] Separately, they may generate duties to cooperate in the same way that resolutions by international organizations, even if not binding, can found future obligations.[8] Furthermore, when UNFCCC developing countries are 'encouraged' to 'contribute to mitigation actions in the forest sector' as part of REDD+ COP decisions,[9] they act not only according to their obligations under treaties (including human rights treaties) and customary international law, but to the Cancun Safeguards, whose adherence may be a precondition for the receipt of REDD+ funds. We have shown how their practice is made transparent through the Warsaw Framework's 'safeguards information system'. Such protections are particularly necessary if local communities assume responsibilities for REDD+ projects (such as financial liability if the permanence of the sequestered carbon is

[5] See further in Chapter 4. [6] See further in Chapter 3 at n 80 and accompanying text.

[7] See on Vienna Convention on the Law of Treaties, Chapter 2 at n 79 and accompanying text.

[8] See, e.g., *Whaling in the Antarctic (Australia v Japan)* ICJ GL No 148 (31 March 2014), para 137 (in that case, the Court found that Japan had failed to pay 'due regard' to resolutions of the International Whaling Commission, notwithstanding that such resolutions were not binding); see further Chapter 3 at n 103 and accompanying text.

[9] The phrase is from the Cancun Agreements: Conference of the Parties, UNFCCC, *Report of the Conference of the Parties on Its Sixteenth Session, Held in Cancun from 29 November to 10 December 2010 – Addendum – Part Two: Action Taken by the Conference of the Parties at Its Sixteenth Session*, UN Doc FCCC/CP/2010/7/Add.1 (15 March 2011), Decision 1/CP.16 ('The Cancun Agreements: Outcome of the Work of the Ad Hoc Working Group on Long-Term Cooperative Action under the Convention') para 70.

disrupted by a natural disaster).[10] Thus, the integration of the rights of indigenous peoples, biodiversity, and social and environmental protections is a requirement for REDD+ recipient states through a combination of law and an experimentalist-orientated benchmarking and transparency provision, which we consider further in Section B.

These issues of regime interaction are not limited to REDD+ project host states. Appropriate regime interaction also flows from obligations of donor countries. Indeed, calling for attention to the human rights obligations of states that are donors, partners or even just members of international financial institutions could improve those frameworks that 'are more usually the purview of broad policy commitments, programs, and process than of treaties'.[11] For example, if there is a principle of customary international law requiring government authorities to consult indigenous peoples on government policies or matters affecting them,[12] this will bind all states in REDD+ implementation. Even stronger protections – on substance if not in prospects for enforcement – flow from the UNDRIP, which is supported by the recognition of FPIC in the UN-REDD Programme Guidelines (but without any specific review procedures for affected communities). Protections such as FPIC and adequate benefit-sharing have been recognised as essential in situations where legal fragmentation implicates indigenous cultures.[13] To counter the suggestion that obligations of states must be tied to the REDD+ projects occurring within their territory, regard could be had to other contexts of public international law, where states have assumed accountability – or at least some form of prescriptive jurisdiction – in faraway places.[14] One of the directions for

[10] See further Chapter 4; see also Irmeli Mustalahti and O Sarobidy Rakotonarivo, 'REDD+ and Empowered Deliberative Democracy: Learning from Tanzania' (2014) 59 *World Development* 199, 208.

[11] Siobhán McInerney-Lankford, 'Human Rights and Development Regime Interaction and the Fragmentation of International Law' in Hassane Cissé et al. (eds), *The World Bank Legal Review: Legal Innovation and Empowerment for Development, Volume 4* (World Bank and Martinus Nijhoff, 2013) 123, 153.

[12] See further Chapter 3, n 98 and accompanying text.

[13] Gunther Teubner and Peter Korth, 'Two Kinds of Legal Pluralism: Collision of Transnational Regimes in the Double Fragmentation of World Society' in Margaret A Young (ed), *Regime Interaction in International Law: Facing Fragmentation* (Cambridge University Press, 2012) 23, 52.

[14] For example, UN peacekeeping usually involves the assumption of any liabilities by the host state (such as where a peacekeeping operation exercised the host government's authority in seizing private property), but in certain circumstances of peacekeeper misconduct, the relevant conduct may be attributed to the troop-contributing state or even the UN itself. A different example can be found in trade policy, where the trade laws

further research suggested by the findings of this book is a consideration of how the emerging international legal concept of 'shared responsibility' – which seeks to allocate responsibility for harms that international law aims to prevent[15] – may have application with respect to any violations to indigenous peoples' rights caused by REDD+. The substance of these responsibilities may also arise in the context of bilateral funding arrangements for REDD projects (although such bilateral arrangements were observed only in the periphery of our country-specific examples,[16] they are an ongoing and growing feature of carbon sequestration activities). In addition, there is an ongoing need to consider how developed countries can better address deforestation drivers, which are often outside of the capabilities of tropical-forested countries themselves.[17]

### 2. *Implied Powers of International Organizations*

This book has demonstrated that rights and accountability mechanisms may be relevant for the activities of international organizations and other implementing agencies (including the FCPF and the UN-REDD Programme). While international organizations do not have the same responsibilities as states, they are still subjects of public international law. The 'learning by doing' approach, articulated expressly by the World Bank's FCPF[18] and apparent also in practices of the UN-REDD Programme, may well be appropriate for an emerging regime such as REDD+ that is faced with existing pluralism, uncertainty and a need for adaptive responses. Yet rights protections may be necessary in

of an importing state are sometimes said to represent an exercise of extraterritorial jurisdiction: see further Margaret A Young, 'Trade Measures to Address Environmental Concerns in Faraway Places: Jurisdictional Issues' (2014) 23 *RECIEL* 302.

15 See André Nollkaemper and Dov Jacobs, 'Shared Responsibility in International Law: A Conceptual Framework' (2013) 34 *Michigan Journal of International Law* 359; see also Catherine Redgwell, 'Shared International Responsibility for Transboundary Harm Arising from Energy Activities' in Lila Barrera-Hernández, Barry Barton, Lee Godden, Alistair Lucas and Anita Rønne (eds), *Sharing the Costs and Benefits of Energy and Resource Activity* (Oxford University Press, 2016), 59.

16 See, e.g., the pilot programmes in Papua New Guinea described in Chapter 8, Section F.

17 Charlotte Streck and Michaela Schwedeler, 'Addressing Drivers of Deforestation and Forest Degradation through International Law' in Christina Voigt (ed), *Research Handbook on REDD+ and International Law* (Edward Elgar, 2016) 213, 225.

18 International Bank for Reconstruction and Development, *Charter Establishing the Forest Carbon Partnership Facility* (Revised 23 November 2015), www.forestcarbonpartnership.org/charter-and-governance-documents, Article 3.1(b).

certain circumstances, particularly on issues affecting indigenous peoples: although rights predominantly are owed by states to individuals, the UNDRIP acknowledges that the 'organs and specialized agencies of the United Nations systems and other intergovernmental organizations' shall contribute to the realisation of its goals, and establish means to ensure participation of indigenous peoples.[19]

Organizations have obligations due to their own constitutive documents and procedures, external agreements and implied powers. For example, we show in Chapter 2 that the UN-REDD Programme operates according to a 'human rights–based approach'.[20] The FCPF is more restrained: we demonstrated that its position relates to a long controversy over whether enhanced recognition of human rights obligations by the World Bank flouts common understandings of its mandate.[21] Adopting or 'mainstreaming' a human rights approach is seen by some as risking a return to didactic conditionality in the bank's lending practices,[22] and by others as risking the co-option of human rights principles.[23] Of equal importance is the institutional dimension which sees the FCPF exposed to a review and complaints procedure of the World Bank's independent Inspection Panel (for REDD+ implementation rather than readiness activities), as compared to a lack of such procedure within the UN-REDD Programme. The 'risks versus rights' difference between the two organizations[24] has a special meaning in this regard. These perspectives

[19] UNDRIP, Article 41.

[20] FAO, UNDP and UNEP, *UN Collaborative Programme on Reducing Emissions from Deforestation and Forest Degradation in Developing Countries (UN-REDD) Framework Document* (20 June 2008) 6, www.un-redd.org/Portals/15/documents/publications/UN-REDD_FrameworkDocument.pdf.

[21] This relates to an asserted prohibition to enter into political decision-making: see Hassane Cissé, 'Should the Political Prohibition in Charters of International Financial Institutions Be Revisited? The Case of the World Bank' in Hassane Cissé, Daniel D Bradlow and Benedict Kingsbury (eds), *The World Bank Legal Review, Volume 3: International Financial Institutions and Global Legal Governance* (The World Bank, 2012) 59. See also aspects of institutional culture mentioned at Chapter 2, n 121 and surrounding text.

[22] David Kinley, 'Human Rights and the World Bank: Practice, Politics, and Law' in Ana Palacio, Caroline Mary Sage and Michael Woolcock (eds), *The World Bank Legal Review: Law, Equity, and Development, Volume 2* (World Bank and Martinus Nijhoff, 2006) 353, 356 and source cited therein.

[23] Martti Koskenniemi, 'Hegemonic Regimes' in Margaret A Young (ed), *Regime Interaction in International Law: Facing Fragmentation* (Cambridge University Press, 2012) 305, 319 (and sources cited therein).

[24] See further in Chapter 4.

underlie the difficulties faced by the REDD+ implementing agencies in incorporating norms and practices from other regimes.

Whether or not the relevant development regimes formally endorse a human rights approach, the phenomenon of fragmentation could be argued to have created additional implied powers for international organizations.[25] Such implied powers build upon international institutional law and provide that, in circumstances where regimes are dependent upon other regimes (such as in setting standards of conduct), intergovernmental organizations have the discretion to apply norms and facts from external sources 'because such collaboration is necessary for their functions'.[26] These implied powers are not boundless, of course. The often-observed risk of managerialism and technocratic decision-making, where unseen experts control global politics, could be exacerbated by an ability to act independently from express mandates.[27] One way to counter this risk is a requirement that such bodies be open and transparent in their practices. While the biases and cultural attitudes within these bodies are not erasable, enhanced transparency can lead to better understanding of the influence of their attitudes on regime interaction. It is encouraging, for example, that the FCPF has instituted regular meetings at which governments, forest-dependent indigenous peoples and forest-dweller organizations, international organizations and NGOs can participate:[28] these fora allow actors to commission and then comment on evaluations of the work of the FCPF, which are themselves made public.[29] A 2011 review commissioned by the Participants Committee (of host and

[25] Margaret A Young, *Trading Fish, Saving Fish: The Interaction between Regimes in International Law* (Cambridge University Press, 2011), 273. This idea is elaborated in 271–278 and draws on jurisprudence of the International Court of Justice such as *Reparations for Injuries Suffered in the Service of the United Nations* [1949] ICJ Reports 174 and *Legality of the Use by a State of Nuclear Weapons in Armed Conflict* [1966] ICJ Reports 66.

[26] Ibid. [27] Ibid. 276–278 and sources cited therein.

[28] The participants committee is made up of donor and recipient states as well as private actors participating in the carbon fund: International Bank for Reconstruction and Development, *Charter Establishing the Forest Carbon Partnership Facility*, above n 18, section 11. A larger 'participants assembly' can include representatives of international organizations, relevant NGOs, forest-dependent indigenous peoples and forest dwellers, and relevant private-sector entities as observers if invited: section 10.1(a).

[29] See Baastel and Nordic Agency for Development and Ecology, 'First Program Evaluation for the Forest Carbon Partnership Facility (FCPF): Evaluation Report (13 June 2011) 30 (noting inter alia that much of the mutual learning taking place between the FCPF and the UNFCCC was due to the fact that many FCPF PC members are also UNFCCC negotiators). See also the Forest Carbon Partnership Facility, FCPF 'Eighteenth Meeting

recipient states as well as potential financers) found that there had been high levels of transparency, 'in large part through the effective functioning of the FCPF website, but also through other processes such as the conduct of PC meetings'.[30] Similarly, the UN-REDD Programme has undergone an external evaluation at the request of the UN-REDD Policy Board.[31] The 2014 evaluation found that recognition of indigenous rights had been supported by its implementation of the Cancun Safeguards, although the social and environmental safeguards were difficult to put into practice due in part to 'lack of clear guidance on how to implement, monitor and enforce these'.[32] There is an ongoing need for assessment on whether the practices of the relevant bodies meet the ideals of transparency and accountability.

The need to act in a context of existing legal regimes and institutions, with due sensitivity to established practices, does not ever seem to have been ignored by the UN-REDD Programme or the FCPF, which have sought to be participatory and consultative in readiness activities. That in itself is a progressive development in international institutional practice and conforms to hopes for appropriate regime interaction.[33] Yet the choices and practices of these bodies – relating to *which* international norm to endorse, or *which* inter-organizational relationship to foster, or *which* indigenous or local community to consult, or *which* capacity to build – have major consequences. On the one hand, one may say that sovereign concerns should be paramount, and the states who are the masters of the international organizations must formally agree on such strategies. When the activities of international organizations relate to strictly domestic constitutionalist understandings about property law or the resolution of indigenous peoples' claims, this seems particularly apt. On the other hand, it is right for international lawyers to call attention to relevant legal mechanisms, powers

of the FCPF Participants Committee (PC18), Seventh Meeting of the FCPF Participants Assembly (PA7), Co-Chairs' Summary' (Arusha, Tanzania, 31 October–2 November 2014) (which noted inter alia that the upcoming review of the FCPF should include evaluation of the FCPF's relation to the frameworks of the UNFCCC: at 3). In addition to the FCPF Participants Committee–commissioned review, an earlier review is the Independent Evaluation Group, *The Forest Carbon Partnership Facility* (Global Program Review vol 6, issue 3, 2011) para 4.12 (noting inter alia that the FCPF 'has been a fair and inclusive program').

30 Baastel and Nordic Agency for Development and Ecology, above n 29, 13.

31 Alain Frechette, Minoli de Bresser and Robert Hofstede, External Evaluation of the United Nations Collaborative Programme on Reducing Emissions from Deforestation and Forest Degradation in Developing Countries (the UN-REDD Programme): *Volume I – Final Report* (July 2014), http://wedocs.unep.org/rest/bitstreams/776/retrieve.

32 Ibid. vi. 33 See, e.g., Young, above n 25, 301.

and duties of international organizations.[34] A future research direction is the application of the concept of shared responsibility to the international organizations that are so demonstrably influential in issues relating to REDD+'s impact on indigenous peoples.

### 3. *Deliberative Credentials and Emerging Duties*

We have demonstrated that nonstate actors such as civil society groups, academic and research institutions, and corporate interests play a major role in regime interaction on REDD+. Nonstate actors have an uncertain place in public international law, with individuals only emerging as rights holders in discrete areas such as human rights regimes and investment treaties that guarantee particular protections; very little law imposes obligations outside of the territorial context of state regulation. Yet, given their involvement and impact on regime interaction on issues relating to REDD+, greater attention to the accountability of nonstate actors is needed. Guiding principles recommended by the UN require business enterprises to avoid causing or contributing to adverse human rights impacts;[35] these principles seem to require that the entities providing private REDD+ funds respect the norms of human rights regimes. This combines with the recognition that private bodies are implicated in the drivers of deforestation and forest degradation, and must take action to address them.[36]

Of course, the limitation of such principles is that they do not necessarily lead to legal avenues for grievance and redress. It is encouraging that the REDD Readiness Preparation Proposal (R-PP) template for states, which seeks consistency with the Cancun Safeguards, requires states seeking REDD+ Readiness funds to propose some form of 'feedback and grievance mechanism'.[37]

[34] See, e.g., International Law Commission, *Responsibility of International Organizations*, UN Doc A/CN.4/L.778 (30 May 2011); see further Nollkaemper and Jacobs, above n 15, and accompanying text.

[35] John Ruggie, 'Report of the Special Representative of the Secretary-General on the Issue of Human Rights and Transnational Corporations and Other Business Enterprises: Guiding Principles on Business and Human Rights: Implementing the United Nations "Protect, Respect and Remedy" Framework, A/HRC/17/31 21 (March 2011).

[36] Conference of the Parties, United Nations Framework Convention on Climate Change, *Report of the Conference of the Parties on Its Nineteenth Session, Held in Warsaw from 11 to 23 November 2013 – Addendum – Part Two: Action Taken by the Conference of the Parties at Its Nineteenth Session*, UN Doc FCCC/CP/2013/10/Add.1 (13 January 2014) Decision 15/CP.19.; see further Streck and Schwedeler, above n 17.

[37] Forest Carbon Partnership Facility (FCPF) and United Nations REDD Programme (UN-REDD), 'Readiness Preparation Proposal (R-PP)' (Template Version 5, Working Draft, 30 October 2010) 16.

This mechanism aims to 'improve responsiveness to citizen concerns, help identify problems early, and foster greater trust and accountability with program stakeholders'.[38] To impose accountability for businesses involved in REDD+, such a mechanism requires broad jurisdictional reach and effective domestic legal processes. As our studies in the domestic contexts of Malaysia, Papua New Guinea, Vanuatu and Australia have demonstrated, such issues are highly variable within states and much depends on established recognition and legal status of indigenous peoples, tenure arrangements and nuanced property law.

Civil society groups active in the sites of interaction between REDD+ and other regimes have much to offer in bridging understanding between regimes. Indeed, they may be essential in assisting lower-level actors with local and contextualised knowledge. Forms of accountability for these actors will recognise that their functions, constituencies and intended beneficiaries are different from states and international organizations.[39] They will also require conceptions of indigenous identity and membership, which, as we show in Chapter 4, are already being promoted by consultative standards of the UN-REDD Programme and the FCPF.[40] Seeking to account for these participants in REDD+ interaction with other regimes requires nonstate actors to be open about their credentials and interests. It also gives rise to ongoing strategic questions about particular forms of engagement – such as whether engaging in regime interaction risks assimilation or loss of identity and critical power.[41] Varied responses to such questions are visible so far, as is evident from the manifold positions taken by indigenous peoples on the REDD+ safeguards.[42] Perspectives such as these have an important deliberative place in the emerging REDD+ regime. How these views traverse the international, domestic and local contexts, in ways that provide solutions to generally agreed problems while avoiding top-down prescriptions, evokes

[38] Ibid.

[39] Sasha Courville, 'Understanding NGO-Based Social and Environmental Regulatory Systems: Why We Need New Models of Accountability' in Michael W Dowdle (ed), *Public Accountability: Designs, Dilemmas and Experiences* (Cambridge University Press, 2006) 271; see further Young, above n 25, 282.

[40] See generally Kirsty Gover, *Tribal Constitutionalism: States, Tribes, and the Governance of Membership* (Oxford University Press, 2010).

[41] Koskenniemi, above n 23, 323.

[42] UN ECOSOC Permanent Forum on Indigenous Issues, *Indigenous People's Rights and Safeguards in Projects Related to Reducing Emissions from Deforestation and Forest Degradation*, UN Doc E/C.19/2013/7 (5 February 2013).

questions about forms of governance, especially as REDD+ is considered to be 'the most important laboratory for expanding environmental democracy.'[43] Section B assesses whether REDD+ is a form of experimentalist governance.

## B. Experimentalist Governance

Finding solutions to climate change – framed as a global problem – that satisfy the demands of a local forest-based populace is highly complex. With no recognised international authority to issue binding rulings on which international, domestic or local laws have priority, there is little hope for clarity or certainty, much less justice. Yet the UNFCCC and associated mechanisms are seeking to achieve climate change mitigation as best they can, and theories of experimentalist governance can help to account for how decentralised efforts can be effective. A whole area of normative development has been concentrated in the area of 'new governance', where sovereignty and hard law give way to non-binding norms (soft law) that offer flexible, dynamic pathways on substance as well as a set of procedural prescriptions. This normative development has been observed in diverse contexts. For example, new governance approaches have been documented from the European Union's efforts to cope with pluralism and diversity,[44] and from the practices of international financial institutions such as the World Bank in seeking to incorporate social protections in funding arrangements.[45] Various forestry policies have been assessed through the changing roles of learning, monitoring and compliance within firms,[46] and experimentalist governance has also been observed in the context of climate change technologies such as geoengineering.[47] Whether governance allows for open participation,

[43] David Takacs, 'Environmental Democracy and Forest Carbon (REDD Plus)' (2014) 44 *Environmental Law* 71.

[44] See, e.g., Joanne Scott and David M Trubek, 'Mind the Gap: Law and New Approaches to Governance in the European Union' (2002) 8 *European Law Journal* 1.

[45] Kerry Rittich, 'The Future of Law and Development: Second-Generation Reforms and the Incorporation of the Social' in David M Trubek and Alvaro Santos (eds), *The New Law and Economic Development: A Critical Appraisal* (Cambridge University Press, 2006) 203, 205.

[46] Christine Overdevest and Jonathan Zeitlin, 'Assembling an Experimentalist Regime: Transnational Governance Interactions in the Forest Sector' (2014) 8 *Regulation & Governance* 22.

[47] Chiara Armeni, 'Global Experimentalist Governance, International Law and Climate Change Technologies' (2015) 64 *International and Comparative Law Quarterly* 875.

empowered local communities and reasoned decision-making and transparency are explored in domestic and transnational contexts.[48] Such questions are increasingly being posed in the climate context, such as the parameters outlined by Sabel and Victor for effective experimentalist, bottom-up climate policy, which includes the use of penalty defaults.[49] In experimentalist governance, '[t]here is no overarching international constitutional framework with institutionalized hierarchical relations between governance units or courts. There are areas of agreed authority, but on many issues authority is overlapping, contested and fluid'.[50] We investigate in the remainder of this section whether the emerging REDD+ regime can be classified as a form of experimentalist governance.

### 1. *Participatory Processes*

An important aspect of experimentalist governance is the openness to participation of relevant entities or stakeholders in a non-hierarchical process of decision-making. While concepts such as FPIC are apparent guarantees of inclusion of the voices of indigenous peoples, our findings show that such inclusion is not straightforward. The different emphasis by the UN-REDD Programme and the FCPF on whether FPIC requires consent or consultation shows that the pursuit of participatory ideals may differ depending on the institutional context in which REDD+ is implemented.[51] Our country-specific examples show that schemes have failed to involve affected communities for a variety of reasons.[52] These findings resonate with similar investigations on other contexts, as exemplified by field studies on REDD+ in Tanzania, which show that outcomes are shaped by 'power relations both between and within communities, central, local, and village governments and the private sector'.[53] To be effective, participation must not only encourage local

[48] See, e.g., de Búrca, Keohane and Sabel, above n 2 (on 'global experimentalist governance'); Archon Fung and Erik Olin Wright 'Deepening Democracy: Innovations in Empowered Participatory Governance' (2001) 29 *Politics & Society* 5.

[49] Charles F Sabel and David G Victor, 'Governing Global Problems under Uncertainty: Making Bottom-Up Climate Policy Work' (2015)(May) *Climatic Change* 1.

[50] de Búrca, Keohane and Sabel, above n 2, 726.

[51] See further Chapter 4.

[52] See further on Malaysia in Chapter 7 and on Papua New Guinea in Chapter 8.

[53] Mustalahti and Rakotonarivo, above n 10, 208. The authors test the concepts of empowered deliberative democracy set out in Fung and Olin Wright, above n 48.

representatives but also explicitly recognise the need for disadvantaged groups to become engaged and informed about the management of forest resources.[54]

Of course, clear markers are available to guide states to achieve basic protections for indigenous and forest communities. These markers are apparent in the Cancun Agreements, especially in paragraph 72's 'request' to developing states to

> address, inter alia, the drivers of deforestation and forest degradation, land tenure issues, forest governance issues, gender considerations and the [Cancun] safeguards ... ensuring the full and effective participation of relevant stakeholders, inter alia indigenous peoples and local communities.[55]

Moreover, as Chapter 2 demonstrates, the seven safeguards agreed upon at Cancun include laudable aspirations for the meeting by UNFCCC parties of a whole range of legal obligations from exogenous regimes.[56] These include reference to the UNDRIP, which itself gives legal expression to the FPIC concept.

Indeed, what is distinctive about these soft-law standards is the ongoing influence and interplay with norms and standards from existing international regimes. The Cancun Agreements are the clearest example of states exhibiting 'systems thinking' in the successful implementation of climate mitigation: demonstrating an awareness of the wider context of REDD+ and the need to have regard for it. This could lead to policy coordination amongst government ministries such as agriculture, energy, forestry, trade and culture. It could lead to state delegates to the UNFCCC being briefed by all these interests, and taking them into account not only in negotiations, but in dealings with other states, with industry and with affected communities. Yet the procedural prescriptions to provide any kind of transparency or oversight for the achievement of this request are vague. There are no requirements for oversight of the ideal behaviour of states encouraged by paragraph 72 of the Cancun Agreements. There is, however, the prospect that states may make their funding of REDD+ contingent on the publication of information on safeguards adherence, and whether this amounts to the

[54] Ibid. See also Taiji Fujisaki et al, 'Does REDD+ Ensure Sectoral Coordination and Stakeholder Participation? A Comparative Analysis of REDD+ National Governance Structures in Countries of Asia-Pacific Region' (2016) 7 *Forests* 195.

[55] *Cancun Agreements*, above n 9. See especially Chapters 2 and 4.

[56] See Chapter 2, Section A3a.

'penalty default' that is central to effective experimentalist governance demands consideration.

In this vein, it is important to note that states are the single most important drivers of REDD+ – a development that was not inevitable given that earlier forestry reforms concentrated on devolution to provinces and local communities.[57] Moreover, in the course of international efforts at conservation and sustainable development of areas that might be conceived as common heritage, national sovereignty over forest resources has been a continual refrain, and this book demonstrates that the refrain has been heard in the emerging REDD+ regime. Yet, a consequentialist assessment that states are unfettered in their decisions about REDD+ and its impact on indigenous and forest communities would be wrong.

### 2. *The Prospect of Penalty Default: The Safeguards Information System*

The UNFCCC Warsaw Framework includes a 'safeguards information system' (SIS), which requires REDD+-recipient states to provide evidence of compliance with the Cancun Safeguards.[58] In addition to the COPs, the SIS is supported by the R-PP documents required by the UN-REDD Programme and the FCPF, which draw on the Cancun Safeguards as well as their own social and environmental standards.[59] In addition, transparency is emphasised by the use of a web portal. At this point, only two countries (Brazil and Democratic Republic of Congo) have uploaded their SIS documentation on the web platform,[60] which may reflect difficulties in capacity and the need for time rather than a refusal to engage with the framework.[61]

Looking to the future, the arrangement does not limit itself to information sharing, but may lead to penalty defaults in the following sense: donor states may choose to make funding conditional on satisfactory review of a recipient states' summary. Questions remain, however, about

[57] Annecoos Wiersema, 'Climate Change, Forests, and International Law: REDD's Descent into Irrelevance' (2014) 47 *Vanderbilt Journal of Transnational Law* 1.

[58] See further Chapter 2, n 65 and accompanying text. [59] See further Chapters 4 and 5.

[60] See http://redd.unfccc.int/submissions.html.

[61] See, e.g., the ongoing work evidenced in D Rey, S Korwin, U Ribet and L Rivera, 'Best Practices and Considerations for the Development of REDD+ Country Approaches to Safeguards and Design of Safeguards Information Systems' (CLP and SNV United Kingdom 2016), uploaded at http://redd.unfccc.int/.

whether such conditionality relates to 'compliance' with safeguards or merely the provision of information. Both situations, however, give rise to the prospect that REDD+ recipient states may be shut out of the REDD+ system for failure to achieve at least some level of expectations about protections for indigenous peoples and other requirements.

A remaining consideration here is that the decision on whether to link safeguards to funding remains the discretion of the donor states. Evidence of compliance through an SIS is highly contingent and only transparent to the broader public if the relevant recipient state chooses to make their records public on the UNFCCC web platform.

Moreover, the safeguards do not encompass an embrace of all forest-relevant international law. For example, we have documented the highly contested and political choices in the COP decisions in response to SFM.[62] This demonstrates a point often observed in studies of fragmentation and regime interaction: differential language in a context of existing laws and policies is often hard fought and meaningful for particular interests and (generally, savvy and well-resourced) states. Outside of the context of the potential for safeguards to exert pressure, prospects for accountability remain in traditional legal avenues.[63]

### 3. *Established Practices for Revision, Involving Peer Review*

Experimentalist governance requires revision by peers in order that best practice is identified and becomes evident to the whole. Whether the rules and practices surrounding REDD+ are open to revision is a matter of debate. Formally, there is peer review for scientists within the IPCC and negotiators within the UNFCCC to revise rules and practices, and indeed much time has been devoted to establishing methodologies for monitoring, verification and reporting. A range of countries are involved in seeking to establish best practices: Papua New Guinea, for example, has sought a leadership role through the facilitation of workshops and the offering of 'south-south advice'.[64] Moreover, the UN-REDD Programme and the FCPF are involved in this peer review process in that they seek to apply best-practice norms in considering funding

[62] See further in Chapter 3.

[63] See, e.g., electoral institutions: Mustalahti and Rakotonarivo, above n 10, 207.

[64] UN-REDD Programme, 'National Programme, Semi-Annual Report: Papua New Guinea' (January to June 2015) 13. See further in Chapter 8. See more generally COP endorsements of step-wise approaches and interim measures: e.g., Decision 12/CP.17, para 10-11.

applications.[65] The experience from Vanuatu in the early pilot programmes, for example, showed that capacity building was necessary, and this led to revisions in the expectations for readiness funding.[66] The endorsement of the UN-REDD Programme of a 'theory of change' is promising in confronting the need for early and intermediate outcomes, as well as a long-term objective of climate mitigation.[67]

Yet critics of REDD+ argue that REDD+ demonstration and pilot activities have already assumed the inevitability of a market response.[68] They argue there is a failure to address the real drivers of deforestation. This relates to whether the common problem has been adequately identified in REDD+, an issue to which we now turn.

## 4. *A Broadly Agreed Common Problem?*

Experimentalist governance is based on responses to a problem that, though complex and difficult, share broadly agreed common agreement as to its definition (although solutions are avowedly open and changing). REDD+'s progression from RED to REDD to REDD+ shows the difficulty in articulating a common problem: Is it climate mitigation or sustainable forests? Is it a system of benefits to indigenous and local forest-dependent communities in recognition of the ecosystem services that they and the forests they inhabit provide? After a series of COPs, the Cancun Agreements and Warsaw Frameworks may be said to have resolved this point, yet the difficulties and complexities that REDD+ has encountered, especially in attending to issues such as tenure and benefit-sharing, show that the problem may yet refuse definition. An uneven distribution of beneficiaries is ever-present in the causes of anthropogenic climate change and in climate change mitigation, and indigenous peoples and forest-dependent local communities are little implicated in either. This is perhaps yet to be fully appreciated by the attempts within international, domestic and local law to address common problems defined by others.

More fundamentally, the idea behind 'carbon offsets' that may develop in the REDD+ context (and that are already apparent in voluntary carbon markets utilising carbon sequestration) is of low-cost mitigation

[65] See further our descriptions of the R-PP template and its associated SESA in Chapter 4.

[66] See further in Chapter 8.

[67] See further in Chapter 2, n 90 and surrounding text.

[68] Julia Dehm, 'Indigenous Peoples and REDD+ Safeguards: Rights as Resistance or as Disciplinary Inclusion in the Green Economy?' (2016) 7 *Journal of Human Rights and the Environment* 170.

regardless of source. Whether the implementation of this idea can ever be localised and contextualised deserves further reflection and evaluation in the light of this book's findings. In the REDD+ model, it is the forested communities that have had minimal or no impact on GHG emissions levels that will be facing the on-the-ground implementation. One may pose the question of whether a true experimentalist model would require the communities most responsible for GHG emissions to experience solutions to climate change in a tangible, place-based way. Whilst mitigation in faraway places may appear to be 'least cost' according to economic analysis, the real, lived, social and legal realities may be far from this.

## C. Conclusion

In this chapter, we argue for better understanding and legal accountability of the actors promoting or obstructing regime interaction on forest-related climate mitigation. This is not merely a question of 'learning lessons' or 'achieving coherence', but goes to more general issues of effectiveness, accountability and even legitimacy for REDD+.

We have demonstrated that regime interaction does and will occur in many ways and for many purposes: not just for coordination but sometimes to prioritise one set of regime preferences over another, and sometimes to co-opt a weaker regime. Calls for 'coherence' can miss these important aspects. We have shown in this chapter that appropriate oversight and review of the activities of actors during regime interaction – which include but are not limited to states – should be maintained using rights or accountability mechanisms. Obligations to consult with affected indigenous peoples, which are held by states and relevant international organizations, are the minimum of such mechanisms. Such issues arise in models of experimentalist governance, which are useful in emphasising concepts of peer review, open participation, revision of rules and bench-marking. While the power disparities among local-level actors as well as between local-level actors and state-level bodies need more visibility in these models, especially when dealing with indigenous peoples and other marginalised groups, accommodating the need for local, contextualised action and feedback to a broader set of aspirations is essential for climate mitigation.

# 11

# Conclusion

This book's primary objective has been to bring the legal interests of indigenous peoples and forest-dependent communities to the foreground of global developments in climate change mitigation. It analyses policy approaches and positive incentives that aim to reduce emissions from deforestation and forest degradation that also engage in positive forest-related activities (REDD+), which have been promoted by countries under the auspices of the UNFCCC. The book assesses the various impacts of REDD+ on indigenous peoples and forest communities and offers a critical view of the relationships between collective indigenous-specific rights (especially with respect to FPIC and tenure) and human rights, as well as background norms existing in contemporary treaties and customary international law. We documented the stances taken on these matters by the REDD+ funding bodies, particularly the UN-REDD Programme and the World Bank's FCPF, and demonstrated that institutions, states, civil society and other actors contribute to diverse framings of REDD+ in national contexts such as in Malaysia, Papua New Guinea and Vanuatu. We examined in depth the history of land law and forest resources laws in these countries and the ways that these laws and policies intersect with legal frameworks for the recognition of the rights of indigenous peoples and local forest communities. At a national and sub-national level, REDD+ is being developed in the context of existing complex legal and cultural interactions. Further, experience from the regional context in Australia shows how indigenous involvement in land-sector carbon sequestration and mitigation schemes depends on the complex interface of federal constitutional arrangements, land-rights legislation, provincial government land laws and natural resource management regimes. In short, the rights and interests of indigenous peoples are distinctive, and, although REDD+ is developing amongst a plethora of laws relating to human rights and social and environmental safeguards, the REDD+ regime has not yet addressed the complexity of the issues raised by indigenous claims to forests and forested lands. In this final chapter, we review our main findings and conclude on the necessity of having REDD+ attend to governance at the international, domestic and local levels.

The book argues that REDD+ can be conceptualised as an emerging international regime, with climate change mitigation as its overarching objective, UNFCCC conferences as its dominant set of decision-making-fora, and implementation by institutions such as the FCPF and the UN-REDD Programme as main behavioural approaches.[1] Existing and potential interactions of REDD+ with other regimes within the international legal system have important implications for indigenous peoples and forest communities. These interactions sometimes develop in surprising ways, such as the UN-REDD Programme's incorporation in its project documentation of norms from ILO Convention 169 even absent the ratification of that convention by many REDD+ participating states.[2]

The book sought to account for the conditions of normative and institutional fragmentation that are an unavoidable aspect of the evolution and development of REDD+ within the system of international law. REDD+ promotes the reduction of emissions from deforestation and forest degradation, but other regimes oblige states to prioritise biodiversity, sustainable forestry management or even food security. We considered whether the interests of local and affected communities have been incorporated in national and international law-making, standard setting and policy implementation when relevant international regimes interact with REDD+, and the extent to which indigenous rights are respected in those processes. How to integrate REDD+ into the complex field of forestry governance and associated legal regimes in a manner that protects and furthers the interests of affected indigenous and local forest-dwelling communities is an ongoing challenge. We show that regime interaction at the international level has extended to outreach by organizations such as the UN-REDD Programme, and is important in forestalling conflicts between norms. Yet regime interaction cannot by itself provide methods with which to resolve disputes over access to land and resources where these involve indigenous peoples' claims. The unresolved legal claims of indigenous peoples are deeply context-specific and must be interrogated at the local level. If REDD+ is to assist in this enterprise, it must include methods for resolving contestation between international norms and commitments where these dictate the trajectory

[1] See especially Chapter 2. [2] See further in Chapter 3.

of domestic disputes involving indigenous claims. Attention to historic injustice, dispossession, the non-recognition of indigenous law, and the social and economic marginalisation of indigenous and forest communities is essential.

Critical reflection on the value and limits of studies on regime interaction is an important aspect of this book. We showed that regime interaction at the international level is but one of several factors that change outcomes for indigenous peoples and local communities in climate mitigation scenarios. A key finding is that the normative and institutional developments that accrue from the interaction between international legal regimes must be attentive to domestic and local-level legal processes: for example, the influence of the CBD on REDD+ safeguards is shaped by the international commitments of states, but outcomes in any given context are also fundamentally determined by the domestic environmental laws or politics of a country receiving REDD+ funding. Or, in another example, since the operational policies of the UN-REDD Programme instruct states to protect and respect all human rights, indigenous rights and property rights, and it remains the case that many conflict with one another in principle as well as in practice, in the absence of any advice on how to balance rights against others or apply proportionality analyses, it will remain for the state to determine which right to prioritise over others in any single instance. This is why the book's attempt to traverse the international, domestic and local contexts is important.

In this vein, we evaluated the efforts of institutions within REDD+ to articulate safeguards on indigenous peoples' rights, as demonstrated perhaps most vividly by the principle of FPIC. We pointed to the divergences between the standards applied by the UN-REDD Programme and FCPF, respectively – where the 'c' in FPIC moves between 'consent' and 'consultation'.[3] We have critically examined the 'human rights–based approach' taken by the UN-REDD Programme and its constituent agencies, arguing that some formulations of this approach can undermine the legitimate claims of indigenous peoples.[4] While such pressures must be assessed in comparison with the threats posed by commercial logging, agricultural expansion and so on, we have shown that REDD+ impacts on indigenous peoples in ways that are not dissimilar to deforestation – such as the loss of livelihoods, the loss of informal title in

[3] See further in Chapter 4. [4] Ibid.

tenure-formalisation processes and increased incentives for states to take public title of forested lands to garner benefits to themselves. Additional risks to indigenous peoples may also include the assumption of liabilities for the permanence of sequestered carbon.

Secure and clear tenure is seen by many as a legal guarantee that must precede REDD+ projects.[5] We discussed current UN-REDD Programme and FCPF standards and advice on tenure arrangements in participant states, and considered the methods used to address indigenous land claims in the domestic courts of North America and Australasia and by international and regional human rights bodies including especially the IACtHR. We emphasised that the policies used by the UN-REDD Programme derive from the FAO's Voluntary Guidelines on the Responsible Governance of Tenure, designed as part of the FAOs food security mandate, and noted that the guidelines provide only limited protection for indigenous communities that are no longer in possession of their traditional lands. We further noted that the World Bank's OP 4.10 on indigenous peoples, which guides the activities of the FCPF, protects dispossessed peoples only where the loss of possession was involuntary and occurred within the lifetime of the affected community members. Accordingly, we noted, REDD+ policy tends to favour current users and occupiers of forested land, and could support the formalisation of these rights at the expense of the claims of indigenous peoples asserting historic and customary title to forested lands. In countries where customary owners form the majority (as in Papua New Guinea), tenure certainty is no longer so pressing, but retaining use and access to land by customary owners remains vital, with FPIC to projects playing an essential role. In all cases, caution is needed if locating the imperative for tenure reform in climate change mitigation obscures the broader need for reform of the position of indigenous peoples within societies and forecloses a more systemic way of achieving it.

The book extended this analysis by examining whether REDD+ incorporates a system of co-benefits, which requires the revisiting of its beginnings as an ecosystem-services model.[6] We traced its evolution into a more sophisticated benefit-sharing system and examined the progressive moves to include biodiversity and socio-economic outcomes within the rubric of efforts to advance carbon sequestration, emissions reduction and avoided-deforestation objectives. In exploring the nature of benefits

[5] See Chapter 5. [6] See Chapter 6.

and examining the articulation between the recognition of the rights of indigenous peoples and local forest communities to forest and land resources and the receipt of benefits, we have canvassed arguments about whether indigenous and local community carbon rights might be better articulated as a resource right rather than by adhering to tenure models.

The country case studies grounded our arguments in real-world scenarios. They also demonstrated the dialogic process between the international and domestic: for example, on the one hand, concerns from local and national contexts become manifest at the international level (such as in human rights standards) but find a different formulation at the international level which is sometimes inadequate to address local concerns. On the other hand, countries may be intransigent in accepting or implementing changed legal norms issued from 'above'. Malaysia – a nation with large tracts of tropical forest and a long history of forest governance and forest certification schemes prior to becoming a UN-REDD Programme participant – has a variable human rights record in respect of indigenous and community land claims. The Malaysian experience, exemplified by relatively large indigenous and 'native' populations in East Malaysia, and a federal government structure with devolved regional systems for forest governance, usefully demonstrates tensions generated by decentralised decision-making authority in the REDD+ context.[7] In contrast, two least developed nation states – Papua New Guinea and Vanuatu – provide examples where customary owners are the majority of the population.[8] Both nations are heavily economically dependent on the forest and land-use sectors, and in both, forests constitute important sources of community subsistence and customary law is constitutionally protected. Points of tension arise from the manner in which third party development interests have sought access to customary held land and forests through long term leasing arrangements, a practice that will influence the ways that REDD+ projects might be structured in these nations as well as the role to be played by FPIC. We found that capacity to effectively govern and distribute benefits to customary owners under REDD+ remains a potential barrier to their effective participation.

The book stepped away from REDD+ itself to examine a closely related model of carbon sequestration and land-sector emissions-reduction programmes developed within Australia. The focus on northern Australia – an area characterised by statutory land rights for Aboriginal peoples

[7] See Chapter 7. [8] See Chapter 8.

and a growing number of native title determinations – provides a longer-term perspective on the connections between climate mitigation policies and indigenous peoples.[9] We have examined Australian federal carbon farming legislation and reciprocal regional government models of carbon rights (severed from landholding) and the extent to which indigenous tenures and resource rights were incorporated within these laws. This also required us to examine the opportunities and challenges for indigenous peoples' involvement in emissions-reduction projects, including innovative savanna-burning modes (this model is based around the reduction of emissions through the adoption of traditional Aboriginal practices for burning the land). The Australian experience highlights the complexity that surrounds the package of laws and policies that must support REDD+-style projects and offers a model that can provide useful information and learning about the implementation of REDD+-style schemes.

On the basis of our analysis of the international, domestic and local contexts, we support efforts within the international institutional framework of REDD+ to facilitate the sharing of local experiences and expertise in REDD+ implementation. We have sought to better understand the legal rights and duties of actors according to an experimentalist governance model, which includes both recipient and donor states, international organizations and nonstate actors.[10] We also encourage the establishment of plausible benchmarks based on real-world state and indigenous responses to REDD+, and analyse whether the SIS might perform this role. Penalty defaults (including the exclusion of participation in REDD+ funds) may be important as a future governance mechanism.[11] At this point, however, it is not clear who will ultimately decide on the trade-offs in values contained in the REDD+ safeguards. Moreover, local-level participation and empowerment may be ineffective or counterproductive without due acknowledgment of power disparities facing indigenous peoples and marginalised or disadvantaged groups.

In sum, this book responded to key questions in climate change mitigation: how a set of international policies and approaches will impact on indigenous peoples and forest-dependent communities, and what role law should play in providing protections, managing risks, promoting hidden perspectives or obstructing dangerous assumptions. This issue required traversing a number of legal disciplines, including public

[9] See further in Chapter 9. [10] See further in Chapter 10. [11] Ibid.

international law and regime interaction, human rights law, the law of indigeneity and land tenure, and environmental law. It also required a grounded understanding of REDD+ policies as they are intended to operate in places such as Malaysia, Papua New Guinea and Vanuatu. Our findings are relevant for governments and policy-makers seeking just and effective ways to achieve climate mitigation through forest policies, indigenous peoples working with or against universalising norms and standards, and international lawyers seeking to develop new norms amongst complex, diverse and non-harmonious legal arrangements.

While REDD+ is an initiative developed within the specific international law framework of the UNFCCC, the resonance or divergence of law and legal principles from other bodies of law will determine the outcomes of REDD+ schemes and the extent to which they further the interests of indigenous and local communities affected by forest governance. REDD+ has the potential to yield material and political benefits for indigenous and forest-dwelling communities in the countries where it currently operates or will one day be implemented. The implementation of REDD+ schemes, however, engages law at multiple levels, including various international law regimes, national legal systems and the law of local communities, and it is not yet clear which legal principles or bodies of law will have ascendancy, or who will ultimately decide between them. Over time, one thing has become clear: REDD+ is *not* a straightforward, 'win-win' approach to climate change mitigation. In seeking to accommodate the needs of indigenous peoples and forest-dwelling communities (as well as forest management, biodiversity protection, food security and so on) through its definitions and safeguards, REDD+ has become increasingly complex and unwieldy. In one sense, this is as it should be, because just and effective climate change mitigation *must* incorporate the perspectives of those who are affected by it and not merely those who have the resources to address the problem of climate change (and may well be most implicated in the causes of that problem). If climate change mitigation is to be global in its orientation, it cannot aspire to be anything but sensitive, time consuming, transparent, pluralist and inclusive. Moreover, it must not obscure or derail efforts to provide justice for indigenous peoples and forest-dwelling communities.

# BIBLIOGRAPHY

## Books and Articles

Abdullah, Saiful Arif and Adnan A Hezri, 'From Forest Landscape to Agricultural Landscape in the Developing Tropical Country of Malaysia: Pattern, Process, and Their Significance on Policy' (2008) 42 *Environmental Management* 907

Abidin, Handa, *The Protection of Indigenous Peoples and Reduction of Forest Carbon Emissions: The REDD-Plus Regime and International Law* (Brill Nijhoff, 2015)

Adcock, Fleur, 'Maori and the Bill of Rights Act: A Case of Missed Opportunities?' (2013) 11 *New Zealand Journal of Public and International Law* 183

Aiken, S Robert and Colin H Leigh, 'Seeking Redress in the Courts: Indigenous Land Rights and Judicial Decisions in Malaysia' (2011) 45 *Modern Asian Studies* 825

Allen, Matthew and Rebecca Monson, 'Land and Conflict in Papua New Guinea: The Role of Land Mediation' (2014) 10(2) *Security Challenges* 1

Altman, Jon and Seán Kerins (eds), *People on Country: Vital Landscapes, Indigenous Futures* (Federation Press, 2012)

Altman, Jon and Francis Markham, 'Burgeoning Indigenous Land Ownership: Diverse Values and Strategic Potentialities' in Sean Brennan et al (eds), *Native Title from Mabo to Akiba: A Vehicle for Change and Empowerment?* (Federation Press, 2015) 126

Amankwah, H A, J T Mugambwa and G Muroa, *Land Law in Papua New Guinea* (LBC Information Services, 2001)

Antonio, Wycliffe, Max Wagi and Lewi Kari, 'Chapter 7: Incorporated Land Groups' in Charles Yala (ed), *The Genesis of the Papua New Guinea Land Reform Program: Selected Papers from the 2005 National Land Summit* (National Research Institute, Papua New Guinea, 2010) 61

Armeni, Chiara, 'Global Experimentalist Governance, International Law and Climate Change Technologies' (2015) 64 *International and Comparative Law Quarterly* 875

Atela, Joanes O, et al, 'Implementing REDD+ at the Local Level: Assessing the Key Enablers for Credible Mitigation and Sustainable Livelihood Outcomes' (2015) 157 *Journal of Environmental Management* 238

Ayari, Inès, 'Commentary: The Dynamics between Indigenous Rights and Environmental Governance: A Preliminary Analysis and Focus on the Impact of Climate Change Governance through the Reducing Emissions from Deforestation and Forest Degradation (REDD) Programme' (2014) 10 *AlterNative: An International Journal of Indigenous Peoples* 81

Aziz, Ammar Abdul, et al, 'Using REDD+ to Balance Timber Production with Conservation Objectives in a Mangrove Forest in Malaysia' (2015) 120 *Ecological Economics* 108

Aziz, Sheema A, et al, 'Why Conservationists Should Be Concerned about Natural Resource Legislation Affecting Indigenous Peoples' Rights: Lessons from Peninsular Malaysia' (2013) 22 *Biodiversity and Conservation* 639

Babon, Andrea, 'Snapshot of REDD+ in Papua New Guinea' (CIFOR *Info Brief* No 40, August 2011) 1

Babon, Andrea and Gae Yansom Gowae, 'The Context of REDD+ in Papua New Guinea: Drivers, Agents and Institutions' (Occasional Paper 89, CIFOR, 2013) 1

Bäckstrand, Karin and Eva Lövbrand, 'Planting Trees to Mitigate Climate Change: Contested Discourses of Market Efficiency, Planetary Carbon Control and Civic Participation' (2006) 6(1) *Global Environmental Politics* 50

Baldwin, P J, et al, 'Vanuatu National Forest Inventory' in G B Wood and B J Turner (eds), *Integrating Forest Information over Space and Time: IUFRO Conference 13–17 January 1992* (ANUTECH, 1992) 295

Barelli, Mauro, 'Free, Prior and Informed Consent in the Aftermath of the UN Declaration on the Rights of Indigenous Peoples: Developments and Challenges Ahead' (2012) 16 *The International Journal of Human Rights* 1

Barrera-Hernández, Lila, et al (eds). *Sharing the Costs and Benefits of Energy and Resources Activity – Legal Change and Impact on Communities* (Oxford University Press, 2016)

Barnett, Katy, 'One Step Forward and Two Steps Back: Native Title and the Bundle of Rights Analysis' (2000) 24 *Melbourne University Law Review* 462

Bavikatte, Kabir Sanjay and Tom Bennett, 'Community Stewardship: The Foundation of Biocultural Rights' (2015) 6(1) *Journal of Human Rights and the Environment* 7

Bigot, Léa, et al, 'Case Study: Vietnam' in Joyeeta Gupta, Nicolien van der Grijp and Onno Kuik (eds), *Climate Change, Forests and REDD: Lessons for Institutional Design* (Routledge, 2013) 99

Birrell, Kathleen, Lee Godden and Maureen Tehan, 'Climate Change and REDD+: Property as a Prism for Conceiving Indigenous Peoples' Engagement' (2012) 3 *Journal of Human Rights and the Environment* 196

Blackstone, William, *Commentaries on the Law of England* (Dublin, 6th ed, 1775) (Eighteenth Century Collections Online) vol. 2

Bolin, Anna and Dominic Taku Tassa, 'Exploring Climate Justice for Forest Communities Engaging in REDD+: Experiences from Tanzania' (2012) 39 *Forum for Development Studies* 5

Boyd, Emily, et al, 'Exploring Development Futures in a Changing Climate: Frontiers for Development Policy and Practice' (2009) 27 *Development Policy Review* 659

Boyd, William, 'Climate Change, Fragmentation, and the Challenges of Global Environmental Law: Elements of a Post-Copenhagen Assemblage' (2010) 32 *University of Pennsylvania Journal of International Law* 457

'Ways of Seeing in Environmental Law: How Deforestation Became an Object of Climate Governance' (2010) 37 *Ecology Law Quarterly* 843

Brookfield, Harold (ed), *Transformation with Industrialization in Peninsular Malaysia* (Oxford University Press, 1994)

Brown, Kenneth, 'Customary Law in the Pacific: An Endangered Species?' (1999) 3 *Journal of South Pacific Law*, www.usp.ac.fj/index.php?id=13145

Bulan, Ramy, 'Native Status and the Law' in Min Aun Wu (ed), *Public Law in Contemporary Malaysia* (Longman, 1999) 248

'Native Customary Land: The Trust as a Device for Land Development in Sarawak' in Fadzilah Majid Cooke (ed), *State, Communities and Forests in Contemporary Borneo* (ANU ePress, 2006) 48, http://press.anu.edu.au/titles/asia-pacific-environment-monographs/borneo_citation

'Native Title in Malaysia: A Complementary Sui Generis Proprietary Right under the Federal Constitution Commentary' (2007) 11(1) *Australian Indigenous Law Review* 54

'Indigenous Land Rights in Malaysia' in J Ronald Engel, Laura Westra and Klaus Bosselmann (eds), *Democracy, Ecological Integrity and International Law* (Cambridge Scholars, 2010) 171

'Ripples of Mabo: Aboriginal and Native Customary Land Rights in Malaysia' in Toni Bauman and Lydia Glick (eds), *The Limits of Change: Mabo and Native Title 20 Years On* (AIATSIS, 2012) 41

Bulan, Ramy and Amy Locklear, *Legal Perspectives on Native Customary Land Rights in Sarawak* (Suruhanjaya Hak Asasi Manusia, 2008)

Cabello, Joanna and Tamra Gilbertson, 'A Colonial Mechanism to Enclose Lands: A Critical Review of Two REDD+-Focused Special Issues' (2012) 12 *Ephemera* 162

Carodenuto, Sophia and Kalame Fobissie, 'Operationalizing Free, Prior and Informed Consent (FPIC) for REDD+: Insights from the National FPIC Guidelines of Cameroon' (2015) 9 *Carbon & Climate Law Review* 156

Cashore, Benjamin and Michael W Stone, 'Can Legality Verification Rescue Global Forest Governance? Analyzing the Potential of Public and Private Policy Intersection to Ameliorate Forest Challenges in Southeast Asia' (2012) 18 *Forest Policy and Economics* 13

Chapman, Sophie, Martijn Wilder and Ilona Millar, 'Defining the Legal Elements of Benefit Sharing in the Context of REDD+' (2014) 8 *Carbon & Climate Law Review* 270

Chapman, Sophie, et al, 'A Legal Perspective of Carbon Rights and Benefit Sharing under REDD+: A Conceptual Framework and Examples from Cambodia and Kenya' (2015) 9 *Carbon & Climate Law Review* 143

Charters, Claire, 'Multi-Sourced Equivalent Norms and the Legitimacy of Indigenous Peoples' Rights under International Law' in Tomer Broude and Yuval Shany (eds), *Multi-Sourced Equivalent Norms in International Law* (Hart Publishing, 2011) 289

Chávez, Luis Enrique, 'The Declaration on the Rights of Indigenous Peoples – Breaking the *Impasse*: The Middle Ground' in Claire Charters and Rodolfo Stavenhagen (eds), *Making the Declaration Work: The United Nations Declaration on the Rights of Indigenous Peoples* (IWGIA, 2009) 96

Chee-Beng, Tan, 'Indigenous People, the State and Ethnogenesis: A Study of the Communal Associations of the "Dayak" Communities in Sarawak, Malaysia' (1997) 28 *Journal of Southeast Asian Studies* 263

Chhatre, Ashwini and Arun Agrawal, 'Trade-Offs and Synergies between Carbon Storage and Livelihood Benefits from Forest Commons' (2009) 106 *PNAS* 17667

Chomba, Susan, et al, 'Roots of Inequity: How the Implementation of REDD+ Reinforces Past Injustices' (2016) 50 *Land Use Policy* 202

Churchill, Robin R and Geir Ulfstein, 'Autonomous Institutional Arrangements in Multilateral Environmental Agreements: A Little-Noticed Phenomenon in International Law' (2000) 94(4) *American Journal of International Law* 623

Cissé, Hassane, 'Should the Political Prohibition in Charters of International Financial Institutions Be Revisited? The Case of the World Bank' in Hassane Cissé, Daniel D Bradlow and Benedict Kingsbury (eds), *The World Bank Legal Review, Volume 3: International Financial Institutions and Global Legal Governance* (The World Bank, 2012) 59

Cohen, Felix S, 'Dialogue on Private Property' (1954) 9 *Rutgers Law Review* 357

Cooney, Rosie, 'CITES and the CBD: Tensions and Synergies' (2001) 10 *Review of European Community & International Environmental Law* 259

Cooperative Research Centre for Tropical Savannas Management (Australia) and Ecological Society of Australia, *Savanna Burning* (Blackwell Science Asia, 2012)

Corbera, Esteve and Heike Schroeder, 'Governing and Implementing REDD+' (2011) 14 *Environmental Science & Policy* 89

Corrin, Jennifer, 'The Root of the Problem: Carbon Rights and Natural Resources Issues in Solomon Islands and Vanuatu' (2014) 26 *New Zealand Universities Law Review* 1

Costanza, Robert and Ralph D'Arge, 'The Value of the World's Ecosystem Services and Natural Capital' (1997) 387 *Nature* 253

Courville, Sasha, 'Understanding NGO-Based Social and Environmental Regulatory Systems: Why We Need New Models of Accountability' in Michael W

Dowdle (ed), *Public Accountability: Designs, Dilemmas and Experiences* (Cambridge University Press, 2006) 271

Cromberg, Marina, Amy E Duchelle and Isa de Oliveira Rocha, 'Local Participation in REDD+: Lessons from the Eastern Brazilian Amazon' (2014) 5 *Forests* 579

Cuthbert, R J, et al, *Analysis of National Circumstances in the Context of REDD+ and Identification of REDD+ Abatement Levers in Papua New Guinea* (report produced by the Wildlife Conservation Society for Papua New Guinea's UN-REDD National Programme, 2016)

Davenport, Deborah, et al, 'Forests and Sustainability' in Jeremy Rayner, Alexander Buck and Pia Katila (eds), *Embracing Complexity: Meeting the Challenges of International Forest Governance. A Global Assessment Report* (Global Forest Expert Panel on the International Forest Regime, IUFRO World Series vol. 28, 2010) 75

de Búrca, Gráinne, Robert O Keohane and Charles Sabel, 'New Modes of Pluralist Global Governance' (2013) 45 *New York University Journal of International Law and Politics* 723

Dehm, Julia, 'Indigenous Peoples and REDD+ Safeguards: Rights as Resistance or as Disciplinary Inclusion in the Green Economy?' (2016) 7 *Journal of Human Rights and the Environment* 170

*Reconsidering REDD+: Authority and Law in the Green Economy* (Cambridge University Press, forthcoming)

de Jong, Wil, 'Developing Swidden Agriculture and the Threat of Biodiversity Loss' (1997) 62(2–3) *Agriculture, Ecosystems & Environment* 187

den Besten, Jan Willem, Bas Arts and Patrick Verkooijen, 'The Evolution of REDD+: An Analysis of Discursive-Institutional Dynamics' (2014) 35 *Environmental Science & Policy* 40

Dickson, Barney, et al, 'REDD+ Beyond Carbon: Supporting Decisions on Safeguards and Multiple Benefits' (Policy Brief, UN-REDD Programme, 2012), http://theredddesk.org/resources/un-redd-policy-brief-redd-beyond-carbon-supporting-decisions-safeguards-and-multiple

Dimitrov, Radoslav S, et al, 'International Nonregimes: A Research Agenda' (2007) 9 *International Studies Review* 230

Dixon, Katherine, 'Working with Mixed Commons/Anticommons Property: Mobilizing Customary Land in Papua New Guinea the Melanesian Way' (2007) 31 *Harvard Environmental Law Review* 219

Doelle, Meinhard, 'The Paris Agreement: Historic Breakthrough or High Stakes Experiment?' (2016) 6 *Climate Law* 1

Doolittle, Amity A, 'Imagining New Environmental Futures: Alternative Strategies for Natural Resource Governance' in *Property and Politics in Sabah, Malaysia: Native Struggles over Land Rights* (University of Washington Press, 2005) 152

*Property and Politics in Sabah, Malaysia: Native Struggles over Land Rights* (University of Washington Press, 2005)

'Redefining Customary Law in Govuton' in *Property and Politics in Sabah, Malaysia: Native Struggles over Land Rights* (University of Washington Press, 2005) 67

Doolittle, Amity, 'Native Customary Land Rights in Sabah, Malaysia 1881–2010' in Marcus Colchester and Sophie Chao (eds), *Divers Paths to Justice: Legal Pluralism and the Rights of Indigenous Peoples in Southeast Asia* (Forest Peoples Programme and Asia Indigenous Peoples Pact, 2011) 81

Dore, Jeremy, et al, 'Carbon Projects and Indigenous Land in Northern Australia' (2014) 36 *The Rangeland Journal* 389

Downing, Thomas E and Martin L Parry, 'Introduction: Climate Change and World Food Security' (1994) 19 *Food Policy* 99

Dudley, Nigel, Jean-Paul Jeanrenaud and Francis Sullivan, *Bad Harvest? The Timber Trade and the Degradation of the World's Forests* (Routledge, 2014)

Dupuy, Pierre-Marie, 'Formation of Customary International Law and General Principles' in Daniel Bodansky, Jutta Brunnée and Ellen Hey (eds), *The Oxford Handbook of International Environmental Law* (Oxford University Press, 2008) 449

Dunoff, Jeffrey L, 'A New Approach to Regime Interaction' in Margaret A Young (ed), *Regime Interaction in International Law: Facing Fragmentation* (Cambridge University Press, 2012) 136

Eckert, Sandra and Richard McKellar, 'Securing Rights to Carbon Sequestration: The Western Australian Experience' (2008) 8(2) *Sustainable Development Law & Policy* 30

Eikermann, Anja, *Forests in International Law: Is There Really a Need for an International Forest Convention?* (Springer International Publishing Switzerland, 2015)

Endicott, Kirk, 'Introduction' in Kirk Endicott (ed), *Malaysia's Original People: Past, Present and Future of the Orang Asli* (NUS Press, 2016) 1

Evans, Kristen, Laura Murphy and Wil de Jong, 'Global Versus Local Narratives of REDD: A Case Study from Peru's Amazon' (2014) 35 *Environmental Science & Policy* 98

Farran, Sue, 'Navigating Changing Land Use in Vanuatu' (2011) 34 *Pacific Studies* 250

Filer, Colin, 'A Serious Case of Conditionality: The World Bank Gets Stuck in the Forests of PNG' (2004) 65 *Development Bulletin* 95

'Local Custom and the Art of Land Group Boundary Maintenance in Papua New Guinea' in James Weiner and Katie Glaskin (eds), *Customary Land Tenure and Registration in Australia and Papua New Guinea: Anthropological Perspectives* (ANU ePress, 2007) 135

'Why Green Grabs Don't Work in Papua New Guinea' (2012) 39 *The Journal of Peasant Studies* 599

'The Double Movement of Immovable Property Rights in Papua New Guinea' (2014) 49 *Journal of Pacific History* 76

Filer, Colin, Navroz K Dubash and Kalit Kilyali, *The Thin Green Line: World Bank Leverage and Forest Policy Reform in Papua New Guinea* (The National Research Institute & Australian National University, 2000)

Filer, Colin, et al, 'Deforestation and Degradation in Papua New Guinea' (2009) 66 *Annals of Forest Science* 813

Filer, Colin and Nikhil Sekhran, *Loggers, Donors and Resource Owners* (International Institute for Environment and Development, 1998)

Fingleton, Jim, *Legal Papers Online No 73: Pacific Land Tenures: New Ideas for Reform* (FAO, 2008)

Fischer-Lescano, Andreas and Gunther Teubner, 'Regime-Collisions: The Vain Search for Legal Unity in the Fragmentation of Global Law' (2004) 25 *Michigan Journal of International Law* 999

Fisher, Brendan, David P Edwards and David S Wilcove, 'Logging and Conservation: Economic Impacts of the Stocking Rates and Prices of Commercial Timber Species' (2014) 38 *Forest Policy and Economics* 65

Fisher, Robert and Rosemary Lyster, 'Land and Resource Tenure: The Rights of Indigenous People and Forest Dwellers' in Rosemary Lyster, Catherine MacKenzie and Constance McDermott (eds), *Law, Tropical Forests and Carbon: The Case of REDD+* (Cambridge University Press, 2013)

Fitzpatrick, Peter, *Modernism and the Grounds of Law* (Cambridge University Press, 2001)

Foucault, Michel, *Security, Territory, Population: Lectures at the Collège de France, 1977–78* (Palgrave Macmillan, 2007)

*Power/Knowledge: Selected Interviews and Other Writings 1972–77*, Colin Gordon (ed) (Pantheon Books, 1980)

Foukona, Joseph D, 'Management of Customary Land as a Form of Community Property in the Solomon Islands, Vanuatu and Fiji' in Lee Godden and Maureen Tehan (eds), *Comparative Perspectives on Communal Lands and Individual Ownership: Sustainable Futures* (Routledge, 2010) 263

Fox, Jefferson, Jean-Christophe Castella and Alan D Ziegler, 'Swidden, Rubber and Carbon: Can REDD+ Work for People and the Environment in Montane Mainland Southeast Asia?' (2014) 29 *Global Environmental Change* 318

Fujisaki, Taiji, et al, 'Does REDD+ Ensure Sectoral Coordination and Stakeholder Participation? A Comparative Analysis of REDD+ National Governance Structures in Countries of the Asia-Pacific Region' (2016) 7 *Forests* 195

Fung, Archon and Erik Olin Wright 'Deepening Democracy: Innovations in Empowered Participatory Governance' (2001) 29 *Politics & Society* 5

Gilberthorp, Emma, 'Fasu Solidarity: A Case Study of Kin Networks, Land Tenure, and Oil Extraction in Kutubu, Papua New Guinea' (2007) 109 *American Anthropologist* 101

Godden, Lee, 'Death, Desire, Modernity and Redemption: Climate Change and Public International Environmental Law' (2009) 10 *Melbourne Journal of International Law* 543

'"Globalized Localisms": Three Phases of International Environmental Governance for Biodiversity Protection' in Shawkat Alam, Natalie Klein and Juliette Overland (eds), *Globalisation and the Quest for Social and Environmental Justice: The Relevance of International Law in an Evolving World Order* (Routledge, 2011)

Godden, Lee, et al, 'Reducing Emissions from Deforestation and Forest Degradation in Developing Countries (REDD): Implementation Issues' (2010) 36 *Monash University Law Review* 139

Godden, Lee and Stuart Cowell, 'Conservation Planning and Indigenous Governance in Australia's Indigenous Protected Areas' (2016) 24 *Restoration Ecology* 692

Godden, Lee and Lily O'Neill, 'Benefit Agreements with Indigenous Communities: The Native Title Act in Australia' in Lila Barrera-Hernández et al (eds), *Sharing the Costs and Benefits of Energy and Resource Activity: Legal Change and Impact on Communities* (Oxford University Press, 2016) 135

Godden, Lee and Maureen Tehan, 'Translating Native Title to Individual "Title" in Australia: Are Real Property Forms and Indigenous Interests Reconcilable?' in Elizabeth Cooke (ed), *Modern Studies in Property Law: Volume 4* (Hart Publishing, 2007)

(eds), *Comparative Perspectives on Communal Lands and Individual Ownership: Sustainable Futures* (Routledge, 2010)

Golub, Alex, 'Ironies of Organization: Landowners, Land Registration, and Papua New Guinea's Mining and Petroleum Industry' (2007) 66 *Human Organization* 38

Goroh, Eleanor, 'Malaysia' in Kathrin Wessendorf (ed), *Yearbook – The Indigenous World* (International Work Group for Indigenous Affairs, 2011) 280, www.iwgia.org/regions/asia/malaysia

Gover, Kirsty, *Tribal Constitutionalism: States, Tribes, and the Governance of Membership* (Oxford University Press, 2010)

'REDD+, Tenure and Indigenous Property: The Promise and Peril of a "Human Rights-Based Approach"' in Christina Voigt (ed), *Research Handbook on REDD+ and International Law* (Edward Elgar, 2016) 249

Graham, Victoria, et al, 'A Comparative Assessment of the Financial Costs and Carbon Benefits of REDD+ Strategies in Southeast Asia' (2016) 11 *Environmental Research Letters* 4022

Gray, Kevin, 'Property in Thin Air' (1991) 50 *Cambridge Law Journal* 252

Grear, Anna, 'Towards "Climate Justice"? A Critical Reflection on Legal Subjectivity and Climate Injustice: Warning Signals, Patterned Hierarchies, Directions for Future Law and Policy' (2014) 5(0) *Journal of Human Rights and the Environment* 103

Gupta, Joyeeta, et al, 'The Future of Forests' in Joyeeta Gupta, Nicolien van der Grijp and Onno Kuik (eds), *Climate Change, Forests and REDD: Lessons for Institutional Design* (Routledge, 2013) 229

Gupta, Joyeeta, Nicolien van der Grijp and Onno Kuik (eds), *Climate Change, Forests and REDD: Lessons for Institutional Design* (Routledge, 2013)

Hall, Anthony, 'Forests, Conservation and Climate Change in Latin America' in *Forests and Climate Change: The Social Dimensions of REDD in Latin America* (Edward Elgar, 2012) 10

Hepburn, Samantha, 'Carbon Rights as New Property: The Benefits of Statutory Verification' (2009) 31 *Sydney Law Review* 239

Hite, Kristen, 'Adjudicating Disputes across Scales: Global Administrative Law Considerations for REDD+' in Christina Voigt (ed), *Research Handbook on REDD+ and International Law* (Edward Elgar, 2016) 408

Hoang, Minh Ha, et al, 'Benefit Distribution Across Scales to Reduce Emissions from Deforestation and Forest Degradation (REDD+) in Vietnam' (2013) 31 *Land Use Policy* 48

Howell, Signe, 'Continuity through Change: Three Decades of Engaging with Chewong: Some Issues Raised by Multitemporal Fieldwork' in Kirk Endicott (ed), *Malaysia's Original People: Past, Present and Future of the Orang Asli* (NUS Press, 2016) 57

Howes, Stephen, 'Cheap But Not Easy: The Reduction of Greenhouse Gas Emissions from Deforestation and Forest Degradation in Papua New Guinea' (2009) 24 *Pacific Economic Bulletin* 130

Humphreys, David, *Logjam: Deforestation and the Crisis of Global Governance* (Earthscan, 2006)

Humphreys, Stephen, 'Editorial: In the Shadow of Paris: Theories of Justice and Principles of Harm' (2016) 7 *Journal of Human Rights and the Environment* 1

Ingram, John, Polly Ericksen and Diana Liverman (eds), *Food Security and Global Environmental Change* (Earthscan, 2010)

Jodoin, Sébastian, *Forest Preservation in a Changing Climate: REDD+ and Community Rights in Indonesia and Tanzania* (Cambridge University Press, 2017)

Jomo, K S and Wee Chong Hui, 'The Political Economy of Malaysian Federalism: Economic Development, Public Policy and Conflict Containment' (2003) 15 *Journal of International Development* 441

Jorgensen, Dan, 'Clan-Finding, Clan-Making and the Politics of Identity in a Papua New Guinea Mining Project' in James F Weiner and Katie Glaskin (eds), *Customary Land Tenure and Registration in Australia and Papua New Guinea: Anthropological Perspectives* (ANU ePress, 2007) 57

Jowitt, Anita, 'Indigenous Land Grievances, Customary Land Disputes and Restorative Justice' (2004) 8(2) *Journal of South Pacific Law* 23

Kalinoe, Lawrence, 'Customary Land Dispute Settlement' in Charles Yala (ed), *The Genesis of the Papua New Guinea Land Reform Program: Selected Papers from the 2005 National Land Summit* (National Research Institute, Papua New Guinea, 2010) 49

Kalinoe, Lawrence and Josepha Kiris, 'Socio-Economic Changes and Their Implications for Customary Land Tenure' in Charles Yala (ed), *The Genesis of the Papua New Guinea Land Reform Program: Selected Papers from the 2005 National Land Summit* (National Research Institute, Papua New Guinea, 2010) 23

Kariyawasam, Kanchana, 'Protecting Biodiversity, Traditional Knowledge and Intellectual Property in the Pacific: Issues and Challenges' (2008) 16 *Asia Pacific Law Review* 73

Karsenty, Alain, Aurélie Vogel and Frédéric Castell, '"Carbon Rights", REDD+ and Payments for Environmental Services' (2014) 35 *Environmental Science & Policy* 20

Kawi, John S and Wilfred W Golman, 'Project Financing in Papua New Guinea' in Eric L Kwa (ed), *Natural Resources Law of Papua New Guinea* (Law Book Company, 2001) 152

Keenan, Rodney J, et al, 'Science and the Governance of Australia's Climate Regime' (2012) 2 *Nature Climate Change* 477

Keohane, Robert O and David G Victor, 'The Regime Complex for Climate Change' (2011) 9 *Perspectives on Politics* 7

Kingsbury, Benedict, 'Operational Policies of International Institutions as Part of the Law-Making Process: The World Bank and Indigenous Peoples' in Guy S Goodwin-Gill and Stefan Talmon (eds), *The Reality of International Law: Essays in Honour of Ian Brownlie* (Oxford University Press, 1999) 323

Kinley, David, 'Human Rights and the World Bank: Practice, Politics, and Law' in Ana Palacio, Caroline Mary Sage and Michael Woolcock (eds), *The World Bank Legal Review: Law, Equity, and Development, Volume 2* (The World Bank and Martinus Nijhoff, 2006) 353

Kleinman, P J A, D Pimentel and R B Bryant, 'The Ecological Sustainability of Slash-and-Burn Agriculture' (1995) 52 *Agriculture, Ecosystems & Environment* 235

Koskenniemi, Martti, 'Hegemonic Regimes' in Margaret A Young (ed), *Regime Interaction in International Law: Facing Fragmentation* (Cambridge University Press, 2012) 305

Krasner, Stephen, 'Structural Causes and Regime Consequences: Regimes as Intervening Variables' in Stephen Krasner (ed), *International Regimes* (Cornell University Press, 1983) 1

Langton, Marcia, et al (eds), *Honour Among Nations?: Treaties and Agreements with Indigenous People* (Melbourne University Press, 2004)

*Settling with Indigenous People: Modern Treaty and Agreement-Making* (Federation Press, 2006)

Langton, Marcia and Judy Longbottom (eds), *Community Futures, Legal Architecture: Foundations for Indigenous Peoples in the Global Mining Boom* (Routledge, 2012)

Langton, Marcia and Odette Mazel, 'The Resource Curse Compared: Australian Aboriginal Participation in the Resource Extraction Industry and Distribution of Impacts' in Marcia Langton and Judy Longbottom (eds), *Community Futures, Legal Architecture: Foundations for Indigenous Peoples in the Global Mining Boom* (Routledge, 2012)

Larson, Anne M, 'Forest Tenure Reform in the Age of Climate Change: Lessons for REDD+' (2011) 21 *Global Environmental Change* 540

La Viña, Antonio G M, Alaya De Leon and Reginald Rex Barrer, 'History and Future of REDD+ in the UNFCCC: Issues and Challenges' in Christina Voigt (ed), *Research Handbook on REDD+ and International Law* (Edward Elgar, 2016) 11

Lawlor, Kathleen, et al, 'Community Participation and Benefits in REDD+: A Review of Initial Outcomes and Lessons' (2013) 4 *Forests* 296

Lawson, Sam, 'Illegal Logging in Papua New Guinea' (Energy, Environment and Resources EER PP 2014/04, Chatham House, April 2014)

Lea, David, 'A Critical Review of the Recent Amendments to the Customary Land Tenure System in Papua New Guinea' (2013) 35(3) *Social Development Issues* 63

Leach, Melissa and Ian Scoones, 'Political Ecologies of Carbon in Africa' in Melissa Leach and Ian Scoones (eds), *Carbon Conflicts and Forest Landscapes in Africa* (Routledge, 2015) 1

Leggett, Matthew and Heather Lovell, 'Community Perceptions of REDD+: A Case Study from Papua New Guinea' (2012) 12 *Climate Policy* 115

Lewis, Robin A and Stacey R Davis, 'Forest Certification, Institutional Capacity, and Learning: An Analysis of the Impacts of the Malaysian Timber Certification Scheme' (2015) 52 *Forest Policy and Economics* 18

Lyster, Rosemary, 'Reducing Emissions from Deforestation and Degradation: The Road to Copenhagen' in Rosemary Lyster (ed), *In the Wilds of Climate Law* (Australian Academic Press, 2010) 97

'REDD+, Transparency, Participation and Resource Rights: The Role of Law' (2011) 14 *Environmental Science & Policy* 118

'International Legal Frameworks for REDD+: Ensuring Legitimacy' in Rosemary Lyster, Catherine MacKenzie and Constance McDermott (eds), *Law, Tropical Forests and Carbon: The Case of REDD+* (Cambridge University Press, 2013) 3

Lyster, Rosemary, Catherine MacKenzie and Constance McDermott (eds), *Law, Tropical Forests and Carbon: The Case of REDD+* (Cambridge University Press, 2013)

Macintyre, Martha and Simon Foale, 'Land and Marine Tenure, Ownership, and New Forms of Entitlement on Lihir: Changing Notions of Property in the Context of a Goldmining Project' (2007) 66(1) *Human Organization* 49

MacKay, Fergus, 'The Draft World Bank Operational Policy 4.10 on Indigenous Peoples: Progress or More of the Same?' (2005) 22 *Arizona Journal of International & Comparative Law* 65

Majid Cooke, Fadzilah, 'Recent Development and Conservation Interventions in Borneo' in Fadzilah Majid Cooke (ed), *State, Communities and Forests in Contemporary Borneo* (Australian National University Press, 2006) 4

'In the Name of Poverty Alleviation: Experiments with Oil Palm Smallholders and Customary Land in Sabah, Malaysia' (2012) 53 *Asia Pacific Viewpoint* 240

'Constructing Rights: Indigenous Peoples at the Public Hearings of the National Inquiry into Customary Rights to Land in Sabah, Malaysia' (2013) 28 *SOJOURN: Journal of Social Issues in Southeast Asia* 512

Majid Cooke, Fadzilah, Sumei Toh and Justine Vaz, 'Community-Investor Business Models: Lessons from the Oil Palm Sector in East Malaysia' (International Institute for Environment and Development, 2011), http://pubs.iied.org/pdfs/12570IIED.pdf

Mathy, Sandrine and Odile Blanchard, 'Proposal for a Poverty-Adaptation-Mitigation Window within the Green Climate Fund' (2016) 16 *Climate Policy* 752

McDermott, Constance, 'REDD+ and Multi-Level Governance: Governing for What and For Whom?' in Rosemary Lyster, Catherine MacKenzie and Constance McDermott (eds), *Law, Tropical Forests and Carbon: The Case of REDD+* (Cambridge University Press, 2013) 84

McDermott, Constance, et al, 'Mapping the Core Actors and Issues Defining International Forest Governance' in Jeremy Rayner, Alexander Buck and Pia Katila (eds), *Embracing Complexity: Meeting the Challenges of International Forest Governance. A Global Assessment Report* (Global Forest Expert Panel on the International Forest Regime, IUFRO World Series vol. 28, 2010) 19

McDermott, Constance L, et al, 'Operationalizing Social Safeguards in REDD+: Actors, Interests and Ideas' (2012) 21 *Environmental Science and Policy* 63

McGregor, Andrew, et al, 'Practical Critique: Bridging the Gap between Critical and Practice-Oriented REDD+ Research Communities' (2014) 55 *Asia Pacific Viewpoint* 277

McInerney-Lankford, Siobhán, 'Human Rights and Development: Regime Interaction and the Fragmentation of International Law' in Hassane Cissé et al (eds), *The World Bank Legal Review, Volume 4 Legal Innovation and Empowerment for Development* (The World Bank and Martinus Nijhoff, 2013) 123

McMorrow, Julia and Mustapa Abdul Talip, 'Decline of Forest Area in Sabah, Malaysia: Relationship to State Policies, Land Code and Land Capability' (2001) 11 *Global Environmental Change* 217

Melick, David, 'Credibility of REDD and Experiences from Papua New Guinea' (2010) 24 *Conservation Biology* 359

Mertz, Ole, et al, 'Swidden Change in Southeast Asia: Understanding Causes and Consequences' (2009) 37 *Human Ecology* 259

'The Last Swiddens of Sarawak, Malaysia' (2013) 41 *Human Ecology* 109

Milne, Robert Stephen, *Malaysia – New States in a New Nation: Political Development of Sarawak and Sabah in Malaysia* (Frank Cass, 1974)

Mohd Shahwahid, H O, 'Forest Certification in Malaysia' in Benjamin Cashore et al (eds), *Confronting Sustainability: Forest Certification in Developing and Transitioning Countries* (Yale School of Forestry and Environmental Studies, 2006) 69

Morgera, Elisa, 'The EU and Environmental Multilateralism: The Case of Access and Benefit-Sharing and the Need for a Good-Faith Test' (2014) 16 *Cambridge Yearbook of European Legal Studies* 109

'The Need for an International Concept of Fair and Equitable Benefit Sharing' (2016) 27 *European Journal of International Law* 353

Morgera, Elisa, Matthias Buck and Elsa Tsioumani (eds), *The 2010 Nagoya Protocol on Access and Benefit-Sharing in Perspective: Implications for International Law and Implementation Challenges* (Brill Nijhoff, 2012)

Morgera, Elisa, Elsa Tsioumani and Matthias Buck, *Unraveling the Nagoya Protocol: A Commentary on the Nagoya Protocol on Access and Benefit-Sharing to the Convention on Biological Diversity* (Brill Nijhoff, 2014)

Mousseau, Frédéric and Peiley Lau, *The Great Timber Heist: The Logging Industry in Papua New Guinea* (The Oakland Institute, February 2016), www.oaklandinstitute.org/great-timber-heist-logging-industry-papua-new-guinea

Mustalahti, Irmeli and O Sarobidy Rakotonarivo, 'REDD+ and Empowered Deliberative Democracy: Learning from Tanzania' (2014) 59 *World Development* 199

Nagendra, Harini and Elinor Ostrom, 'Polycentric Governance of Multifunctional Forested Landscapes' (2012) 6 *International Journal of the Commons* 104

Newing, H S, 'Traditional Knowledge in International Forest Policy: Contested Meanings and Divergent Discourses' (2009) 6 *Journal of Integrative Environmental Sciences* 175

Niall, Stephanie, et al, 'Climate Change and REDD+: Integrating Customary Fire-Management Schemes in East Malaysia and Northern Australia' (2013) 28 *SOJOURN: Journal of Social Issues in Southeast Asia* 538

Nicholas, Colin, Jenita Engi and Teh Yen Ping, *The Orang Asli and the UNDRIP: From Rhetoric to Recognition* (Center for Orang Asli Concerns, 2010)

Nollkaemper, André and Dov Jacobs, 'Shared Responsibility in International Law: A Conceptual Framework' (2013) 34 *Michigan Journal of International Law* 359

Oberborbeck Andersen, Astrid, et al, 'Negotiating Development Narratives within Large-Scale Oil Palm Projects on Village Lands in Sarawak, Malaysia' (2016) 182 *The Geographical Journal* 364

O'Connor, Pamela, et al, 'From Rights to Responsibilities: Reconceptualising Carbon Sequestration Rights in Australia' (2013) 30 *Environmental and Planning Law Journal* 403

O'Donnell, Michael, 'Native Title – A Right to Burn and Fire the Land? Savanna Burning and the Carbon Farming Initiative in Northern Australia' (2013) 30 *Environmental and Planning Law Journal* 553

Oels, Angela, 'Rendering Climate Change Governable: From Biopower to Advanced Liberal Government?' (2005) 7 *Journal of Environmental Policy & Planning* 185

O'Faircheallaigh, Ciaran, 'Native Title and Mining Negotiations: A Seat at the Table, But No Guarantee of Success' (2007) 6(26) *Indigenous Law Bulletin* 18

Okereke, Chukwumerije, 'Climate Justice and the International Regime' (2010) 1 *WIREs Climate Change* 462

Ostrom, Elinor, 'Polycentric Systems for Coping with Collective Action and Global Environmental Change' (2010) 20 *Global Environmental Change* 550

Overdevest, Christine and Jonathan Zeitlin, 'Assembling an Experimentalist Regime: Transnational Governance Interactions in the Forest Sector' (2014) 8 *Regulation & Governance* 22

Parrotta, John A, Christoph Wildburger and Stephanie Mansourian (eds), *Understanding Relationships between Biodiversity, Carbon, Forests and People: The Key to Achieving REDD+ Objectives – A Global Assessment Report* (International Union of Forest Research Organizations (IUFRO), IUFRO World Series vol. 31, 2012)

Peel, Jacqueline, Lee Godden and Rodney J Keenan, 'Climate Change Law in an Era of Multi-Level Governance' (2012) 1 *Transnational Environmental Law* 245

Perez, Oren, *Ecological Sensitivity and Global Legal Pluralism: Rethinking the Trade and Environment Conflict* (Hart Publishing, 2004)

Peskett, Leo, 'REDD+ and Development' in Rosemary Lyster, Catherine MacKenzie and Constance McDermott (eds), *Law, Tropical Forests and Carbon* (Cambridge University Press, 2013) 230

Petsonk, Annie, 'Legal Obligations and Institutions of Developing Countries: Rethinking Approaches to Forest Governance' in Hassane Cissé, Daniel D Bradlow and Benedict Kingsbury (eds), *The World Bank Legal Review:*

*International Financial Institutions and Global Legal Governance, Volume 3* (The World Bank; Martinus Nijhoff, 2012) 293

Phua, Mui-How, et al, 'Estimating Logged-Over Lowland Rainforest Aboveground Biomass in Sabah, Malaysia Using Airborne LiDAR Data' (2016) 27 *Terrestrial, Atmospheric & Oceanic Sciences* 481

Pistorius, Till, 'From RED to REDD+: The Evolution of a Forest-Based Mitigation Approach for Developing Countries' (2012) 4 *Current Opinion in Environmental Sustainability* 638

Power, Tony, 'Reconciling Customary Ownership and Development: Incorporated Land Groups in Papua New Guinea' in *Making Land Work* (Australian Agency for International Development (AusAID), 2008) vol. 2 (Case Studies on Customary Land and Development in the Pacific) 3, http://dfat.gov.au/about-us/publications/Pages/making-land-work.aspx

Rae, Jessica, Mahala Gunther and Lee Godden, 'Governing Tropical Forests: REDD+, Certification and Local Forest Outcomes in Malaysia' (2011) 7(2) *Macquarie Journal of International and Comparative Environmental Law* 40

Rajamani, Lavanya, 'The Making and Unmaking of the Copenhagen Accord' (2010) 59 *International and Comparative Law Quarterly* 824

Rayner, Jeremy, 'Preface' in Jeremy Rayner, Alexander Buck and Pia Katila (eds), *Embracing Complexity: Meeting the Challenges of International Forest Governance. A Global Assessment Report* (Global Forest Expert Panel on the International Forest Regime, IUFRO World Series vol. 28, 2010)

Rayner, Jeremy, et al, 'Introduction' in Jeremy Rayner, Alexander Buck and Pia Katila (eds), *Embracing Complexity: Meeting the Challenges of International Forest Governance. A Global Assessment Report* (Global Forest Expert Panel on the International Forest Regime, IUFRO World Series vol. 28, 2010)

Redgwell, Catherine, 'Shared International Responsibility for Transboundary Harm Arising from Energy Activities' in Lila Barrera-Hernández et al (eds), *Sharing the Costs and Benefits of Energy and Resource Activity* (Oxford University Press, 2016) 59

Reeves, John, *Building on Land Rights for the Next Generation: Report of the Review of the Aboriginal Land Rights (Northern Territory) Act 1976* (Commonwealth of Australia, 1998)

Regenvanu, Ralph, 'Issues with Land Reform in Vanuatu' (2008) 12(1) *Journal of South Pacific Law*, www.paclii.org/journals/fJSPL/vol12no1/pdf/regenvanu.pdf

Ritter, David, *The Native Title Market* (UWA Publishing, 2009)

Rittich, Kerry, 'The Future of Law and Development: Second-Generation Reforms and the Incorporation of the Social' in David M Trubek and Alvaro Santos (eds), *The New Law and Economic Development: A Critical Appraisal* (Cambridge University Press, 2006) 203

Robinson, Catherine J, et al, 'Indigenous Benefits and Carbon Offset Schemes: An Australian Case Study' (2016) 56 *Environmental Science and Policy* 129

Robinson, Cathy J, et al, 'Australia's Indigenous Carbon Economy: A National Snapshot' (2014) 52 *Geographical Research* 123

Robinson, Daniel F, *Biodiversity, Access and Benefit-Sharing: Global Case Studies* (Taylor and Francis, 2014)

Robiou du Pont, Yann, et al, 'National Contributions for Decarbonizing the World Economy in Line with the G7 Agreement' (2016) 11 *Environmental Research Letters* 054005

Rodman, Margaret, 'Breathing Spaces: Customary Land Tenure in Vanuatu' in R Gerard Ward and Elizabeth Kingdon (eds), *Land, Custom and Practice in the South Pacific* (Cambridge University Press, 1995) 65

Ruhl, J B, 'The Twentieth Annual Lloyd K Garrison Lecture: In Defense of Ecosystem Services' (2015) 32 *Pace Environmental Law Review* 306

Russell-Smith, Jeremy, Peter Whitehead and Peter Cooke (eds), *Culture, Ecology and Economy of Fire Management in North Australian Savannas: Rekindling the Wurrk Tradition* (CSIRO, 2009)

Sabel, Charles F and David G Victor, 'Governing Global Problems under Uncertainty: Making Bottom-Up Climate Policy Work' (2015) (May) *Climatic Change*, doi:10.1007/s10584-015-1507-y

Sajeva, Giulia, 'Rights with Limits: Biocultural Rights – Between Self-Determination and Conservation of the Environment' (2015) 6 *Journal of Human Rights and the Environment* 30

Sands, Philippe and Jacqueline Peel, *Principles of International Environmental Law* (Cambridge University Press, 3rd edn, 2012)

Santosa, Mas Achmad, Josi Khatarina and Aldilla Stephanie Suwana, 'Indonesia REDD+: Beyond Carbon, More Than Just Forest' in Paul Martin et al (eds), *The Search for Environmental Justice* (Edward Elgar, 2015) 158

Sarfaty, Galit A, 'Why Culture Matters in International Institutions: The Marginality of Human Rights at the World Bank' (2009) 103 *American Journal of International Law* 647

Savaresi, Annalisa, 'The Human Rights Dimension of REDD' (2012) 21 *Review of European Community & International Environmental Law* 102

'REDD+ and Human Rights: Addressing Synergies between International Regimes' (2013) 18(3) *Ecology and Society* 149

'The Role of REDD in the Harmonisation of Overlapping International Obligations' in Erkki J Hollo, Kati Kulovesi and Michael Mehling (eds), *Climate Change and the Law*, Ius Gentium: Comparative Perspectives on Law and Justice, vol. 21 (Springer, 2013) 391

'The Legal Status and Role of Safeguards' in Christina Voigt (ed), *Research Handbook on REDD+ and International Law* (Edward Elgar, 2016) 126

Schlager, Edella and Elinor Ostrom, 'Property-Rights Regimes and Natural Resources: A Conceptual Analysis' (1992) 68 *Land Economics* 249

Schroeder, Heike, 'Agency in International Climate Negotiations: The Case of Indigenous Peoples and Avoided Deforestation' (2010) 10(4) *International Environmental Agreements: Politics, Law & Economics* 317

Scott, Joanne, 'The Multi-Level Governance of Climate Change' in Paul Craig and Gráinne de Búrca (eds), *The Evolution of EU Law* (Oxford University Press, 2nd edn, 2011) 805

Scott, Joanne and David M Trubek, 'Mind the Gap: Law and New Approaches to Governance in the European Union' (2002) 8 *European Law Journal* 1

Seymour, Frances J and Navroz K Dubash, *The Right Conditions: The World Bank, Structural Adjustment and Forest Policy Reform* (World Resources Institute, 2000)

Shearman, Phil L, et al, *The State of the Forests in Papua New Guinea: Mapping the Extent and Condition of Forest Cover and Measuring the Drivers of Forest Change in the Period 1972–2002* (University of Papua New Guinea, 2008)

Shelton, Dinah, 'Principle 22: Indigenous People and Sustainable Development' in Jorge E Viñuales (ed), *The Rio Declaration on Environment and Development: A Commentary* (Oxford University Press, 2015) 541

Shrestha, Shanti, Bhaskar Singh Karky and Seema Karki, 'Case Study Report: REDD+ Pilot Project in Community Forests in Three Watersheds of Nepal' (2014) 5 *Forests* 2425

Stern, Nicholas, *The Economics of Climate Change: The Stern Review* (Cambridge University Press, 2007)

Stewart, Richard B, Bryce Rudyk and Kiri Mattes, 'Governing a Fragmented Climate Finance Regime' in Hassane Cissé, Daniel D Bradlow and Benedict Kingsbury (eds), *The World Bank Legal Review: International Financial Institutions and Global Legal Governance, Volume 3* (The World Bank and Martinus Nijhoff, 2012) 363

Stocks, Sarah N, 'Review of Forest and REDD+ Related Policy and Legislation in Papua New Guinea' (United States Agency for International Development (USAID) and Lowering Emissions in Asia's Forests (LEAF) Program, 2014)

Streck, Charlotte, 'Financing Mechanisms for Climate Change Mitigation' in Thomas Hale and David Held (eds), *The Handbook of Transnational Governance: Institutions and Innovations* (Polity Press, 2011) 377

Streck, Charlotte and Michaela Schwedeler, 'Addressing Drivers of Deforestation and Forest Degradation through International Law' in Christina Voigt (ed), *Research Handbook on REDD+ and International Law* (Edward Elgar, 2016) 213

Subramaniam, Yogeswaran, 'Orang Asli, Land Rights and the Court Process: A "Native Title" Lawyer's Perspective' in Kirk Endicott (ed), *Malaysia's Original People: Past, Present and Future of the Orang Asli* (NUS Press, 2016) 423

Takacs, David, 'Environmental Democracy and Forest Carbon (REDD Plus)' (2014) 44 *Environmental Law* 71

Tararia, Almah and Lisa Ogle, 'Incorporated Land Groups and the Registration of Customary Lands: Recent Developments in PNG' in Tim Anderson and Gary Lee (eds), *In Defense of Melanesian Customary Land* (Aidwatch, 2010) 21

Tehan, Maureen, 'A Hope Disillusioned, An Opportunity Lost? Reflections on Common Law Native Title and Ten Years of the *Native Title Act*' (2003) 27 *Melbourne University Law Review* 523

'Customary Land Tenure, Communal Titles and Sustainability: The Allure of Individual Title and Property Rights in Australia' in Lee Godden and Maureen Tehan (eds), *Comparative Perspectives on Communal Lands and Individual Ownership: Sustainable Futures* (Routledge, 2010) 354

Tehan, Maureen and Lee Godden, 'Legal Forms and Their Implications for Long-Term Relationships and Economic, Cultural and Social Empowerment: Sturcturing Agreements in Australia' in Marcia Langton and Judy Longbottom (eds), *Community Futures, Legal Architecture: Foundations for Indigenous Peoples in the Global Mining Boom* (Routledge, 2012) 111

Teubner, Gunther and Peter Korth, 'Two Kinds of Legal Pluralism: Collision of Transnational Regimes in the Double Fragmentation of World Society' in Margaret A Young (ed), *Regime Interaction in International Law: Facing Fragmentation* (Cambridge University Press, 2012) 23

Terrill, Leon, 'The Days of the Failed Collective: Communal Ownership, Individual Ownership and Township Leasing in Aboriginal Communities in the Northern Territory' (2009) 32 *University of New South Wales Law Journal* 814

*Beyond Communal and Individual Ownership: Indigenous Land Reform in Australia* (Routledge, 2016)

Teubner, Gunther and Peter Korth, 'Two Kinds of Legal Pluralism: Collision of Transnational Regimes in the Double Fragmentation of World Society' in Margaret Young (ed), *Regime Interaction in International Law: Facing Fragmentation* (Cambridge University Press, 2012) 23

Thompson, Mary C, Manali Baruah and Edward R Carr, 'Seeing REDD+ as a Project of Environmental Governance' (2011) 14 *Environmental Science & Policy* 100

van Asselt, Harro, 'Managing the Fragmentation of International Environmental Law: Forests at the Intersection of the Climate and Biodiversity Regimes' (2012) 44 *New York University Journal of International Law and Politics* 1205

van Asselt, Harro and Constance L McDermott, 'The Institutional Complex for REDD+: A "Benevolent Jigsaw"?' in Christina Voigt (ed), *Research Handbook on REDD+ and International Law* (Edward Elgar, 2016) 63

van Trease, Howard, 'The History of Land Alienation' in Peter Larmour (ed), *Land Tenure in Vanuatu* (Institute of Pacific Studies, 1984)

Vijay, Varsha, et al, 'The Impacts of Oil Palm on Recent Deforestation and Biodiversity Loss' (2016) 11(7) *PLoS ONE* e0159668, doi:10.1371/journal.pone.0159668

Visseren-Hamakers, Ingrid J, et al, 'Trade-Offs, Co-Benefits and Safeguards: Current Debates on the Breadth of REDD+' (2012) 4 *Current Opinion in Environmental Sustainability* 646

Voigt, Christina, 'Introduction: The Kaleidoscopic World of REDD+' in Christina Voigt (ed), *Research Handbook on REDD+ and International Law* (Edward Elgar, 2016) 1

(ed), *Research Handbook on REDD+ and International Law* (Edward Elgar, 2016)

Voigt, Christina and Felipe Ferreira, 'Differentiation in the Paris Agreement' (2016) 6 *Climate Law* 58

'The Warsaw Framework for REDD+: Implications for National Implementation and Results-Based Finance' in Christine Voigt (ed), *Research Handbook on REDD+ and International Law* (Edward Elgar, 2016) 30

Wagner, John, 'Conservation as Development in Papua New Guinea: The View from Blue Mountain' (2007) 66(1) *Human Organization* 28

Wallbott, Linda, 'Indigenous Peoples in UN REDD+ Negotiations: "Importing Power" and Lobbying for Rights through Discursive Interplay Management' (2014) 19(1) *Ecology and Society* 166

Walton, Nerissa and James Fitzsimons, 'Payment for Ecosystem Services in Practice: Savanna Burning and Carbon Abatement at Fish River, Northern Australia' in Penelope Figgis et al (eds), *Valuing Nature: Protected Areas and Ecosystem Services* (Australian Committee for IUCN Inc, 2015) 78

Weiss, Meredith L and Saliha Hassan, 'Introduction' in Meredith L Weiss and Saliha Hassan (eds), *Social Movements in Malaysia: From Moral Communities to NGOs* (Routledge, 2002) 1

West, Simon, '"Command Without Control": Are Market Mechanisms Capable of Delivering Ecological Integrity to REDD?' (2010) 6 *Law, Environment & Development Journal* 298

Westholm, Lisa, et al, 'REDD+ and Tenure: A Review of the Latest Developments in Research, Implementation and Public Policy Debates' in Randall A Bluffstone and Elizabeth J Z Robinson (eds), *Forest Tenure Reform in Asia and Africa: Local Control for Improved Livelihoods, Forest Management, and Carbon Sequestration* (Taylor and Francis, 2015) 221

Wilder, Martijn, Sophie Chapman and Arjuna Dibley, 'Special Issue: The Legal Aspects of REDD+ Implementation: Translating the International Rules into Effective National Frameworks' (2015) 9 *Carbon & Climate Law Review* 99

White, Andy and Alejandra Martin, *Who Owns the World's Forests? Forest Tenure and Public Forests in Transition* (Forest Trends and Center for International Environmental Law, 2002), www.era-mx.org/biblio/politica/ForestTenure.pdf

Whiting, Amanda, 'In the Shadow of Developmentalism: The Human Rights Commission of Malaysia at the Intersection of State and Civil Society Priorities' in C Raj Kumar and D K Srivastava (eds), *Human Rights and Development: Law, Policy and Governance* (LexisNexis, 2006) 324

Wiersema, Annecoos, 'Climate Change, Forests, and International Law: REDD's Descent into Irrelevance' (2014) 47 *Vanderbilt Journal of Transnational Law* 1

Wyn, Lyn Teck, 'Critical Review of the Forest Regulatory Framework and Its Implementation in Malaysia' in Henry Scheyvens (ed), *Critical Review of Selected Forest-Related Regulatory Initiatives: Applying a Rights Perspective* (Institute for Global Environmental Strategies, 2011) 51

Yagi, Sylver, 'PNG First in the Region with Climate Change Law', *Loop*, 5 September 2015, www.looppng.com/content/png-first-region-climate-change-law

Yala, Charles, 'Overview' in Charles Yala (ed), *The Genesis of the Papua New Guinea Land Reform Program: Selected Papers from the 2005 National Summit* (National Research Institute, 2010)

Yapao, George, Lee Godden and Steven Pettigrove, 'Papua New Guinea: Conflicts, Customary Landholding and Resource Exploitation' in Marcia Langton and Judy Longbottom (eds), *Community Futures, Legal Architecture: Foundations for Indigenous Peoples in the Global Mining Boom* (Routledge, 2012) 77

Young, Margaret A, 'Climate Change Law and Regime Interaction' (2011) 5 *Carbon & Climate Law Review* 147

*Trading Fish, Saving Fish: The Interaction between Regimes in International Law* (Cambridge University Press, 2011)

'Introduction: The Productive Friction between Regimes' in Margaret A Young (ed), *Regime Interaction in International Law: Facing Fragmentation* (Cambridge University Press, 2012) 1

'Regime Interaction in Creating, Implementing and Enforcing International Law' in Margaret A Young (ed), *Regime Interaction in International Law: Facing Fragmentation* (Cambridge University Press, 2012) 85

'Fragmentation' in Anthony Carty (ed), *Oxford Bibliographies in International Law* (Oxford University Press, 2014), www.oxfordbibliographies.com

'Trade Measures to Address Environmental Concerns in Faraway Places: Jurisdictional Issues' (2014) 23 *Review of European Community & International Environmental Law* 302

'Fragmentation, Regime Interaction and Sovereignty' in Christine Chinkin and Freya Baetens (eds), *Sovereignty, Statehood and State Responsibility* (Cambridge University Press, 2015) 71

'Principle 12: Trade and the Environment' in Jorge E Vinuales (ed), *The Rio Declaration on Environment and Development: A Commentary* (Oxford University Press, 2015) 325

'REDD+ and Interacting Legal Regimes' in Christina Voigt (ed), *Research Handbook on REDD+ and International Law* (Edward Elgar, 2016) 89

Young, Margaret A and Cobi Smith, *International Laws for Forest Governance* (University of Melbourne, 2016) (Dataset), doi.org/10.4225/49/5567C4E16BB15

Young, Margaret A and Sebastian R Sullivan, 'Evolution through the Duty to Cooperate: Implications of the *Whaling* Case at the International Court of Justice' (2015) 16 *Melbourne Journal of International Law* 310

Zahar, Alexander, *International Climate Change Law and State Compliance* (Taylor and Francis, 2014)

Zahar, Alexander, Jacqueline Peel and Lee Godden, *Australian Climate Law in Global Context* (Cambridge University Press, 2013)

## Papers Submitted to or Produced By International Organizations

Aboriginal and Torres Strait Islander Social Justice Commissioner, *Native Title Report 2007* (Australian Human Rights Commission, 2008), www.humanrights.gov.au/our-work/aboriginal-and-torres-strait-islander-social-justice/publications/native-title-report-2007

'Achieving the ITTO Objective 2000 and Sustainable Forest Management in Papua New Guinea: Report of the Diagnostic Mission' (ITTC(XLII)/7, International Tropical Timber Council, 4 April 2007)

Allen, Cameron and Stuart Clouth, 'A Guidebook to the Green Economy' (UN Division for Sustainable Development, August 2012)

Anaya, James, Special Rapporteur, *Report of the Special Rapporteur on the Rights of Indigenous Peoples*, 21st sess, UN Doc A/HRC/21/47 (6 July 2012)

Ancrenaz, Marc, et al, *Strategic Plan of Action (Sabah): The Heart of Borneo Initiative (2014–2020)* (Sabah Forestry Department & WWF-Malaysia, 2013)

Australian Government, *Carbon Farming Initiative (CFI): Savanna Burning* (2014) Department of the Environment and Energy, www.environment.gov.au/climate-change/emissions-reduction-fund/cfi/publications/factsheet-savanna-burning

Climate Change Authority, 'Australia's Climate Policy Options' (Special Review, Second Draft Report, November 2015)

Australian Government, Department of the Environment, 'Emissions Reduction Fund' (Methodology Determination Guidelines, January 2015), www.environment.gov.au/climate-change/emissions-reduction-fund/publications/methodology-determination-guidelines

Baastel and Nordic Agency for Development and Ecology (NORDECO), *First Program Evaluation for the Forest Carbon Partnership Facility (FCPF)* (Evaluation Report, 13 June 2011)

Bank Information Centre (BiC), 'SESA, Safeguards and the FCPF: A Guide for Civil Society' (January 2013), www.bankinformationcenter.org/wp-content/uploads/2013/01/SESA+informational+doc+for+PC6.pdf

Canada, 'Preliminary Comments, Part VI' (1995) doCip Doc 023a (on file with author)

'Statement on Lands and Resources' (21 October 1999) doCip Doc 014a (on file with author)

Canada on Articles 19 and 20, 'Articles 19, 20, 22 & 23' (28 October 1996) doCIP Doc 228a (on file with author)

Chao, Sophie, 'Forest Peoples: Numbers Across the World' (Forest Peoples Programme, 2012), www.forestpeoples.org/sites/fpp/files/publication/2012/05/forest-peoples-numbers-across-world-final_0.pdf

Cheney, Emelyne, et al, 'REDD+ Safeguards Information Systems: Practical Design Considerations' (UN-REDD Programme Technical Brief 01, October 2015)

Commonwealth of Australia, 'Reducing Emissions from Deforestation and Forest Degradation in Developing Countries (REDD)' (Submission to the AWG-LCA, AWG-KP and SBSTA, 2008), http://climatechange.gov.au/sites/climatechange/files/files/Reducing-emissions-deforestation-forest-degradation-developing-countries-Nov-08.pdf

Corrin, Jennifer, 'REDD+ and Forest Carbon Rights in Vanuatu: Background Legal Analysis' (Report prepared for Secretariat of the South Pacific Community and the Federal Ministry for the Environment, Nature Conservation and Nuclear Safety, November 2012)

Cotula, Lorenzo and James Mayers, 'Tenure in REDD: Start-Point or After-thought?' (International Institute for Environment and Development (IIED), 2009), pubs.iied.org/13554IIED/?k=tenure+in+redd

Daes, Erica I, *Working Paper by the Chairperson-Rapporteur on the Concept of 'Indigenous People'*, UN Doc E/CN.4/Sub.2/AC.4/1996/2 (10 June 1996)

Davis, Crystal, Ruth Nogueron and Anne-Gaelle Javelle, 'Analysis of Institutional Mechanisms for Sharing REDD+ Benefits: Property Rights and Resource Governance Project (PRRGP)' (United States Agency for International Development (USAID), March 2012), www.land-links.org/research-publication/analysis-of-institutional-mechanisms-for-sharing-redd-benefits

Denman, K L, et al, 'Couplings between Changes in the Climate System and Biogeochemistry' in S. Solomon et al (eds), Climate Change 2007: The Physical Science Basis. Contribution of Working Group I to the Fourth Assessment Report of the Intergovernmental Panel on Climate Change (Cambridge University Press, 2007)

Department of Forests, 'Vanuatu Forest Policy: 2013–2023' (2011), http://theredddesk.org/countries/policies/vanuatu-forest-policy-2013–2023

DoCIP, United Nations Declaration on the Rights of Indigenous Peoples: Comparative Table (2011) (on file with author)

Dooley, Kate, et al, *Smoke and Mirrors: A Critical Assessment of the Forest Carbon Partnership Facility* (FERN and Forest Peoples Programme, February 2011)

FAO, *FAO Policy on Indigenous and Tribal Peoples* (Food and Agriculture Organization, 2010), www.fao.org/docrep/013/i1857e/i1857e00.htm

'Land Tenure and Rural Development' (FAO Land Tenure Series 3, Food and Agriculture Organization of the United Nations, 2002), www.fao.org/docrep/005/y4307e/y4307e00.htm

'Towards Food Security and Improved Nutrition: Increasing the Contribution of Forests and Trees' (Food and Agriculture Organization of the United Nations, 2013), www.fao.org/docrep/018/i2969e/i2969e.pdf

*Voluntary Guidelines on the Responsible Governance of Tenure of Land, Fisheries and Forests in the Context of National Food Security* (Food and Agriculture Organization of the United Nations, 2012), www.fao.org/nr/tenure/voluntary-guidelines/en

FAO, UNDP and UNEP, *UN Collaborative Programme on Reducing Emissions from Deforestation and Forest Degradation in Developing Countries (UN-REDD): Framework Document* (20 June 2008)

Fey, Chip, 'Review of Legal Frameworks for Community-Based Natural Resource Management in Selected Asian Countries (Draft)' (World Agroforestry Centre, September 2007)

Food and Agriculture Organization of the United Nations (FAO), *Rome Declaration on World Food Security and the World Food Summit Plan of Action* (World Food Summit, Rome, Italy, 13–17 November 1996)

'Forest Resources Assessment 2000: On Definitions of Forest and Forest Change' (Working Paper 33, FAO, 2000)

*Voluntary Guidelines to Support the Progressive Realization of the Right to Adequate Food in the Context of National Food Security* (2005)

Food and Agriculture Organization of the United Nations, 'Global Forest Resources Assessment 2010: Country Report – Malaysia' (2010), www.fao.org/docrep/013/al558E/al558e.pdf

'Constitution' in *Basic Texts of the Food and Agriculture Organization of the United Nations (Volumes I and II)* (FAO, 2013), www.fao.org/docrep/meeting/022/k8024e.pdf

Food and Agriculture Organization of the United Nations (FAO), 'Global Forest Resources Assessment 2015: Country Report – Malaysia' (2014), www.fao.org/documents/card/en/c/08bc6d0d-bb0f-4313-ba3b-056fc307e79f

*Respecting Free, Prior* and *Informed Consent: Practical Guidance for Governments, Companies, NGOs, Indigenous Peoples and Local Communities in Relation to Land Acquisition* (2014), www.fao.org/documents/card/en/c/e5e9c08e-ff02-5509–8929-6f6537883f7b

*Free Prior and Informed Consent: An Indigenous Peoples' Right and a Good Practice for Local Communities – Manual for Project Practitioners* (2016), www.fao.org/documents/card/en/c/5202ca4e-e27e-4afa-84e2-b08f8181e8c9

Forest Carbon Partnership, 'REDD+ Annual Country Progress Reporting: Vanuatu' (January 2015–August 2015), http://forestcarbonpartnership.org/sites/fcp/files/2015/September/FCPF_August2015_Vanuatu.pdf

Forest Carbon Partnership Facility (FCPF), 'Information Memorandum' (13 June 2008), www.forestcarbonpartnership.org/readiness-fund

Forest Carbon Partnership Facility, 'A Guide to the FCPF Readiness Assessment Framework' (June 2013)

*Carbon Fund Methodological Framework* (22 June 2016), www.forestcarbonpartnership.org/carbon-fund-methodological-framework

'Eighteenth Meeting of the FCPF Participants Committee (PC18), Seventh Meeting of the FCPF Participants Assembly (PA7), Co-Chairs' Summary' (Arusha, Tanzania, 31 October–2 November 2014), www.forestcarbonpartnership.org/PA7/PC18

Forest Carbon Partnership Facility (FCPF) Carbon Fund, 'World Bank Safeguard Policies and the UNFCCC REDD+ Safeguards' (FMT Note CF-2013-3, 28 August 2013)

Forest Carbon Partnership Facility (FCPF) Readiness Fund, 'Incorporating Environmental and Social Considerations into the Process of Getting Ready for REDD Draft,' (Note FMT 2009–6, 15 October 2009), www.forestcarbonpartnership.org/sites/fcp/files/Documents/tagged/FCPF_en_soc_guidelines_10-15-09.pdf

'Incorporating Environmental and Social Considerations into the Process of Getting Ready for REDD' (Note FMT 2010–16, 23 June 2010), www.forestcarbonpartnership.org/sites/fcp/files/2014/June/FMT%20Note%202010-16%20SESA%20Mainstreaming.pdf

'Common Approach to Environmental and Social Safeguards for Multiple Delivery Partners' (Revised 9 August 2012)

Forest Carbon Partnership Facility (FCPF) and The United Nations Collaborative Programme on Reducing Emissions from Deforestation and Forest Degradation in Developing Countries (UN-REDD), 'Readiness Preparation Proposal (R-PP)' (Template Version 6, Working Draft, 4 April 2012)

'Readiness Preparation Proposal (R-PP)' (Template Version 6, 20 April 2012)

Forest Carbon Partnership Facility (FCPF) and United Nations REDD Programme (UN-REDD), 'Readiness Preparation Proposal (R-PP)' (Template Version 5, Working Draft, 30 October 2010)

Forest Peoples Programme, 'The Palangka Raya Declaration on Deforestation and the Rights of Forest Peoples' (Declaration of Looking Ahead: International Workshop on Deforestation Drivers and the Rights of Forest

Peoples, Palangka Raya, Central Kalimantan, Indonesia, 18 May 2014), www.forestpeoples.org/sites/fpp/files/news/2014/03/declaration_english.pdf

Forest Trends, *Logging, Legality and Livelihoods in Papua New Guinea: Synthesis of Official Assessments of the Large-Scale Logging Industry: Volume 1* (2006), www.forest-trends.org/publication_details.php?publicationID=105

*Forests: Action Statements and Action Plans* (Climate Summit 2014, New York, 23 September 2014)

The Forests Dialogue, 'Giving REDD+ Life: Integrating REDD+ with Broader Development Goals' (TFD Review, No 6, 2012), http://theforestsdialogue.org/sites/default/files/tfdreview_givingredd-pluslife_en.pdf

Friends of the Earth Japan and Global Environment Forum, 'Evaluation of Social and Environmental Risks Accompanying the Procurement of Timber from Papua New Guinea' (June 2011)

Global Environment Facility (GEF) and United Nations Development Programme (UNDP), 'Biodiversity Conservation in Multiple-Use Forest Landscapes in Sabah, Malaysia' (United Nations Development Programme County: Regional (Sabah, Malaysia) – Project Document, rev 14 March 2012), http://info.undp.org/docs/pdc/Documents/MYS/00060158_4186_FINAL_Sabah_Prodoc.pdf

Global Environmental Justice Think Tank, 'Second Norwich Declaration on Environmental Justice: From Safeguarding to Mainstreaming Local Rights: Democratic Forest Governance as Pre-Requisite for a Just and Effective REDD+' (Declaration of Global Environmental Justice Think Tank, University of East Anglia, Norwich, United Kingdom, 14–16 May 2014), www.uea.ac.uk/documents/439774/6205791/Second+Norwich+Declaration+on+Global+Environmental+Justice.pdf/8c39f1a2-456f-4f68-9ee4-76f25b282251

Global Expert Workshop on Biodiversity, *Outcomes of the Global Expert Workshop on Biodiversity Benefits of Reducing Emissions from Deforestation and Forest Degradation in Developing Countries*, UN Doc UNEP/CBD/WS-REDD/1/3 (23 September 2010)

Global Witness, 'Understanding REDD+: The Role of Governance, Enforcement and Safeguards in Reducing Emissions from Deforestation and Forest Degradation' (November 2010)

Government of Papua New Guinea, Office of Climate Change and Development, 'Interim Action Plan for Climate-Compatible Development (Draft Document for Public Consultation)' (August 2010), www.occd.gov.pg/images/stories/documents/PNG_Interim_Action_Plan.pdf

'Readiness Preparation Proposal (R-PP) for Papua New Guinea (Final)' (9 December 2013)

Government of Papua New Guinea, et al, 'National Programme Document' (UN Collaborative Programme on Reducing Emissions from Deforestation and Forest Degradation in Developing Countries, 2011)

Government of the Republic of Vanuatu, 'Disaster Risk Reduction and Disaster Management National Action Plan 2006–2016' (2007)

Government of Vanuatu, 'Readiness Preparation Proposal (R-PP) for Vanuatu (Final)' (7 October 2013)

Green Climate Fund, *Decisions of the Board – Eighth Meeting of the Board, 14–17 October 2014* (GCF/B.08/45, 3 December 2014)

Green Climate Fund, 'Initial Logic Model and Performance Measurement Framework for REDD+ Results-Based Payments' (GCF/B.08/08/Rev.01, 17 October 2014)

Greenpeace, *Up for Grabs: Millions of Hectares of Customary Land in PNG Stolen for Logging* (Greenpeace Asia Pacific, 2012)

Guidelines for the Development of Terms of Reference (ToRs) for an Environmental and Social Management Framework (ESMF) in the Context of the REDD-Plus Readiness Preparation Process Supported by the FCPF', (Annex C to Forest Carbon Partnership Facility (FCPF) and The United Nations Collaborative Programme on Reducing Emissions from Deforestation and Forest Degradation in Developing Countries (UN-REDD), 'Readiness Preparation Proposal (R-PP), Template Version 6, 20 April 2012, rev 9 August 2012)

HRBA Portal, 'The Human Rights Based Approach to Development Cooperation: Towards a Common Understanding Among UN Agencies' (2003), http://hrbaportal.org/the-human-rights-based-approach-to-development-cooperation-towards-a-common-understanding-among-un-agencies

Human Rights Commission of Malaysia (SUHAKAM), *Report of the National Inquiry into the Land Rights of Indigenous Peoples* (2013)

'National Inquiry: Suhakam Inquiry Highlights Issues Faced by Sabah's Indigenous People' (11 January 2014), www.suhakam.org.my/national-inquiry

Independent Evaluation Group, *The Forest Carbon Partnership Facility* (Global Program Review, vol. 6, Issue 3, 2011)

Indufor, *Second Program Evaluation of the Forest Carbon Partnership Facility (FCPF)* (Inception Report, The World Bank, 13 November 2015)

Institute for Global Environmental Strategies (IGES), 'Papua New Guinea REDD+ Readiness: State of Play' (August 2012) www.inapng.com/pdf_files/PNG_12-09–2012.pdf

International Bank for Reconstruction and Development, 'Management Report and Recommendation in Response to the Inspection Panel Investigation Report: Independent State of Papua New Guinea – Smallholder Agriculture Development Project (IDA Credit 4374-PNG)' (Report No INSP/64458-PG, 31 October 2011)

*Charter Establishing the Forest Carbon Partnership Facility* (Revised 23 November 2015), www.forestcarbonpartnership.org/charter-and-governance-documents

International Bar Association, Climate Change Justice and Human Rights Task Force, *Achieving Justice and Human Rights in an Era of Climate Disruption* (2014)

International Labour Organization, *Ratifications of C169 – Indigenous and Tribal Peoples Convention, 1989 (No 169)*, www.ilo.org/dyn/normlex/en/f?p=NORMLEXPUB:11300:0::NO::P11300_INSTRUMENT_ID:312314

International Law Association, *Sofia Conference (2012): Rights of Indigenous Peoples (Final Report)* (2012)

International Tropical Timber Organization, 'Revised ITTO Criteria and Indicators for the Sustainable Management of Tropical Forests Including Reporting Format' (ITTO Policy Development Series No 15, 2005), www.itto.int/direct/topics/topics_pdf_download/topics_id=9630000&no=1&disp=inline

IPCC, Climate Change 2013: The Physical Science Basis. Contribution of Working Group I to the Fifth Assessment Report of the Intergovernmental Panel on Climate Change (Cambridge University Press, 2013)

Kugan, Frederick and Heidi William, 'The Long and Bumpy Road to REDDINESS – Why Bother?' (Sabah Forestry Headquarters, Sandakan Sabah, Malaysia, 2011) (copy on file with author)

Larsen, Gaia and Florence Daviet, 'Safeguarding Forests and People: A Framework for Designing a National System to Implement REDD+ Safeguards' (World Resources Institute, November 2012), www.wri.org/publication/safeguarding-forests-and-people

Le Groupe-conseil baastel ltée and Nordic Agency for Development and Ecology, 'First Program Evaluation for the Forest Carbon Partnership Facility (FCPF): Evaluation Report' (2011)

Madsen, Becca, Nathaniel Carroll and Kelly Moore Brands, *State of Biodiversity Markets Report: Offset and Compensation Programs Worldwide* (Forest Trends, 2010), www.ecosystemmarketplace.com/publications/state-of-biodiversity-markets

Martínez Cobo, José R, Special Rapporteur, *Study of the Problem of Discrimination against Indigenous Populations – Volume V: Conclusions, Proposals and Recommendations*, UN Doc E/CN.4/Sub.2/1986/7 (March 1987)

Mikkelsen, Cæcilie (ed), *The Indigenous World 2013* (The International Work Group for Indigenous Affairs (IWGIA), 2013)

National Human Rights Commission of Malaysia, 'Background Paper: SUHAKAM's National Inquiry into the Land Rights of Indigenous Peoples in Malaysia' (May 2011), www.suhakam.org.my/wp-content/uploads/2014/03/BACKGROUND-PAPER-4-May-2011.pdf

Nelson, Anjali, et al, *Project Idea Note: Loru Forest Project* (Live & Learn International, 9 October 2014), www.ekos.org.nz/uploads/8/2/1/0/8210062/loru_pin_d3.3_v1.0_20140606.pdf

*Loru Forest Project – Project Description (PD), Part A: General Description:* (Live & Learn Environmental Education, 2015), www.nakau.org/uploads/5/2/2/5/52251233/loru_pd_part_a_d3.2a_v1.0_20151009b.pdf

'New York Declaration on Forests and Action Agenda (Section 1)' in *Forests: Action Statements and Action Plans* (Climate Summit 2014, New York, 23 September 2014)

Numapo, John, *Commission of Inquiry into the Special Agriculture and Business Lease (SABL): Final Report* (24 June 2013), http://pngexposed.files.wordpress.com/2013/12/numapo-sabl-final-report.pdf

O'Brien, Stephen, 'REDD+ and Forest Carbon Rights in Papua New Guinea: Background Legal Analysis' (SPC/GIZ Regional REDD+ Project: Climate Protection through Forest Conservation in Pacific Island Countries, November 2012), http://theredddesk.org/resources/redd-and-forest-carbon-rights-papua-new-guinea-background-legal-analysis

Office of Climate Change and Development (OCCD), 'PNG REDD+ Readiness: Status and Funding Needs' (2011)

Office of Climate Change and Development, 'Guidelines on FPIC for REDD+ in Papua New Guinea' (Version 4, August 2014)

Office of the United Nations High Commissioner for Human Rights (OHCHR) and the United Nations Environment Programme (UNEP), *Human Rights and the Environment: Rio+20A – Joint Report OHCHR and UNEP* (2012)

Ooft, Max, *UNDP and Indigenous Peoples: Towards Effective Partnerships for Human Rights and Development* (United Nations Development Programme, Oslo Governance Centre, November 2006)

Pacific Forest Alliance Limited, 'April Salumei Sustainable Forest Management Project' (VCS Feasibility Assessment, 4 May 2012), http://theredddesk.org/countries/initiatives/april-salumei-sustainable-forest-management-project-east-sepik

Papua New Guinea, 'Readiness Preparation Proposal (R-PP) – Papua New Guinea' (PNG R-PP Final Document, 9 December 2013), www.unredd.net/documents/un-redd-partner-countries-181/asia-the-pacific-333/a-p-partner-countries/papua-new-guinea-188/r-pp-fcpf-1906.html

Permanent Forum on Indigenous Issues, *Report of the International Workshop on Methodologies Regarding Free, Prior and Informed Consent and Indigenous Peoples*, UN Doc E/C.19/2005/3 (17 February 2005)

REDD+ Papua New Guinea, 'Situation Analysis and Recommendations for Improving Gender Inclusiveness and Participation in Papua New Guinea's National REDD+ Strategy and Policies' (Draft), www.pg.undp.org/content/dam/papua_new_guinea/FCPF/ROAR%20REports/Gender%20assessment%20report.pdf?download

*Resolution H1 – General Guidelines for the Sustainable Management of Forests in Europe* (Second Ministerial Conference on the Protection of Forests in Europe, Helsinki, Finland, 16–17 June 1993)

Rey, D, et al, *A Guide to Understanding and Implementing the UNFCCC REDD+ Safeguards* (ClientEarth, 2013)

Rights and Resources, 'Toward a Global Baseline of Carbon Storage in Collective Lands: An Updated Analysis of Indigenous Peoples' and Local Communities' Contributions to Climate Change Mitigation' (Rights and Resources Initiative, Woods Hole Research Center, World Resources Institute, 1 November 2016), http://rightsandresources.org/wp-content/uploads/2016/10/Toward-a-Global-Baseline-of-Carbon-Storage-in-Collective-Lands-November-2016-RRI-WHRC-WRI-report.pdf

Ruggie, John, *Report of the Special Representative of the Secretary-General on the Issue of Human Rights and Transnational Corporations and Other Business Enterprises: Guiding Principles on Business and Human Rights: Implementing the United Nations 'Protect, Respect and Remedy' Framework*, UN Doc A/HRC/17/31 (21 March 2011)

Sabah Forestry Department and WWF (Malaysia), 'REDD+: Developing Sub-National Strategy in the State of Sabah, Malaysia' (Sabah Forestry Department, Sandakan, Malaysia, 2011) (copy on file with author)

Scientific Certification Systems, 'Final CCBA Project Validation Report, April Salumei, East Sepik, Papua New Guinea' (Validation Report, 30 May 2011), http://s3.amazonaws.com/CCBA/Projects/April_Salumei_Sustainable_Forest_Management_Project/CCB_RMA_AprilSalumei_Validation_RPT_Final_Signed_062411.pdf

Secretariat of the Pacific Island Community, 'Pacific Islands Regional Policy Framework for REDD+' (Supported by the SPC/GIZ Regional Project, 'Climate Protection through Forest Conservation in Pacific Island Countries', 2013), http://theredddesk.org/countries/policies/pacific-islands-regional-policy-framework-redd

Smith, Carsten, Member of the Permanent Forum on Indigenous Issues, *Draft General Comment No 1 (2009): Article 42 of the Declaration on the Rights of Indigenous Peoples*, 8th sess, UN Doc E/C.19/2009/CRP.12 (5 May 2009)

South East Asia National Human Rights Institutions Forum (SEANF) and Human Rights Commission of Malaysia (SUHAKAM), *Project on Good Practices in Promoting and Protecting the Rights of Indigenous Peoples: Good Practices in Malaysia* (2014)

Subsidiary Body for Scientific and Technological Advice, United Nations Framework Convention on Climate Change, *Report of the Subsidiary Body for Scientific and Technological Advice on Its Thirty-Eighth Session, Held in Bonn from 3 to 14 June 2013*, UN Doc FCCC/SBSTA/2013/3 (29 August 2013)

*Views on Experiences and Lessons Learned from the Development of Systems for Providing Information on How All the Safeguards Are Being Addressed and Respected and the Challenges Faced in Developing Such Systems – Submissions from Developing Country Parties*, UN Doc FCCC/SBSTA/2014/MISC.6 (6 November 2014)

Tauli-Corpuz, Victoria, United Nations Special Rapporteur on the Rights of Indigenous Peoples, *Expert Testimony of the UN Special Rapporteur on the Rights of Indigenous Peoples, Victoria Tauli-Corpuz, before the Inter-American Court of Human Rights (IACHR) on the Case of Kaliña and Lokono Peoples vs the Government of Surinam* (3 February 2015), http://unsr.vtaulicorpuz.org/site/index.php/en/statements/56-testimony-case-surinam

'Technical Notes on Implementation of the Montréal Process Criteria and Indicators: Criteria 1–7' (3rd edn, 2009, rev July 2014), www.montrealprocess.org/documents/publications/techreports/MontrealProcessTechnicalNotes3rdEditionRevisedJuly2014.pdf

Terrestrial Carbon Group Project, 'Legal and Institutional Foundations for the National Implementation of REDD: Lessons from Early Experience in Developing and Developed Countries' (Policy Brief 4, Terrestrial Carbon Group and UN-REDD Programme, 2009)

*UNDG Human Rights Mainstreaming Mechanism: Operational Plan 2011–2013* (United Nations Development Group, November 2011)

UNDP, '*UNDP and Indigenous Peoples: A Policy of Engagement*' (June 2001), www.undp.org/content/undp/en/home/librarypage/environment-energy/local_development/undp-and-indigenous-peoples-a-policy-of-engagement.html

UNFCCC, 'Tenth Session of the United Nations Permanent Forum on Indigenous Issues (UNPFII)' (February 2011), www.un.org/esa/socdev/unpfii/documents/session_10_UNFCCC.pdf

*United Nations Development Group (UNDG) Guidelines on Indigenous Peoples' Issues* (United Nations Development Group, February 2008)

United Nations Development Programme, 'UNDP and Indigenous Peoples: A Practice Note on Engagement' (2001), www.undp.org/content/dam/undp/documents/partners/civil_society/publications/policies_and_strategic_documents/UNDP_and_Indigenous_Peoples_A_Practice_Note_On_Engagement_2001.pdf

United Nations Environment Programme (UNEP), 'UNEP and Indigenous Peoples: A Partnership in Caring for the Environment' (Policy Guidance, November 2012)

United States, 'Preliminary Comments, Part VI (Land and Resources)' (28 November 1995) doCip Doc 030a (on file with author)

UN-REDD Programme, 'Operational Guidance: Engagement of Indigenous Peoples and Other Forest Dependent Communities' (Working Document, 25 June 2009)

'Interim REDD+ Partnership Document' (27 May 2010)

'UN-REDD Programme Social and Environmental Principles and Criteria' (Version 3, Draft for Consultation, September 2011)

*UN-REDD Programme Strategy 2011–2015* (February 2011)

*UN-REDD Programme Handbook for National Programmes and Other National-Level Activities* (2012)

'UN-REDD Programme SEPC: Supporting Document' (2012)

'UN-REDD Programme Social and Environmental Principles and Criteria (SEPC)' (UNREDD/PB8/2012/V/1, Asunción, Paraguay, 25–26 March 2012)

'Background Paper for the Expert Meeting: Options for Addressing Tenure under REDD+' (25–27 February 2013)

'Draft Social and Environmental Principles and Criteria – Benefit and Risks Tool (SEPC-BeRT): UN-REDD Programme 2013' (2013), www.unredd.net/index.php?option=com_docman&Itemid=134&view=document&alias=6352-draft-sepc-bert-un-redd-6352&category_slug=latest-version-of-the-draft-sepc-1266

*Guidelines on Free, Prior and Informed Consent* (January 2013)

'Legal Companion to the UN-REDD Programme Guidelines on Free, Prior and Informed Consent (FPIC): International Law and Jurisprudence Affirming the Requirement of FPIC' (January 2013)

'Review of Tenure Issues and Needs Identified by UN-REDD Countries' (9 January 2013), www.unredd.net/index.php?option=com_docman&Itemid=134&view=document&alias=9124-review-of-tenure-issues-and-needs-identified-by-un-redd-countries-9124&category_slug=tenure-and-redd-2647

'National Programme Semi Annual Report: Papua New Guinea' (September 2013), www.unredd.net/index.php?option=com_docman&task=doc_download&gid=11966&Itemid=53

'Report of the Thirteenth Policy Board Meeting' (6–7 November 2014), www.unredd.net/documents/policy-board-86/thirteenth-policy-board-meeting-arusha-tanzania-5-7-november-2014–3553/final-meeting-report-3577.html

'National Programme, Semi-Annual Report: Papua New Guinea' (January to June 2015)

'National Programme Annual Report' (Draft, January–December 2015)

*UN-REDD Programme Launches Benefits and Risks Tool (BeRT): Supporting Countries to Address and Respect the Cancun Safeguards* (19 May 2015), www.unredd.net/announcements-and-news/2131-un-redd-programme-launches-benefits-and-risks-tool-bert-supporting-countries-to-address-and-respect-the-cancun-safeguards.html

*UN-REDD Programme Strategic Framework 2016–2020* (UNREDD/PB14/2015/III/3, Revised Draft, 7 May 2015)

UN-REDD Programme and Forest Carbon Partnership Facility (FCPF), 'Joint Guidelines on Stakeholder Engagement in REDD+ Readiness with a Focus on Indigenous Peoples and Other Forest-Dependent Communities' (20 April 2012), www.undp.org/content/undp/en/home/ourwork/democraticgovernance/focus_areas/focus_human_rights/empowering_indigenous_peoples/UNREDD.html

'Country Needs Assessment: A Report on REDD+ Readiness among UN-REDD Programme and Forest Carbon Partnership Facility Member Countries' (October 2012)

UN Committee on the Elimination of Racial Discrimination, *Decision 1(66): New Zealand Foreshore and Seabed Act 2004*, UN Doc CERD/C/66/NZL/Dec. 1 (27 April 2004)

*General Recommendation No 23: Indigenous Peoples*, 52nd sess, UN Doc A/52/18 (18 August 1997) annex V at 122

*General Recommendation No 32: The Meaning and Scope of Special Measures in the International Convention on the Elimination of All Forms [of] Racial Discrimination*, UN Doc CERD/C/GC/32 (24 September 2009)

UN ECOSOC Commission on Human Rights, *Report of the Sub-Commission on Prevention of Discrimination and Protection of Minorities on Its Forty-Sixth Session*, UN Doc E/CN.4/1995/2 (28 October 1994)

UN ECOSOC Permanent Forum on Indigenous Issues, *Indigenous People's Rights and Safeguards in Projects Related to Reducing Emissions from Deforestation and Forest Degradation*, UN Doc E/C.19/2013/7 (5 February 2013)

United Nations Conference on Trade and Development, *International Tropical Timber Agreement 2006*, UN Doc TD/TIMBER.3/12 (1 February 2006, entered into force 7 December 2011)

UN Secretariat of the Permanent Forum on Indigenous Issues, 'The Concept of Indigenous Peoples' (Background Paper to Workshop on Data Collection and Disaggregation for Indigenous Peoples, New York, United States, 19–21 January 2004) UN Doc PFII/2004/WS.1/3

Vanuatu, 'Readiness Preparation Proposal (R-PP): Vanuatu' (7 October 2013), http://theredddesk.org/resources/readiness-preparation-proposal-r-pp-vanuatu

Weaver, Sean, Robbie Henderson and Anjali Nelson, *Nakau Methodology Framework: General Methodology for the Nakau Programme: An Indigenous Forest Conservation Programme through Payments for Ecosystem Services* (April 2014), http://theredddesk.org/sites/default/files/nakau_methodology_framework_d2.1_v1.0_20140428_0.pdf

Weaver, Sean, Martin Herold and Murray Ward, 'The Forest Carbon Partnership Facility (FCPF) Readiness Plan Idea Note (R-PIN) Template' (Vanuatu National Advisory Committee on Climate Change, 29 July 2008)

Weaver, Sean, Anjali Nelson and Robbie Henderson, *Loru Forest Project Monitoring Report 1* (Live and Learn Environmental Education, October 2015), www.nakau.org/uploads/5/2/2/5/52251233/loru_monitoring_report_1_d3.3__1__v1.0_20151009b.pdf

Weaver, Sean and Brian Phillips, 'Vanuatu Carbon Credits Project: Phase 2 Roadmap Workshop, 11–13 February, 2008' (Workshop Report, 2008)

World Bank, *Environmental and Social Framework (Proposed Third Draft)* (Review and Update of The World Bank's Safeguard Policies, 4 August 2016)

*Environmental and Social Framework: Setting Environmental and Social Standards for Investment Project Financing* (4 August 2016)

World Bank, 'Environmental and Social Standard 7: Indigenous Peoples/ Sub-Saharan African Historically Underserved Traditional Local Communities' in *Environmental and Social Framework: Setting Environmental and Social Standards for Investment Project Financing* (World Bank, 4 August 2016)

World Bank Inspection Panel, 'Report and Recommendation, Papua New Guinea: Smallholder Agriculture Development Project (SADP) (IDA Credit No 43740-PNG)' (Inspection Report No 53280-PG, 10 March 2010), http://documents.worldbank.org/curated/en/456211468285074449/pdf/532800IPR0P079101Official0Use0Only1.pdf

'Investigation Report, Independent State of Papua New Guinea: Small Agriculture Development Project (IDA Credit No 4374-PNG)' (Inspection Report No 64458-PG, World Bank Group, 19 September 2011), http://ewebapps.worldbank.org/apps/ip/PanelCases/62-Investigation%20Report%20%28English%29.pdf

World Bank Legal Department, 'Legal Note on Indigenous Peoples' (8 April 2005), http://siteresources.worldbank.org/INTINDPEOPLE/Publications/20571167/Legal%20Note.pdf

World Bank, 'Operational Manual: OP 4.10 – Indigenous Peoples' (July 2005, rev April 2013), http://policies.worldbank.org/sites/ppf3/PPFDocuments/090224b0822f89d5.pdf

Yong, Carol, 'Forest Governance in Malaysia: An NGO Perspective' (JOANGOHutan, 2006), www.fern.org/sites/fern.org/files/media/documents/document_3754_3755.pdf

'Logging in Sarawak and the Rights of Sarawak's Indigenous Communities' (JOANGOHutan, 2010), www.pro-regenwald.de/docs/logginginsarawak_joangohutanreport.pdf

## Policy Papers, News Sources and Unpublished Works

Aikman, Amos, 'Aboriginal Land Rights Out of Step?' *The Australian*, 31 July 2014

Babon, Andrea, 'Snapshot of REDD+ in Papua New Guinea' (CIFOR *Infobrief*, No 40, August 2011)

Babon, Andrea and Gae Yansom Gowae, 'The Context of REDD+ in Papua New Guinea: Drivers, Agents and Institutions' (Occasional Paper 89, CIFOR, 2013)

Bauman, Toni, Chris Haynes and Gabrielle Lauder, 'Pathways to the Co-Management of Protected Areas and Native Title in Australia' (AIATSIS Research Discussion Paper No 32, May 2013), http://aiatsis.gov.au/sites/default/files/products/discussion_paper/pathways-to-comanagement-of-protected-areas-and-native-title.pdf

Bingeding, Nalau, 'Reducing Emissions from Deforestation and Degradation of Forests in Papua New Guinea: Issues and Options' (Discussion Paper No 119, National Research Institute, Papua New Guinea, 2011)

Bird, Neil, et al, 'The Current Legal and Institutional Framework of the Forest Sector in Papua New Guinea: Papua New Guinea Forest Studies 2' (Overseas Development Institute, January 2007)

Bradley, Amanda, 'Tenure and REDD+: Update and Key Messages' (Presentation at the 13th Policy Board Meeting, UN-REDD Programme, Arusha, Tanzania, 5 November 2014)

Brown, Carmen, 'Credits Roll in for Territory Carbon Project' (23 April 2013) *ABC Rural News*, www.abc.net.au/site-archive/rural/news/content/201304/s3743095.htm

Bulan, Ramy, 'Indigenous Peoples and the Right to Participate in Decision Making in Malaysia' (Paper prepared for International Expert Seminar on Indigenous Peoples and the Right to Participate in Decision-Making, Chang Mai, Thailand, 20–22 January 2010)

Calí Tzay, Jose Fransisco, 'Letter from the Committee on the Elimination of Racial Discrimination to the Government of Papua New Guinea Re Special Agricultural Business Letters', 17 February 2016, http://tbinternet.ohchr.org/Treaties/CERD/Shared%20Documents/PNG/INT_CERD_ALE_PNG_7926_E.pdf

CCDA Media Release 'GoPNG Seeks to Strengthen the Community's Voice on Climate Change and Sustainable Management of Forests' (10 February 2017), www.ccda.gov.pg/images/REDD+%20GRM.pdf

Dickson, Barney, et al, 'REDD+ Beyond Carbon: Supporting Decisions on Safeguards and Multiple Benefits' (UN-REDD Programme Policy Brief, Issue No 2, November 2012)

Email from Dan Nekitel, OCCD, to Kirsty Gover, 20 February 2017

Filer, Colin, 'The New Land Grab in Papua New Guinea' (paper presented at International Conference on Global Land Grabbing, Future Agricultures Consortium at the Institute of Development Studies, University of Sussex, 6–8 April 2011)

Fincke, Annelie, 'Indigenous Peoples and Climate Change/REDD: An Overview of Current Discussions and Main Issues' (IUCN Briefing Document, 21 July 2010)

'Fire Brings Back Country' (2009) (January–July) *Savanna Links*, www.environorth.org.au/environorth/teach/downloads/Fire_brings_back_mosaic.pdf

'First Sales of Rainforest Carbon Credits Sold in the Pacific – Vanuatu Leading the Charge', *Vanuatu Daily Post*, 29 November 2016, http://dailypost.vu/news/first-sales-of-rainforest-carbon-credits-sold-in-the-pacific/article_97eb0395-87cb-5c75-ae47-a2b70b3f1f31.html

Forest Stewardship Council, 'Forest Stewardship Council Opens Doors to Sustainable Forest Management in Malaysia' (19 November 2014), http://ic.fsc.org/fsc-press-releases.638.979.htm

Fortmann, Lea, et al, 'Incentive Contracts for Environmental Services and Their Potential in REDD' (Policy Research Working Paper 6829, World Bank, April 2014)

Fox, Liam, 'NGO Raises Concern Over PNG's Climate Change Legislation', *ABC News* (9 December 2015), www.abc.net.au/news/2015-12-09/ngo-raises-concerns-over-pngs-climate-change/7014980

Frechette, Alain, Minoli de Bresser and Robert Hofstede, *External Evaluation of the United Nations Collaborative Programme on Reducing Emissions from Deforestation and Forest Degradation in Developing Countries (the UN-REDD Programme): Volume I – Final Report* (July 2014), http://wedocs.unep.org/rest/bitstreams/776/retrieve

Goers Williams, Lauren, 'Putting the Pieces Together for Good Governance of REDD+: An Analysis of 32 REDD+ Country Readiness Proposals' (Working Paper, World Resources Institute, March 2013), www.wri.org/sites/default/files/pdf/putting_the_pieces_together_for_good_governance_of_redd.pdf

Greenpeace, 'PNG: Not Ready for REDD' (Media Release, 25 October 2010), www.greenpeace.org/australia/en/mediacentre/media-releases/forests-media-releases/png-not-ready-for-redd

Greer, Guy and Claudia Henfry, 'Roe 8 Delay: The Decision of *Save Beeliar Wetlands v Jacob*', *Gilbert & Tobin: Insights – WA Resources Update*, 23 March 2016, www.gtlaw.com.au/wa-resources-update-march-2016

Haccius, Justin, 'The Interaction of Modern and Custom Land Tenure Systems in Vanuatu' (Discussion Paper 2011/1, School of International, Political and Strategic Studies, Australian National University, 2011)

Hajjar, Reem, et al, 'Criteria and Indicators for Sustainable Forest Management in the Face of Decentralization: Are They Still Relevant in Their Current Form' (XIII World Forestry Conference, Buenos Aires, Argentina, 18–23 October 2009)

Hansen, Linda, 'Carbon in Country: Legal Pathways and Barriers to Indigenous Participation in Australia's Carbon Market through Savanna Fire Management under the Carbon Farming Initiative' (Working Paper No 3, Centre for Resources, Energy and Environmental Law, Melbourne Law School, September 2015)

Hoare, Alison, 'Illegal Logging and Related Trade: The Response in Malaysia – A Chatham House Assessment' (Research Paper, Energy, Environment and Resources, January 2015), www.chathamhouse.org/publication/illegal-logging-and-related-trade-response-malaysia#

Jones, Natalie, 'WA Group Represents Indigenous Australia at UN Climate Change Conference', *ABC News*, 25 November 2015, www.abc.net.au/news/2015-11-24/wa-groups-heads-to-climate-change-conference/6971072

Jotzo, Frank, 'Australia's 2030 Climate Target Puts Us in the Race, But at the Back', *The Conversation*, 12 August 2015, http://theconversation.com/australias-2030-climate-target-puts-us-in-the-race-but-at-the-back-45931

Kalinoe, Lawrence, 'Incorporated Land Groups in Papua New Guinea' (paper presented at Australasian Law Teachers Association Conference, University of the South Pacific, Port Vila, Vanuatu, 1–4 July 2001)

Kimberley Land Council, 'Traditional Owners to Lose Millions of Dollars under Emission Reduction Fund' (Media Release, 9 July 2014), www.klc.org.au/docs/default-source/Media-Releases/traditional-owners-to-lose-millions-of-dollars-under-emission-reduction-fund?sfvrsn=0

'KLC Puts Aboriginal Fire Management on the World Stage' (28 October 2016), www.klc.org.au/news-media/newsroom/news-detail/2016/10/28/klc-puts-aboriginal-fire-management-on-the-world-stage

Kugan, Frederick, 'Sabah's Roadmap on REDD+ Strategy' (paper presented at National Inception Workshop on REDD+: Building Capacity for Decision Makers and the Way Forward, Sunway Putra Hotel, Kuala Lumpur, 13–14 February 2012)

Land Rights Now, Kimberley Land Council and Oxfam, 'Making Rights a Reality: The Ongoing Struggle for Land Justice in the Kimberley Region, Australia' (Briefing Note, 26 September 2016), www.oxfam.org.au/wp-content/uploads/2016/09/bn-australia-kimberley-land-rights-260916-en.pdf

Lang, Chris, *The Margarita Declaration on Climate Change: 'We Reject the Implementation of False Solutions to Climate Change Such As: Carbon Markets and Other Forms of Privatization and Commodification of Life'* (8 August 2014) REDD-Monitor.org, www.redd-monitor.org/2014/08/08/the-margarita-declaration

Lawson, Sam, 'Illegal Logging in Papua New Guinea' (Energy, Environment and Resources EER PP 2014/04, Chatham House, April 2014)

Leong, Joe, 'Sabah Moves Forward with REDD+ Road Map', *Borneo Post*, 24 August 2011, www.theborneopost.com/2011/08/24/sabah-moves-forward-with-redd-road-map-latest

Letter to Mario Boccucci, UN-REDD Programme Secretariat from Varigini Badira, UN-REDD Programme Papua New Guinea and Goodwill Amos, PNG Forest Activity, 'Request for No-Cost Extension of Papua New Guinea's UN-REDD National Programme', 15 July 2015, www.unredd.net/documents/policy-board-86/intersessional-decisions-829/14750-no-cost-extension-request-and-justification-from-papua-new-guinea-national-programme.html?path=policy-board-86/intersessional-decisions-829

Locke, Sarina, 'Red Meat Producers Biggest Winners from Government's Multi-Billion Dollar Emission Reduction Fund', *ABC News*, 1 December 2016, www.abc.net.au/news/2016-12-01/carbon-auctions-deliver-funds-to-beef-cattle-producers/8081904

Markham, Francis and Jon Altman, 'Values Mapping Indigenous Lands: An Exploration of Development Possibilities' (National Native Title Conference 2013, Alice Springs, Australia, 3 June 2013) (copy on file with author)

Marlay, Sarah, 'Clarifying the Role of Non-Carbon Benefits in REDD+' (Policy Paper, Environmental Defense Fund, July 2013), http://blogs.edf.org/climatetalks/files/2013/08/Non-Carbon-Benefits-Policy-Paper.pdf

Mayers, James, et al, 'Forest-Based Climate Strategies: Making REDD and Other Initiatives Work' (International Institute for Environment and Development (IIED), March 2008), www.fern.org/publications/recommended-reading/forest-based-climate-strategies

McDonnell, Siobhan, Joseph Foukona and Alice Pollard, 'Building a Pathway for Successful Land Reform in Solomon Islands' (2015), www.academia.edu/15598962/Building_a_pathway_for_successful_land_reform_in_Solomon_Islands

McIntyre, Greg, 'Native Title Rights after Yorta Yorta' (paper presented at Native Title Conference 2003: Native Title on the Ground, Alice Springs, Australia, 3–5 June 2003)

Ministry of Natural Resources and Environment (NRE), 'Malaysia Hosted REDD+ Day at COP-22, Marrakech, Morocco' (Media Statement, 15 November 2016), www.nre.gov.my/ms-my/PustakaMedia/Kenyataan%20Media/Media%20Statement%20Malaysia%20Hosted%20REDD+%20Day%20at%20COP%2022,%20Morocco%2014%20November%202016.pdf

Morpi, Murib, 'Ecological Linkages to Connect National Parks', *Borneo Post*, 22 July 2011, www.theborneopost.com/2011/07/22/ecological-linkages-to-connect-national-parks

Nekitel, Danny, 'Draft Guidelines on FPIC for REDD+ in Papua New Guinea' (Policy, Office of Climate Change and Development, August 2014)

O'Faircheallaigh, Ciaran, 'Native Title and Agreement Making in the Mining Industry: Focusing on Outcomes for Indigenous Peoples' (Issues Paper No 25, Native Title Research Unit, Australian Institute of Aboriginal and Torres Strait Islander Studies, 2004)

Ogle, Lisa, 'Free, Prior and Informed Consent for REDD+ in the Asia-Pacific Region: Lessons Learned, Challenges and Recommendations' (UN-REDD Programme, August 2012, 1 October 2013)

*Vanuatu: Legal Framework for REDD+* (SPC/GIZ Regional Project 'Climate Protection through Forest Conservation in Pacific Island Countries', August 2014, http://lrd.spc.int/spcgiz-qclimate-protection-through-forest-conservation-in-the-pacific-islandsq/studies-and-reports/cat_view/137-all/133-climate-change/374/375/501

Ogle, Lisa and Celina Yong, 'What Does It Take to Make a Local Consultation a Success? Input Paper III: Asia Pacific Region' (BMZ, FCPF, UN-REDD Joint Expert Workshop)

Parrotta, John, 'Social Safeguards: Protecting the Rights and Interests of Indigenous Peoples and Forest-Dependent Communities in REDD+' (Keynote Remarks, Forest Day 5, Discussion Forum, 4 December 2011)

Paterson, Don, 'Conflicts Arising from Management and Use of Customary Land' (paper presented at South Pacific Land Tenure Conflict Symposium, University of the South Pacific, 10–12 April 2002)

Peskett, Leo and Gernot Brodnig, 'Carbon Rights in REDD+: Exploring the Implications for Poor and Vulnerable People' (Working Paper 65864, World Bank and REDD-net, 2011)

'Qantas Backs North Kimberley Carbon Project' (2014) (September) *Kimberley Land Council Newsletter* 10

Rothe, Ann-Kristin and Paul Munro-Faure, 'Tenure and REDD+: Developing Enabling Tenure Conditions for REDD+' (Policy Brief, Issue No 6, UN-REDD Programme, 2013), http://theredddesk.org/resources/tenure-and-redd-developing-enabling-tenure-conditions-redd

Samsudin, M, et al, 'Reducing Emissions from Deforestation and Degradation (REDD+) – Malaysia' (paper presented at International Technical Seminar on REDD+, Tokyo, 16–17 February 2011)

Sheehan, John, 'Indigenous Carbon Property Rights' (paper presented at the 16th Annual Pacific Rim Real Estate Society (PRRES) Conference, Wellington, New Zealand, 25 January 2010), www.prres.net/papers/Sheehan_Indigenous_Carbon_Property_Rights.pdf

Simo, Joel, 'Land and the Traditional Economy: "Your Money, My Life" Hu I Kakae Long Basket Blong Laef' (Aid/Watch, April 2010)

Sommerville, Matthew, 'Land Tenure and REDD+: Risks to Property Rights and Opportunities for Economic Growth' (Issue Brief, USAID, 2013), www.land-links.org/issue-brief/land-tenure-and-redd-risks-to-property-rights-and-opportunities-for-economic-growth

Statement by Bill Barker on behalf of the Australian Delegation, Working Group on a Draft Declaration on the Rights of Indigenous Peoples (21 November 1995) (on file with author)

Statement by the Chinese Delegation to the Third Session of the Working Group Established in Accordance with the Commission on Human Rights Resolution 1995/32 (28 October 1997) (on file with author)

Stefanova, Milena, Raewyn Porter and Rod Nixon, 'Towards More Equitable Land Governance in Vanuatu: Ensuring Fair Land Dealings for Customary Groups' (Discussion Note, World Bank, May 2012)

Thuy, Pham Thu, et al, 'Approaches to Benefit Sharing: A Preliminary Comparative Analysis of 13 REDD+ Countries' (Working Paper 108, Center for International Forestry Research, 2013)

'UN-REDD Highlights Tenure as a Success Factor for REDD+' (2013) (36) *UN-REDD Programme Newsletter*

UN-REDD Programme, 'Indigenous Peoples and the UN-REDD Programme: An Overview' (Paper delivered at the International Expert Group Meeting on Indigenous Peoples and Forests, 12–14 January 2011)

'Malaysia and Uganda Join the UN-REDD Programme' (June 2012) *UN-REDD Programme Newsletter*, www.un-redd.org/Newsletter29/Malaysia_Uganda_Join_UN-REDD/tabid/104318/Default.aspx

'Reports & Analysis: Committee on World Food Security Launches Groundbreaking Guidelines on Land Tenure' (June 2012) *UN-REDD Programme Newsletter*, www.un-redd.org/Newsletter28/Impact_of_NewLandTenureGuidelines/tabid/104166/Default.aspx

Walker, Daniel and Allan Dale, 'Land Tenure in Northern Australia: Opportunities and Challenges for Investment' (CSIRO Report, 2013)

Weaver, Sean and Brian Phillips, 'Vanuatu Carbon Credits Project: Phase 2 Roadmap Workshop, 11–13 February, 2008' (Workshop Report, 2008)

Wells, Adrian, Thang Hooi Chiew and Chen Hin Keong, 'Systems for Verification of Legality in the Forest Sector, Malaysia: Domestic Timber Production and Timber Imports' (Country Case Study 8, VERIFOR, February 2008)

World Bank, 'Leases in Vanuatu: Key Data from the World Bank Jastis Blong Evriwan Vanuatu National Leasing Profile' (December 2010), http://siteresources.worldbank.org/INTJUSFORPOOR/Resources/Vanuatu_Leasing_Data_Summary.pdf

'World Bank Board Approves New Environmental and Social Framework' (Press Release, 4 August 2016), www.worldbank.org/en/news/press-release/2016/08/04/world-bank-board-approves-new-environmental-and-social-framework

The World Bank and The Inspection Panel, 'Joint Statement from World Bank Management and Inspection Panel' (News Release, 2014/143/ECR, 2 October 2014), http://ewebapps.worldbank.org/apps/ip/PanelCases/84%20-%20Press%20Release%20on%20Board%20Discussion%20%28English%29.pdf

## Internet Materials

'The Anchorage Declaration' (Indigenous Peoples' Global Summit on Climate Change, 24 April 2009), http://unfccc.int/resource/docs/2009/smsn/ngo/168.pdf

'Are We on Track for 2°C?' Mitigation-Contributions.org, www.mitigation-contributions.org

'Australia', Climate Action Tracker, www.climateactiontracker.org/countries/australia.html

Australian Government, 'About the Emissions Reduction Fund', Department of the Environment and Energy, www.environment.gov.au/climate-change/emissions-reduction-fund/about

'About Us', Indigenous Land Corporation, www.ilc.gov.au/Home/About-Us

'Carbon Farming Initiative Project Transition into the Emissions Reduction Fund', Department of the Environment and Energy, www.environment.gov.au/climate-change/emissions-reduction-fund/carbon-farming-initiative-project-transition

'Emissions Reduction Fund Methods,' Department of the Environment and Energy, www.environment.gov.au/climate-change/emissions-reduction-fund/methods

'Emissions Reduction Fund Project Register' (10 February 2017) Clean Energy Regulator, www.cleanenergyregulator.gov.au/ERF/project-and-contracts-registers/project-register

'Fish River Fire Project', Clean Energy Regulator, www.cleanenergyregulator.gov.au

'Fish River Fire Project: Northern Territory' (2015) Indigenous Land Corporation, www.ilc.gov.au/Home/What-We-Do/Project-Profiles/Fish-River-Fire-Project

'Indigenous Protected Areas: Balanggarra', Department of the Environment, www.environment.gov.au/indigenous/ipa/declared/balangarra.html

'Savanna Fire Management', Department of the Environment and Energy, www.environment.gov.au/climate-change/emissions-reduction-fund/methods/savanna-burning

'West Arnhem Land Fire Abatement (WALFA) Project' (27 January 2017) Clean Energy Regulator, www.cleanenergyregulator.gov.au

BeneLex, 'About the Project', www.benelex.ed.ac.uk/about_us

The Board of Trustees of the Sabah Parks, 'Kinabalu Ecolinc Project', www.sabahparks.org.my/index.php/discover-us-3/kinabalu-ecolinc-project

Butler, Rhett A, 'Malaysia Has the World's Highest Deforestation Rate, Reveals Google Forest Map' (15 November 2013) Mongabay.com, http://news.mongabay.com/2013/1115-worlds-highest-deforestation-rate.html

CAIT Climate Data Explorer: 'Papua New Guinea' (2013) World Resources Institute, http://cait2.wri.org/profile/Papua%20New%20Guinea

The Climate, Community and Biodiversity Alliance, 'Climate Community and Biodiversity Project Design Standards (2nd edn)' (December 2008), http://s3.amazonaws.com/CCBA/Upload/ccb_standards_second_edition_december_2008+%281%29.pdf

'Climate Summit 2014: Catalyzing Action' (23 September 2014), www.un.org/climatechange/summit/about

Dore, Jeremy, 'Is There a Native Title Right to Burn Country?' (27 August 2014) Aboriginal Carbon Fund, http://aboriginalcarbonfund.com.au/blog/2014/8/21/is-there-a-native-title-right-to-burn-country

'ERF and Indigenous Savanna Entwined' (29 November 2016) Aboriginal Carbon Fund, http://aboriginalcarbonfund.com.au/blog/2016/11/29/erf-and-indigenous-savanna-entwined

Fagan, Emma, 'New Methods and Greater Price Certainty Makes the Second ERF Reverse Auction More Attractive' (1 May 2015) Energetics, www.energetics.com.au/resources/latest-news/climate-change-matters/new-methods-and-greater-price-certainty-makes-the

'First Forest Project on Sabah, Borneo Registered under the VCS' (6 September 2011) Forest Carbon Portal: Tracking Terrestrial Carbon, www.forestcarbonportal.com/news/first-forest-project-sabah-borneo-registered-under-vcs

'FLEGT Independent Market Monitoring (IMM)', International Tropical Trade Organization (ITTO), www.itto.int/imm

Food and Agriculture Organization of the United Nations (FAO), 'Global Forest Resources Assessment 2015: How Are the World's Forests Changing?' (2015), www.fao.org/resources/infographics/infographics-details/en/c/325836

'Global Forest Resources Assessments' (2015), www.fao.org/forest-resources-assessment/en

Forest Carbon Partnership Facility (2015), www.forestcarbonpartnership.org

Forest Carbon Partnership Facility REDD+ Readiness Project, UNDP in Papua New Guinea, www.pg.undp.org/content/papua_new_guinea/en/home/operations/projects/environment_and_energy/forest-carbon-partnership-facility-reddt-readiness-project.html

The Forest Research Institute Malaysia, 'FRIM, FFPRI Share Research Findings REDD+' (6 February 2013) Ministry of Natural Resources and Environment, www.frim.gov.my/frim-ffpri-share-redd-research-findings

'Funding Support', Climate Change and Development Authority, www.ccda.gov.pg/Funding%20Support.html

'Governors' Climate and Forests Task Force', www.gcftaskforce.org/about

Global Forest Watch, 'Country Profile: Malaysia', www.globalforestwatch.org/country/MYS

Government of Malaysia, 'Intended Nationally Determined Contribution of the Government of Malaysia', 27 November 2015, available from *INDC – Submissions*, www4.unfccc.int/submissions/indc/Submission%20Pages/submissions.aspx

Hunt, Colin, et al, 'Asking the Question: Is Papua New Guinea REDD-Ready?' (2012), www.colinhunt.com.au/index.php/download_file/view/59/139

International Indigenous Peoples' Forum on Climate Change, 'Resources – 2016: COP 22', www.iipfcc.org/resources

'Malaysia – Country Profile', Convention on Biological Diversity, www.cbd.int/countries/default.shtml?country=my

'The Montréal Process: Member Countries', www.montrealprocess.org/The_Montreal_Process/Member_Countries/index.shtml

'National REDD+ Readiness in Malaysia' (2015) The REDD Desk: A Collaborative Resource for REDD Readiness, http://theredddesk.org/countries/initiatives/national-redd-readiness-malaysia

North Australian Land and Sea Management Alliance (NAILSMA), 'WALFA-West Arnhem Land Fire Abatement Project' (2012), www.nailsma.org.au/walfa-west-arnhem-land-fire-abatement-projecthtml.html

Northern Land Council, 'Fire Management', www.nlc.org.au/articles/info/fire-management

Office of the United Nations High Commissioner for Human Rights, *Frequently Asked Questions on a Human Rights-Based Approach to Development Cooperation* (2006), www.ohchr.org/Documents/Publications/FAQen.pdf

Papua New Guinea, 'Intended Nationally Determined Contribution (INDC) under the United Nations Framework Convention on Climate Change', submitted 24 March 2016, available at NDC Registry (Interim), www4.unfccc .int/ndcregistry/Pages/All.aspx

'Paris Agreement – Status of Ratification', United Nations Framework Convention on Climate Change, http://unfccc.int/paris_agreement/items/9444.php

'Parties to the Nagoya Protocol', Convention on Biological Diversity, www.cbd.int/ abs/nagoya-protocol/signatories

*People's Agreement of Cochabamba* (24 April 2010) World People's Conference on Climate Change and the Rights of Mother Earth, http://pwccc.wordpress .com/2010/04/24/peoples-agreement

The REDD Desk, 'Papua New Guinea: Financing' (Global Canopy Programme), http://theredddesk.org/countries/papua-new-guinea/financing

'Pilot Effective Models for Governance and Implementation of REDD+ in Small Island Developing States to Provide Equitable Benefits for Forest-Dependent Local and Indigenous People', http://theredddesk.org/countries/initiatives/ pilot-effective-models-governance-and-implementation-redd-small-island

'REDD in Malaysia' (2017) The REDD Desk: A Collaborative Resource for REDD Readiness, http://theredddesk.org/countries/malaysia

'REDD+ Countries' The Forest Carbon Partnership Facility', www.forestcarbon partnership.org/redd-countries-1

REDD+ Web Platform, 'Submissions: Malaysia', United Nations Framework Convention on Climate Change, http://redd.unfccc.int/submissions.html?country=mys

Republic of Vanuatu, 'Intended Nationally Determined Contribution (INDC)', submitted 21 September 2016, available at NDC Registry (Interim), www4.unfccc.int/ndcregistry/Pages/All.aspx

Rey, et al, 'Best Practices and Considerations for the Development of REDD+ Country Approaches to Safeguards and Design of Safeguards Information Systems' (CLP and SNV, United Kingdom 2016), http://redd.unfccc.int/ uploads/3539_1_best_practices_and_considerations_for_the_development_ of_redd_country_approaches_to_safeguards_and_design_of_safeguard_ information_systems.pdf.

Russell-Smith, Jeremy, 'Fire Agreement to Strengthen Communities' (1 November 2013) North Australian Land Manager, http://savanna.cdu.edu.au/view/ 250363/fire-agreement-to-strengthen-communities.html

Shihata, Ibrahim F I, 'Prohibition of Political Activities in the Bank's Work: Legal Opinion by the Senior Vice President and General Counsel' (12 July 1995) Transnational Dispute Management, www.transnational-dispute-manage ment.com/article.asp?key=981

State Government Victoria, 'Agreements with Traditional Owners', Environment, Land, Water and Planning, www.depi.vic.gov.au/forestry-and-land-use/managing-land/indigenous-land-management/agreements-with-traditional-owners#Yorta_Yorta_Agreements

'Status of the Doha Amendment', United Nations Framework Convention on Climate Change, http://unfccc.int/kyoto_protocol/doha_amendment/items/7362.php

'Sustainable Forest Management (SFM) Toolbox', Food and Agriculture Organization of the United Nations, www.fao.org/sustainable-forest-management/toolbox/en

'Tackling Climate Change through Sustainable Forest Management and Community Development in Malaysia' (2012) The REDD Desk: A Collaborative Resource for REDD Readiness, http://theredddesk.org/countries/initiatives/tackling-climate-change-through-sustainable-forest-management-and-community

The World Bank, 'Papua New Guinea: Projects & Programs – All Projects', www.worldbank.org/en/country/png/projects/all

The World Bank and The Inspection Panel, 'Kenya: Natural Resource Management Project' (14 January 2013), http://ewebapps.worldbank.org/apps/ip/Pages/ViewCase.aspx?CaseId=89

'Tracking (I)NDCs', Climate Action Tracker, www.climateactiontracker.org

Tropical Savannas CRC, 'West Arnhem Land Fire Abatement Project – What's Been Achieved?' (2007), http://savanna.cdu.edu.au/information/walfa_achievements.html

UNDP, 'Human Rights and UNDP' (Fast Facts, United Nations Development Programme, November 2011), www.undp.org/content/dam/undp/library/corporate/fast-facts/english/FF-Human-Rights.pdf

UNDP and Forest Carbon Partnership Facility (FCPF), 'Papua New Guinea REDD+ Readiness Project', www.pg.undp.org/content/dam/papua_new_guinea/docs/environment%20and%20energy/FCPF%20project/FCPF%20REDD+%20Readiness%20Project%20Brief_CORR.pdf

UNESCO: World Heritage Centre, 'The States Parties: Malaysia' (2015), http://whc.unesco.org/en/statesparties/my

United Nations Development Programme, 'Human Development Report 2015' (2015), http://hdr.undp.org/sites/default/files/2015_human_development_report_1.pdf

'Papua New Guinea: Project Document – Forest Carbon Partnership Facility' (2015) ('UNDP Project Document'), www.pg.undp.org/content/papua_new_guinea/en/home/operations/projects/environment_and_energy/forest-carbon-partnership-facility-reddt-readiness-project.html

United Nations Development Programme (UNDP), 'Biodiversity Conservation in Multiple-Use Forest Landscapes in Sabah, Malaysia' (2012) UNDP in Malaysia, Singapore & Brunei Darussalam, www.my.undp.org/content/

malaysia/en/home/operations/projects/environment_and_energy/80468_forestlandscapes.html

'UN REDD and Indigenous Peoples', www.undp.org/content/undp/en/home/ourwork/democraticgovernance/focus_areas/focus_human_rights/empowering_indigenous_peoples/UNREDD

United Nations Office of the High Commissioner for Human Rights, 'Declaration on the Rights of Indigenous Peoples', www.ohchr.org/EN/Issues/IPeoples/Pages/Declaration.aspx

'Status of Ratification Interactive Dashboard', http://indicators.ohchr.org

United Nations Permanent Forum on Indigenous Peoples, 'Who Are Indigenous Peoples?' www.un.org/esa/socdev/unpfii/documents/5session_factsheet1.pdf

UN Data, 'Country Profile – Vanuatu' (2016), http://data.un.org/CountryProfile.aspx?crName=Vanuatu#Environment

UN-REDD Programme, 'Community-Based REDD+', www.unredd.net/community-based-redd.html

'Partner Countries: Malaysia' (2017), www.unredd.net/index.php?option=com_docman&Itemid=134&view=list&slug=malaysia-2398

*Stakeholder Engagement: Engaging Indigenous Peoples and Other Forest Dependent Communities*, www.unredd.net/knowledge/redd-plus-technical-issues/stakeholder-engagement.html

'13th Policy Board Meeting, Arusha (Tanzania), 5–7 November 2014', www.unredd.net/documents/policy-board-86/thirteenth-policy-board-meeting-arusha-tanzania-5–7-november-2014–3553.html

*Tenure Security*, www.unredd.net/knowledge/redd-plus-technicalissues/tenure-security.html

van Paddenburg, Annawati, et al, 'Heart of Borneo Investing in Nature for a Green Economy', WWF HoB Global Initiative (2012), http://awsassets.wwf.org.au/downloads/fl013_heart_of_borneo_investing_in_nature_for_a_green_economy_jan12.pdf

West, Maddie, 'UN-REDD Programme (2016–2020) Establishes Executive Board, First Meeting to Take Place at FAO Headquarters in June', UN-REDD Programme: Collaborative Online Workspace (5 April 2016), www.unredd.net/announcements-and-news/2390-un-redd-programme-2016–2020-establishes-executive-board-first-meeting-to-take-place-at-fao-headquarters-in-june.html

Wiseman, Virginia, 'UNFCCC Publishes Country Views on REDD+ Safeguard Information Systems' (12 November 2014) IISD/SDG Knowledge Hub, sdg.iisd.org/news/unfccc-publishes-country-views-on-redd-safeguard-information-systems/?rdr=climate-l.iisd.org

World Resources Institute, 'Risk Tool: Malaysia', Forest Legality Initiative, www.forestlegality.org/risk-tool/country/malaysia

## Other

International Union of Forest Research Organizations (IUFRO), United Nations Development Programme (UNDP/UN-REDD), International Union for the Conservation of Nature (IUCN), 'Social Safeguards: Protecting the Rights and Interests of Indigenous Peoples and Forest-Dependent Communities in REDD+' (Programme, Forest Day, Durban, 4 December 2011), www.iufro.org/download/file/7930/4887/redd-social-safeguards-forum_doc

Papua New Guinea, *First INDC* (submitted to UNFCCC 24 March 2016)

Rahman, Dato' Dr Abdul, E Philip and K A Pan, 'REDD Plus Framework for Malaysia', www.cbd.int/doc/meetings/for/wscbredd-apac-01/other/wscbredd-apac-01-malaysia-en.pdf

Republic of Vanuatu, *First INDC* (submitted to UNFCCC 21 September 2016)

# INDEX